Semiparametric Regression

Semiparametric regression is concerned with the flexible incorporation of nonlinear functional relationships in regression analyses. Any application area that uses regression analysis can benefit from semiparametric regression. Assuming only a basic familiarity with ordinary parametric regression, this user-friendly book explains the techniques and benefits of semiparametric regression in a concise and modular fashion. The authors make liberal use of graphics and examples plus case studies taken from environmental, financial, and other applications. They include practical advice on implementation and pointers to relevant software.

This book is suitable as a textbook for students with little background in regression as well as a reference book for statistically oriented scientists – such as biostatisticians, econometricians, quantitative social scientists, and epidemiologists – with a good working knowledge of regression and the desire to begin using more flexible semiparametric models. Even experts on semiparametric regression should find something new here.

David Ruppert is the Andrew Schultz, Jr., Professor of Engineering (School of Operations Research and Industrial Engineering) and Professor of Statistical Science at Cornell University. He has served as editor for a number of prestigious series and journals and has published some 80 articles of his own as well as co-authoring two popular books, *Transformation and Weighting in Regression* and *Measurement Error in Nonlinear Models*. He is also winner of the Wilcoxon Prize for best practical applications paper in technometrics and an elected Fellow of the American Statistical Association and the Institute of Mathematical Statistics.

M. P. Wand is Professor of Statistics at the University of New South Wales in Sydney, Australia. He has held faculty appointments at Harvard University, Rice University, and Texas A&M University. Dr. Wand is a Fellow of the American Statistical Association and has served as an associate editor for the *Journal of the American Statistical Association* and *Biometrika*. He is winner of the P. A. P. Moran Medal for statistical research.

R. J. Carroll is Distinguished Professor of Statistics, Nutrition and Toxicology at Texas A&M University. Among his many honors are the COPSS Presidents' Award, the Fisher Lecture, the Snedecor Award, and the Wilcoxon Prize. He is an elected Fellow of the American Statistical Association and the Institute of Mathematical Statistics as well as an elected member of the International Statistical Institute.

Semiparametric Regression

DAVID RUPPERT

Cornell University

M. P. WAND

Harvard University

R. J. CARROLL

Texas A&M University

CAMBRIDGE
UNIVERSITY PRESS

CAMBRIDGE UNIVERSITY PRESS
Cambridge, New York, Melbourne, Madrid, Cape Town, Singapore,
São Paulo, Delhi, Dubai, Tokyo, Mexico City

Cambridge University Press
32 Avenue of the Americas, New York, NY 10013-2473, USA

www.cambridge.org
Information on this title: www.cambridge.org/9780521785167

First published 2003
Reprinted 2005, 2006, 2008, 2009

A catalog record for this publication is available from the British Library

Library of Congress Cataloging in Publication data

Ruppert, David, 1948–
Semiparametric regression / David Ruppert, M. P. Wand, R. J. Carroll.
p. cm.
Includes bibliographical references and index.
ISBN 0-521-78050-0 – ISBN 0-521-78516-2 (pb.)
1. Regression analysis.   2. Nonparametric statistics.   I. Wand, M. P. (Matthew P.).
II. Carroll, Raymond J.   III. Title.
QA278.2.R87   2003
519.5'36 – dc21   2002041460

ISBN 978-0-521-78050-6 Hardback
ISBN 978-0-521-78516-7 Paperback

To Anne, with love
— David

To my wife's parents, Ayhan and Recep
— Matt

To Brett and Jeb
— Raymond

Contents

Preface

The primary aim of this book is to guide researchers needing to flexibly incorporate nonlinear relationships into their regression analyses. Flexible nonlinear regression is traditionally known as *nonparametric regression*; it differs from parametric regression in that the shape of the functional relationships are not predetermined but can adjust to capture unusual or unexpected features of the data.

Almost all existing regression texts treat either parametric or nonparametric regression exclusively. The level of exposition between books of either type differs quite alarmingly. In this book we argue that nonparametric regression can be viewed as a relatively simple extension of parametric regression and treat the two together. We refer to this combination as *semiparametric regression*. Our approach to semiparametric regression is based on penalized regression splines and mixed models. Indeed, every model in this book is a special case of the linear mixed model or its generalized counterpart. This makes the methodology modular and is in keeping with our general philosophy of *minimalist statistics* (see Section 19.2), where the amount of methodology, terminology, and so on is kept to a minimum. This is the first smoothing book that makes use of the mixed model representation of smoothers.

Unlike many other texts on nonparametric regression, this book is very much problem-driven. Examples from our collaborative research (and elsewhere) have driven the selection of material and emphases and are used throughout the book.

The book is suitable for several audiences. One audience consists of students or working scientists with only a moderate background in regression, though familiarity with matrix and linear algebra is assumed. Marginal notes and the appendices are intended for beginners, especially those from interface disciplines. We make liberal use of graphics because visualization is a particularly effective tool for acquiring intuition in a new subject.

Another audience that we are aiming at consists of statistically oriented scientists (e.g., biostatisticians, econometricians, quantitative social scientists, and epidemiologists) who have a good working knowledge of linear models and the desire to begin using more flexible semiparametric models. There are many connections between linear and nonparametric regression. Our goal is to exploit them and the reader's knowledge of linear models to provide a foundation for understanding nonparametric modeling.

There is enough new material to be of interest even to experts on smoothing, and they are a third possible audience.

There are several competing approaches to nonparametric modeling: smoothing splines (e.g., Eubank 1988, 1999; Wahba 1990; Green and Silverman 1994); series-based smoothers, including wavelets (Tarter and Lock 1993; Ogden 1996); kernel methods, including local regression (Wand and Jones 1995; Fan and Gijbels 1996); and regression splines (Friedman 1991; Stone et al. 1997; Hansen and Kooperberg 2002). All four approaches can be used effectively and have their devotees. We believe that the nature of the data should play a role in the choice among them. For example, wavelets are more suited to highly oscillatory functions. Apart from this, the choice of a nonparametric regression method is a matter somewhat of individual taste and background. Based on our motivating applications and personal tastes, the approach to nonparametric regression used throughout this book is what we call *penalized splines*, although they are also labeled as *P-splines, pseudosplines,* and *low-rank spline smoothers* in the literature. Penalized splines are quite similar to smoothing splines; in fact, they are a generalization of smoothing splines that allow more flexible choices of the spline model, the basis functions for that model, and the penalty.

Penalized splines have close ties with ridge regression, mixed models, and Bayesian statistics, ties that were discovered by researchers working on smoothing splines. These ties allow techniques from mixed models – for example, (restricted) maximum likelihood estimation and likelihood ratio tests – to be added to penalized spline methodology. Similarly, Bayesian techniques based on Markov chain Monte Carlo provide what we believe to be the most satisfactory approach to fitting complex semiparametric models as well as the direction that semiparametric regression is most likely to take in the future. This book includes introductions to mixed models and to Bayesian modeling.

Acknowledgments

We are especially grateful to Ciprian Crainiceanu and Bhaswati Ganguli for their assistance in the preparation of this book. Ciprian wrote the WinBugs program in Appendix B and wrote the programs used for simulations-based *p*-values for likelihood ratio tests. Several other of our colleagues and collaborators have contributed to the book in various ways. We would like to thank Marc Aerts, Babette Brumback, Tianxi Cai, Gerda Claeskens, Brent Coull, Maria Durban, Garrett Fitzmaurice, Jonathan French, Robert Gentleman, Bob Gray, Nick Horton, Joe Ibrahim, Erin Kammann, Göran Kauermann, Robert Kohn, Nan Laird, Nick Lange, Mary Lindstrom, Long Ngo, Doug Nychka, Michael O'Connell, Helen Parise, José Pinheiro, Louise Ryan, Misha Salganik, Joel Schwartz, John Staudenmayer, Sally Thurston, Carrie Wager, Naisyin Wang, Jim Ware, Antonella Zanobetti, and Yihua Zhao for their collaboration, interest, and comments.

We thank Lauren Cowles for being a very supportive and patient editor. We would like to express our gratitude to Tom Ryan and Misha Salganik for sending us errata.

The second author lovingly acknowledges the support of his wife, Handan, and children, Declan and Jaida, throughout this project. Support of the Department of Biostatistics, Harvard University, is also gratefully acknowledged.

Guide to Notation

This chapter gives a brief overview of notational conventions used in the book. Please see the Notation Index for more specialized notation.

The symbol "$\equiv$" means "equal by definition".

We use both lower- and uppercase letters (e.g., x, X, and λ) to denote scalar quantities, either fixed or random. Lowercase bold letters (e.g., $\mathbf{x}$ and $\boldsymbol{\lambda}$) will be used for vectors. Uppercase bold fonts (e.g., $\mathbf{X}$ and $\boldsymbol{\Lambda}$) will denote matrices. The entries of a vector or matrix use the same letter and case as the vector or matrix itself but are not bold. Thus,

$$\mathbf{x} = \begin{bmatrix} x_1 \\ \vdots \\ x_n \end{bmatrix}$$

and

$$\mathbf{A} = \begin{bmatrix} A_{11} & A_{12} \\ A_{21} & A_{22} \end{bmatrix}.$$

If a matrix is partitioned then the submatrices are in bold; for example,

$$\mathbf{A} = \begin{bmatrix} \mathbf{A}_{11} & \mathbf{A}_{12} \\ \mathbf{A}_{21} & \mathbf{A}_{22} \end{bmatrix}.$$

We will indicate the row index of a matrix to the right and the column index below, as in:

$$\mathbf{C} = \begin{bmatrix} c_{ik} \end{bmatrix}_{\substack{1 \le k \le K \\ 1 \le i \le n}}.$$

The transpose of $\mathbf{A}$ is denoted by $\mathbf{A}^\mathsf{T}$. If $\mathbf{A}$ is an invertible square matrix, then $\mathbf{A}^{-1}$ denotes its inverse. Any vector is assumed to be a column, so its transpose is a row.

The *norm* of a vector $\mathbf{x}$ is denoted by $\|\mathbf{x}\|$; that is,

$$\|\mathbf{x}\| \equiv \sqrt{\mathbf{x}^\mathsf{T}\mathbf{x}}.$$

The real line will be denoted by $\mathbb{R}$, and d-dimensional space will be denoted by $\mathbb{R}^d$.

For a function $f(x)$ of a scalar x,

$$f^{(r)}(x) \equiv (d^r/dx^r)f(x),$$

the rth derivative of $f(x)$.

If $f(\mathbf{x})$ is a function from $\mathbb{R}^d$ to $\mathbb{R}$ then the *derivative vector* is a $1 \times d$ row vector with jth entry equal to $(\partial/\partial x_j) f(\mathbf{x})$, the partial derivative of $f(\mathbf{x})$ with respect to x_j, and is denoted by

$$\mathsf{D}f(\mathbf{x}).$$

The *Hessian matrix* is a $d \times d$ matrix whose (i, j) entry is equal to

$$\frac{\partial^2}{\partial x_i \partial x_j} f(\mathbf{x});$$

it is denoted by

$$\mathsf{H}f(\mathbf{x}).$$

If x and y are random variables, then $\mathsf{E}(x)$, $\text{var}(x)$, and st.dev.(x) are the mean, variance, and standard deviation of x, and $\text{cov}(x, y)$ is the covariance between x and y. $\text{Cov}(\mathbf{x})$ is the covariance matrix of a random vector $\mathbf{x}$; see Appendix A for its definition.

1

Introduction

Semiparametric regression can be of substantial value in the solution of complex scientific problems. The real world is far too complicated for the human mind to comprehend in great detail. Semiparametric regression models reduce complex data sets to summaries that we can understand. Properly applied, they retain essential features of the data while discarding unimportant details, and hence they aid sound decision-making.

Figure 1.1 depicts a complex data set corresponding to a cancer study in the Upper Cape Cod region of Massachusetts. Apart from the geographical location of cancer occurrences, there are data on age and smoking status. These data are for females.

One question of interest is whether there are elevated lung cancer rates, relative to all cancers and after adjustment for confounders, in any particular geographical locations. There is clearly a lot of relevant information represented by the one thousand points in this plot. However, it is very difficult to draw any conclusions from this alone. A semiparametric regression analysis leads to Figure 1.2.

Each of the graphics in Figure 1.2 displays an easy-to-comprehend estimate of the effect of smoking status, age, and geographical location on the occurrence of

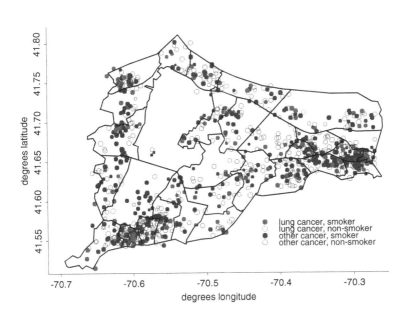

Figure 1.1 One thousand randomly chosen occurrences of female cancer in Upper Cape Cod, Massachusetts, for the period 1986–1994. The data are categorized according to lung cancer (red) or other (blue) and smoker (closed circle) or nonsmoker (open circle). The size of the circle is proportional to age. For confidentiality reasons, the data have been jittered.

Figure 1.2 Graphical outcomes from a semiparametric regression analysis of Upper Cape Cod lung cancer data: top panel, point estimate and approximate 95% confidence interval for the odds ratio of lung cancer among smokers who have some type of cancer; middle panel, estimated odds ratio as function of age; bottom panel, estimated odds ratio as function of geographic location. Higher values correspond to high estimated probabilities of lung cancer, given cancer, measured through the odds ratio.

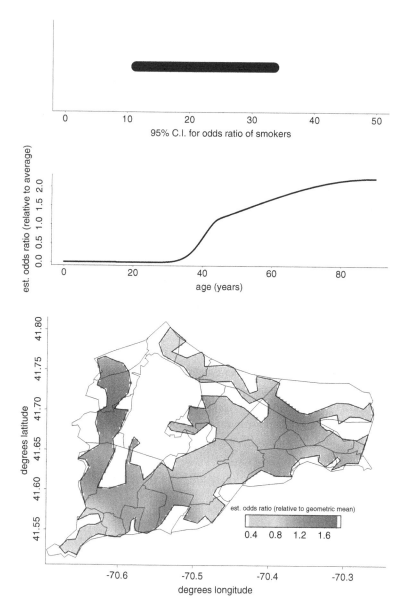

The *odds ratio* of an event *A*, relative to an event *B*, is defined to be the ratio of the odds of *A* to the odds of *B*. The *odds of A* is the probability of *A* occurring divided by the probability of *A* not occurring.

lung cancer, relative to cancer, while controlling for each of the other two variables. Smoking status is a binary variable, so its effect can be modeled through a single parameter. This the simplest type of parametric modeling. The graphic shows an odds ratio estimated to be in the range 11 to 33. Age is a continuous variable and, in this instance, its effect can be modeled reasonably well using parametric regression techniques. However, the nonparametric estimate shown in the middle panel suggests an unusual type of nonlinearity and so nonparametric regression techniques may lead to an improved fit. The effect of geography is difficult to model using traditional parametric models, and the map in Figure 1.2 is the result of a bivariate nonparametric regression technique. It clearly shows

0 ppm			25,000 ppm		
Tumor rates		Mean body weight[b]	Tumor rates		Mean body weight
Overall	Terminal[a]		Overall	Terminal	
32/50	25/30	287	17/50	15/32	254

[a] Tumors found at terminal sacrifice time.
[b] Average body weight at 12 months.

Table 1.1 Observed mammary tumor rates with phenolphthalein. For example, 32 of the 50 animals exposed at 25,000 ppm had tumors at the time of death. Of these, 18 died during the experiment and 32 were sacrificed at the end of the experiment, with 15 of the sacrificed animals being among the 17 with tumors.

regions with elevated lung cancer levels, something that is not easy to discern in Figure 1.1. Since the effects of smoking, age, and location have been modeled using a combination of parametric and nonparametric regression techniques, we call this a *semiparametric regression* analysis.

In the next sections we look at other important scientific investigations where semiparametric regression can play a useful role. We give detailed analyses of these studies (or at least references to where careful analyses can be found) in Chapter 18, after we have developed methodology to tackle them; Chapters 2–17 will be spent describing this methodology.

1.1 Assessing the Carcinogenicity of Phenolphthalein

The U.S. National Toxicology Program (NTP) routinely conducts animal experiments to measure the toxicity of certain foods and drugs. One such example is the assessment of the possible carcinogenicity of *phenolphthalein,* an ingredient of over-the-counter laxatives that was recently withdrawn by the U.S. Food and Drug Administration.

A topic of recent interest in the analysis of carcinogenicity data is how to deal with body weight. A recent editorial in *Science* magazine was highly critical of risk assessment agencies for not controlling for the possible confounding effect of weight, since weight loss caused by a toxic substance might protect against cancer and mask a carcinogenic effect (Abelson 1995). It is not uncommon for control animals to weigh substantially more than the treated animals throughout the course of an experiment owing to toxic effects of the chemical. Several sources have reported a lower incidence of tumors corresponding to lower body weights (Hart et al. 1995; Haseman, Bourbina, and Eustis 1994; Seilkop 1995). Thus, dose-related differences in body weights could affect the conclusions drawn from these studies. Indeed, many studies conducted by the NTP have shown protective effects of the chemical being tested on certain tumor incidences. These apparent reductions in tumor incidence across dose may be due to differences in body weight (Hart et al. 1995). This phenomenon is illustrated in Table 1.1, taken from the NTP study in phenolphthalein.

Figure 1.3 shows nonparametric estimates of the probabilities of four carcinogenic outcomes as a function of weight based on a large NTP set of data on

Figure 1.3
Estimated probability
of mammary tumor,
leukemia, pituitary
tumor, and thyroid
tumor as a function
of weight for a set
of NTP historical
controls. The shaded
region represents plus
and minus twice the
estimated (pointwise)
standard error.

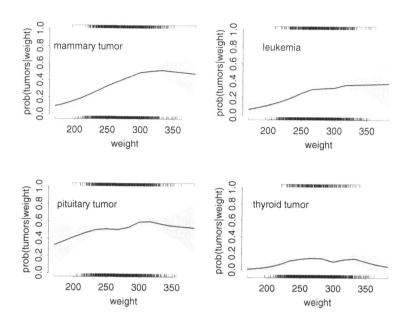

controls. It is apparent from these plots that nonlinear relationships exist and that semiparametric models for incorporation of weight data would be beneficial.

1.2   Salinity and Fishing in North Carolina

This example comes from a larger project to predict the annual shrimp (or prawn) harvest in Pamlico Sound, North Carolina, where shrimping occurs in the summer and autumn. It was believed that low salinity in the sound was detrimental to the shrimp harvest and that salinity values during certain crucial springtime periods would be useful predictors.

Salinity values were not measured regularly during the years prior to the project. However, discharges from rivers that empty into Pamlico Sound were known. The goal of the project was to develop a prediction model that could be used during the spring, early enough to help the fishing industry decide whether to rig for shrimp or instead to harvest some other species such as bluefish.

The data set has 28 cases taken from the spring periods of years 1972 to 1977. In each case, salinity was measured at the current time period and two weeks earlier, giving the variables salinity and lagged.sal. Two other variables were measured, discharge and trend. The variable trend indicated which of six biweekly periods during March to May a case came from. It was felt that trend might model the effects of increasing evaporation as the weather warmed, but no effect of trend was detected and so that variable will be ignored.

Figure 1.4 is a scatterplot matrix of the salinity data. One can see the strong, seemingly linear, relationship between salinity and lagged.sal. The relationship between salinity and discharge is somewhat weaker and possibly nonlinear. There is not a strong relationship between lagged.sal and

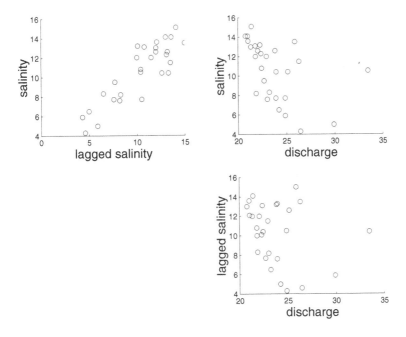

Figure 1.4
Scatterplot matrix of
the salinity data.

discharge, so their effects upon salinity should be individually estimable
with good precision.

The relationship between salinity and discharge is easier to see if we
remove the effects of lagged.sal. To do this, we regressed salinity on
lagged.sal using a straight line model (see Section 2.2). The residuals (i.e.,
the differences between salinity and the predicted values) are plotted against
discharge in Figure 1.5. The nonlinearity is now more evident, especially because a *scatterplot smooth* has been added. This suggests that a semiparametric
regression approach will be beneficial. The observation with discharge equal
to nearly 34 is a "high leverage point," meaning that it has a potentially high influence on the fitted curve. In fact, the fitted curve bends upward in the figure but
would not do so if the leverage point were excluded. However, unlike a linear
fit, the curved fit is only influenced locally – that is, on the right. We will discuss
this point further when we return to this example in Chapter 18.

The notion of
smoothing a
scatterplot will be
described extensively
in Chapters 3 and 5.

1.3 Management of a Retirement Fund

Bryant and Smith (1995) describe a managerial problem based on a real data
set, but with names changed to protect confidentiality. It concerns a company,
Best Retirement Inc. (BRI), that sells retirement plans to corporations around the
United States. To capture a market niche, it has decided to target smaller firms:
those with 500 or fewer employees. The major portion of their revenue comes
from retirement packages.

For a particular type of retirement plan known as 401(k), data are available
on several attributes of the firms from the previous year. It is advantageous that

Figure 1.5
Scatterplot of
residuals from
the regression
of salinity on
lagged.sal. A
scatterplot smooth
has been added. Note
the effect of the high
leverage point on the
extreme right.

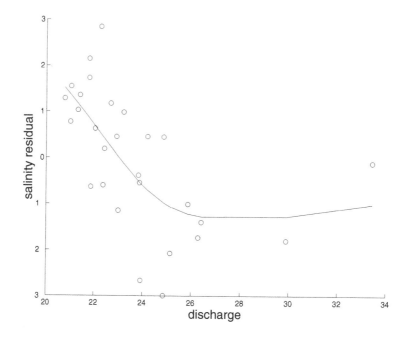

Figure 1.6
Estimated effect
of salary on
contribution to the
logarithm of year-end
contributions in
a semiparametric
regression analysis.
The shaded region
represents plus and
minus twice the
estimated (pointwise)
standard error.

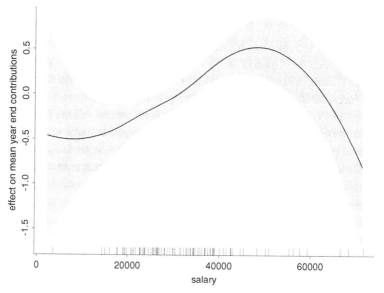

BRI be able to estimate the year-end dollar amount contributed to each plan in
advance so that it can make internal revenue and cost projections.

Apart from building a prediction model for year-end contributions, there are
some other managerial questions that can be addressed using these data. For example, BRI has a sales representative who has been specifically trained to deal
exclusively with 401(k) retirement plans. The company would like to know if her
expertise is a factor that influences contributions to such retirement plans.

Figure 1.6 shows the effect of salary (average salary of each firm) on the
logarithm of year-end contributions as estimated by a semiparametric regression

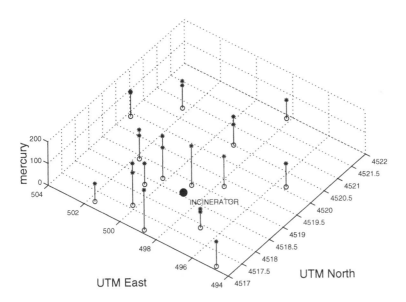

Figure 1.7 Plot of biomonitoring data. Open circles show sampling locations, and asterisks mark the single or replicate values of mercury measured at each sampling location. The large solid circle marks the location of the incinerator.

analysis. There is a pronounced nonlinearity here, which suggests that better predictions and managerial decisions can be realized through the use of semi-parametric regression.

1.4 Biomonitoring of Airborne Mercury

Waste incineration is a major source of environmental mercury. As part of an environmental monitoring program in Warren County, New Jersey, pots of sphaghum moss were placed at 15 sampling locations about a solid waste incinerator and exposed to ambient conditions between July 9 and July 23, 1991. The moss was then collected, dried, and assayed for mercury. The resultant data are shown in Figure 1.7.

The goals of the study include estimating the distribution of mercury about the incinerator and testing the null hypothesis that the mean mercury concentration is constant.

Figure 1.8 shows estimated levels of mercury concentration that were obtained using nonparametric methods described in this book. The plot indicates that mercury concentration peaks north of the incinerator. There are only 15 sampling locations, with replicate moss pots at 7 of these sites, for a total of 22 observations. With so few data, only gross features of mercury deposition can be resolved, but the nonparametric fit provides a pleasing image of these features.

1.5 Term Structure of Interest Rates

Corporations, municipalities, the U.S. Treasury, and other entities raise money by issuing bonds. The purchase price of the bond is a loan to the issuing entity and the bond is a contract requiring that entity to pay to the bond holder both principal

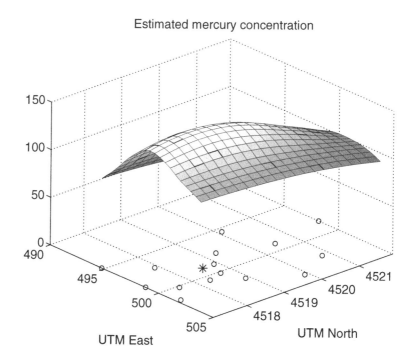

Estimated mercury concentration

and interest according to a schedule. At the time of expiration of the bond, which
is called the *maturity,* the bond holder receives a payment call the *par value.*
There are two general classes of bonds, coupon bonds and zero-coupon bonds.
At fixed periods, often every six months, the holder of a coupon bond receives
a *coupon payment.* Generally, coupon bonds sell at a price near their par value.
The par payment at maturity is a repayment of principal while the coupon payments are interest. Zero-coupon bonds have no coupon payments and sell below
par. The par payment at maturity represents principal and interest.

Frequently, the initial owner of the bond will sell the bond to another investor.
The current price at which bonds trade depends upon the current interest rates.
For example, suppose a corporate coupon bond with a 5% coupon rate is issued
with the initial price equal to par, so that the coupon payments are 5% of the initial price. If the prevailing interest rate increases to 6% then the price of the bond
will drop, so that a new purchaser of the bond will in effect receive a 6% rate.

The interest rates on bonds depend upon their maturities, with long-term bonds
frequently (though not always) paying higher rates than short-term bonds. For example, on January 26, 2001, the rate on a 1-year Treasury bill was 4.83% whereas
the rate on a 30-year Treasury bond was 6.11%. The term structure of interest
rates is a quantitative description of the dependency of rate upon maturity. The
estimation of term structure is essential for financial analysts working, for example, with credit derivatives.

Interest rates not only depend upon the maturity, but for any fixed maturity, the
interest rate on bonds with that maturity will change over time. In this case study,
we are not concerned with such changes. Rather, we will only be concerned with

A financial derivative
is a security whose
value depends on
the value of other
underlying securities.
As an example of a
derivative, consider a
call option on a stock.
A call option gives the
owner the right, but
not the obligation, to
purchase a share of
stock at a fixed price
on a given date, called
the expiration date.
The value of the call
option depends on the
price of the underlying
stock and on such

how interest rates on a given day depend on maturity. Specifically, in our example, we will model bond interest rates on December 31, 1995.

We will work with continuously compounded interest rates. As an illustration, we will start with an unrealistic assumption that the interest rate is constant, that is, not dependent on maturity. If a bond is worth $P(t)$ dollars at time t and is continuously compounded at a constant rate r, then $P(t)$ satisfies the simple differential equation

$$P'(t) = rP(t) \qquad (1.1)$$

and so, at maturity T,

$$P(T) = P(0) \exp(rT). \qquad (1.2)$$

The rate r is called the *forward rate*. It is the rate agreed upon at present for interest in the future, that is, *forward* in time.

Interest rates must be inferred from bond prices. Recall that the bond's value at maturity, $P(T)$, is called the par value. Hence, from (1.2) we have

$$P(0) = \text{par} \exp(-rT), \qquad (1.3)$$

where par is the par value. Suppose a 1-year, par $100 zero-coupon bond is selling now for $92. This means we can buy the bond now for $92 and receive $100 exactly one year from now. Recall that zero-coupon means the bond holder receives no interest payments until maturity. The $8 difference between the present price and the par value is the only interest payment. Here we have $T = 1$, $P(1) = \text{par} = 100$, $P(0) = 92$, and, from (1.2),

$$92 = 100 \exp(-r)$$

or

$$r = \log(100/92) = 0.0834.$$

Thus, the annual continuously compounded interest rate over the next year is 8.34%.

Suppose, in addition, that a 2-year, par $100 zero-coupon bond sells for $85. We assume that this bond pays the just-determined rate of 8.34% the first year but a different interest rate the next year. The rate for the second year, call it r_2, solves

$$83 = 100 \exp\{-(0.0834 + r_2)\}$$

or

$$r_2 = \log(100/83) - 0.0834 = 0.1029.$$

Table 1.2 gives the prices on December 31, 1995, of five bonds previously issued by the U.S. communications company AT&T and maturing at some time after that date. These are the prices at which the bonds were traded – that is, purchased by one investor from another. Each bond price is expressed as a percentage of par, the amount AT&T will pay the bond owner at maturity. The maturity is given in years from December 31, 1995. The bonds make semiannual interest payments called coupons. The time in years of the next coupon and the coupon payments are given in the table. The aim is to determine the forward rate of AT&T bonds from these data.

other variables as the time left until expiration. An example of a interest rate derivative is a *cap.* If an interest rate exceeds the cap, then the owner of the cap is paid the difference between the interest rate and the cap. Clearly, the value of the cap depends on the underlying interest rate. A company paying interest at a floating rate might purchase a cap as insurance against rate increases.

Table 1.2 AT&T
bond prices on
December 31, 1995.
Issue, maturity, and
next coupon dates
are in years from
December 31, 1995.

| | | Next | | |
Issue	Maturity	coupon	Coupon	Price
−3.9644	5.9781	0.0356	7.1250	109.4580
−1.7726	8.1890	0.2274	6.7500	106.2840
−1.5836	10.3562	0.4164	7.5000	111.4360
−0.8384	11.1041	0.1616	7.7500	115.5090
−0.6384	9.3096	0.3616	7.0000	107.6590

We have been assuming that the forward interest rate is constant over each
year. Clearly, this is an oversimplification. Financial analysts model the forward
interest rate as a continuous function of time, $r(t)$. If $P(T)$ is the par value of a
zero-coupon bond maturing at time T and if $P(0)$ is the current price of the bond,
then (1.1) is replaced by

$$P'(t) = r(t)P(t),$$

with solution

$$P(0) = P(T)\exp\left(-\int_0^T r(x)\,dx\right). \tag{1.4}$$

A forward price is a
price negotiated at
the present for the
future delivery of
some commodity. A
forward interest rate
means an interest rate
that is agreed upon
now for a loan in the
future.

The problem is to estimate $r(t)$ from bond prices, such as those shown in Table
1.2. A further complication is that many bonds, including those in the table, have
coupons. A coupon bond can be modeled as a bundle of zero-coupon bonds, one
for each coupon payment and one for the final payment at maturity of the par
value. The bond price is the aggregate price of all of these coupon bonds. Bond
prices such as in Table 1.2 have some random "error" since, for example, they are
really prices at last transaction, not exactly at the current time. Therefore, the estimation of the forward rate curve is a statistical problem. Fisher, Nychka, and
Zervos (1994) have developed a very elegant spline method for estimating the forward rate curve. Their method works well for Treasury bond data because there
are enough Treasury bonds to estimate a continuous forward rate.

For corporate bonds, there is often a paucity of data and so the method of Fisher
and colleagues cannot be applied directly. Jarrow, Ruppert, and Yu (2001) extend
the model of Fisher et al. by assuming that the forward rate for a corporation such
as AT&T differs from the Treasury forward rate by a constant or, perhaps, by a
low-degree polynomial function of time. The corporate forward rate is greater
than the Treasury rate, since Treasury bonds have no risk of default; the U.S. Treasury can always raise money by taxation. The difference between the two rates
is called the *risk premium* or *spread* and reflects the extra interest that investors
demand when buying corporate bonds (which may default) rather than risk-free
Treasury bonds. The model of Jarrow and colleagues is semiparametric in that
the Treasury forward rate is modeled as a spline, but the risk premium is modeled parametrically. This case study is typical of semiparametric models in that
parts of the model for which there is much data are modeled nonparametrically
while parts that are not well supported by data are modeled parametrically.

Figure 1.9 shows the prices of U.S. STRIPS (Separate Trading of Registered
Interest and Principal of Securities), a type of zero-coupon Treasury bond. The

Figure 1.9 U.S. STRIPS prices as a percentage of the par value.

prices are expressed as a percentage of the par value and are plotted against time to maturity. If $r(x)$ is constant, say $r(x) = r_0$ for all x, then by (1.4) we have

$$y_i = 100 \exp(-r_0 T_i) \qquad (1.5)$$

and

$$\log(y_i) = \log(100) - r_0 T_i. \qquad (1.6)$$

Here $P(T_i)$ is the par, $P(0)$ is the present price, $y_i = 100P(0)/P(T_i)$ is the "response," and T_i is the maturity for the the ith U.S. STRIPS.

The rough exponential shape in Figure 1.9 suggests that model (1.5) is at least approximately correct. However, in Figure 1.10 we see $\log(y_i)$ plotted against T_i, and the plot is not quite the straight line that (1.6) suggests. In fact, we fit a straight line to $\{T_i, \log(y_i)\}_{i=1}^{n}$ and plotted the "residuals," which are the differences between the $\log(y_i)$ and the fitted line. This plot, shown as Figure 1.11, shows an obvious deviation from the random cloud that we would expect if the model (1.5) fit the data, thus indicating the need for a nonparametric model. The fitting of straight line models and residual analysis will be discussed in Chapter 2.

1.6   Air Pollution and Mortality in Milan: The Harvesting Effect

In the last decade, a good deal of literature has been published concerning the short-term effect of air pollution on health. Daily mortality counts and hospital admissions have been associated with daily air pollution levels, correcting for several time-dependent confounders. From the public health point of view, the significance of air pollution's short-term effects corresponds to an increase in mortality or morbidity among individuals who would otherwise die much later, not among those who could have died within a few days.

Figure 1.10
Logarithms of U.S.
STRIPS prices as a
percentage of the par
value.

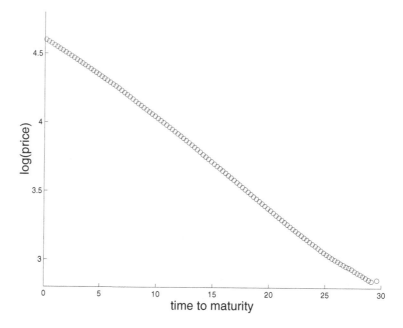

Figure 1.11
Residuals from a
straight line fit to the
logarithms of the U.S.
STRIPS prices as a
percentage of the par
value.

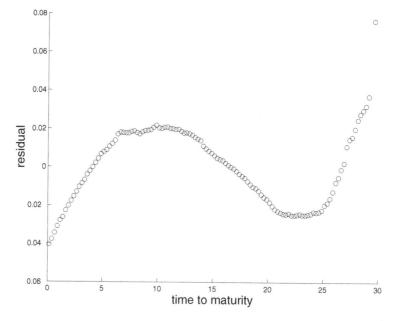

Figure 1.12 is a schematic representation of the dynamics that arise when air pollution has an impact on mortality. The *risk pool* consists of sick and elderly people. Transitions between this state and the general population are affected by air pollution levels.

Consider the following lagged regression model of air pollution and generic mortality:

$$\log\{\mathtt{E}(\mathtt{mortality}_t)\} = \alpha + \beta_0\mathtt{pollution}_t + \cdots + \beta_q\mathtt{pollution}_{t-q} + \varepsilon_t,$$

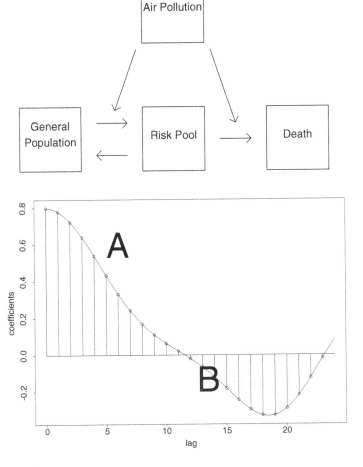

Figure 1.12
Schematic
representation
of the dynamics
that arise when air
pollution has an
impact on mortality.

Figure 1.13
Lag structure
corresponding to the
harvesting effect.

where mortality$_t$ and pollution$_t$ are (respectively) the mortality count and pollution level for day t. The lag structure in Figure 1.13 describes the so-called *harvesting effect*. The horizontal axis is the lag number and the vertical axis shows the coefficients β_ℓ. Each β_ℓ has this interpretation: net effect of pollution level ℓ days ago on mortality.

In the figure, A is the sum of the positive coefficients for low lags and represents the fact that pollution levels in the past few days or weeks have a positive effect on mortality. However, the negative coefficients in B mean that pollution levels a longer period ago have a negative effect. This is due to "depletion of the risk pool," normally made up of elderly and sick people whose deaths have been hastened a few days or weeks by episodes of high pollution; this is known as "harvesting." Here A overestimates the public health significance of pollution, since it is really A + B (where B is negative) that represents deaths induced by a noticeable amount of time.

Daily data over 10 years are available on mortality, air pollution, and several meteorological variables for the city of Milan, Italy. It is of interest to use these to

Figure 1.14
Estimates of the
coefficients of the
lags of sulphur
dioxide on mortality
in Milan, Italy. The
shaded points are
plus and minus 2
times the estimated
standard error of each
coefficient estimate.

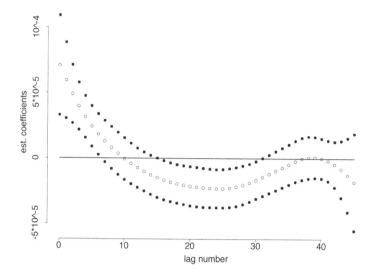

quantify the public health significance of air pollution, incorporating the harvesting effect. By constraining the lag coefficients to be on a smooth (but otherwise
flexible) curve, we obtained Figure 1.14. This suggests some evidence of harvesting. The construction of this result required some nonstandard semiparametric
regression techniques that allowed for the lag coefficients to lie on a smooth curve
and also be influenced by data on daily weather conditions.

Chapter 18 provides much fuller analyses and solutions for a selection of the problems presented in this chapter. Between now and then we will need to describe
techniques for performing semiparametric regression analysis. The next chapter
signals the start of this journey.

2

Parametric Regression

2.1 Introduction

Each of the problems described in the previous chapter can benefit from regression analysis. In this book we focus on the combination of classical parametric regression techniques and modern nonparametric regression techniques to develop useful models for such analyses. Therefore, it is essential to have a good grounding in the principles of *parametric regression* before proceeding to the more complicated semiparametric regression chapters. In particular, some of the theoretical aspects of regression should be well understood since these are important in extensions to semiparametric regression. The present chapter can serve as either a brief introduction to parametric regression for readers without a background in that field or as a refresher for those with a working knowledge of parametric regression but who could benefit from a review. If you are very familiar with parametric regression methodology and theory, then this chapter could be skimmed. Of course, this brief introduction can only cover the main concepts and a few special models. Many widely used parametric models are not discussed. This chapter provides sufficient background in parametric regression for the chapters to follow. However, readers wishing to apply parametric regression models may consult a textbook on parametric regression such as Weisberg (1985), Neter et al. (1996), or Draper and Smith (1998).

Note, moreover, that Section 2.5 contains some new perspectives on parametric regression that are relevant to later chapters on semiparametric models, so this is worth covering regardless of experience.

Toward the end of the chapter we describe some limitations of parametric regression. Most of the remainder of the book is concerned with extensions of parametric regression that have much more flexibility.

The theory in this book makes extensive use of matrix notation and results, a summary of which is given in Appendix A.

2.2 Linear Regression Models

Figure 2.1 shows a scatterplot of 55 months of data from a house in Westchester County, New York. The horizontal variable is temp, the average temperature (in degrees Fahrenheit) for the month; the vertical variable is logkwh, the logarithm of electricity usage in kilowatt-hours (kwh).

The electricity usage data are from the textbook *A Casebook for a First Course in Statistics and Data Analysis* by Chatterjee, Handcock, and Simonoff. The analysis here differs somewhat from the analysis in that textbook – there are many valid ways of analyzing any given set of data.

Figure 2.1
Scatterplot of `logkwh`
versus `temp` from
the electricity usage
study.

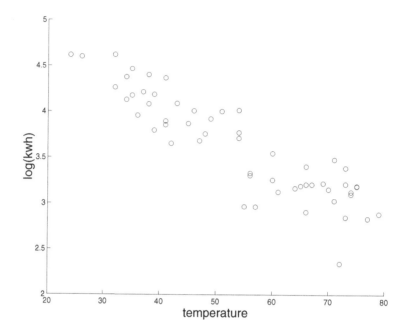

One aim of the study was to determine the relationship between `temp` and `logkwh`. The scatterplot suggests that a plausible relationship is one of the form

$$\texttt{logkwh}_i = \beta_0 + \beta_1 \texttt{temp}_i + \texttt{error}_i, \tag{2.1}$$

where $\texttt{temp}_i$ and $\texttt{logkwh}_i$ are the value of each variable for the ith month. The $\texttt{error}_i$ term accounts for the fact that the points in the scatterplot are randomly scattered about the line, so this term is assumed to be independent realizations of a random variable with mean equal to zero. Hence an alternative formulation is

$$\mathsf{E}(\texttt{logkwh}|\texttt{temp}) = \beta_0 + \beta_1\texttt{temp},$$

which says that the mean `logkwh` value is a linear function of the corresponding `temp` value.

It is common to call `logkwh` the *response variable* and `temp` the *predictor variable*. Many alternative names are commonly used, although we will use this terminology throughout the book.

The model can be fit using *least squares,* which corresponds to finding the line that best fits the scatterplot in terms of minimizing the sum of the squared vertical distances between the points and the line; that is, minimizing over (β_0, β_1) the quantity

$$\sum_{i=1}^{n}\{\texttt{logkwh}_i - (\beta_0 + \beta_1\texttt{temp}_i)\}^2.$$

Algebraically, this can be done by setting up the matrices

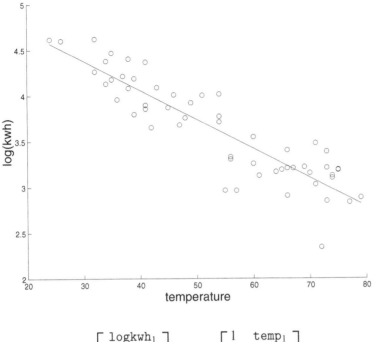

$$\mathbf{y} = \begin{bmatrix} \texttt{logkwh}_1 \\ \vdots \\ \texttt{logkwh}_{55} \end{bmatrix}, \quad \mathbf{X} = \begin{bmatrix} 1 & \texttt{temp}_1 \\ \vdots & \vdots \\ 1 & \texttt{temp}_{55} \end{bmatrix},$$

$$\boldsymbol{\beta} = \begin{bmatrix} \beta_0 \\ \beta_1 \end{bmatrix}, \quad \boldsymbol{\varepsilon} = \begin{bmatrix} \texttt{error}_1 \\ \vdots \\ \texttt{error}_{55} \end{bmatrix},$$

so that (2.1) can be written as

$$\mathbf{y} = \mathbf{X}\boldsymbol{\beta} + \boldsymbol{\varepsilon}.$$

Then the *least-squares estimator* of $\boldsymbol{\beta}$ is calculated as

$$\hat{\boldsymbol{\beta}} = \begin{bmatrix} \hat{\beta}_0 \\ \hat{\beta}_1 \end{bmatrix} = (\mathbf{X}^\mathsf{T}\mathbf{X})^{-1}\mathbf{X}^\mathsf{T}\mathbf{y}. \tag{2.2}$$

This leads to the estimated coefficients

$$\hat{\boldsymbol{\beta}} = \begin{bmatrix} 5.34 \\ -0.0319 \end{bmatrix}$$

and the fitted line

$$\widehat{\texttt{logkwh}} = 5.34 - 0.0319\texttt{temp},$$

which is shown in Figure 2.2.

   This is an example of the *simple linear regression model,* the simplest regression model of all. We use the word "simple" because the model has only one predictor variable, `temp`. The *slope coefficient* $\hat{\beta}_1 = -0.0319$ has the interpretation:

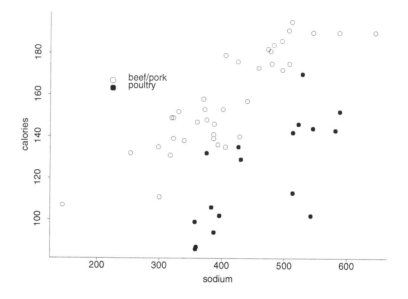

a 1-degree increase in average temperature corresponds to a 0.0319-unit
decrease in *mean* logkwh.

The sausage data are
from the textbook
*Introduction to the
Practice of Statistics*
by Moore and
McCabe (1998).

Figure 2.3 depicts a slightly more complicated data set. In this case, data on
the calorie count for a sample of 54 sausages are plotted against sodium content.
However, the data are categorized according to meat type: beef/pork and poultry.
The plot shows that there is evidence of a linear relationship between calories
and sodium, but the relationship seems to be different according to meat type.
The difference is not so much in the slope as in the *intercept*. In other words,
it seems reasonable to model the mean calorie content in terms of two parallel
lines, one for each type. This can be done through the model:

$$\text{calories}_i = \beta_0 + \beta_1\text{sodium}_i + \beta_2\text{pork.beef}_i + \text{error}_i, \qquad (2.3)$$

where

$$\text{pork.beef}_i = \begin{cases} 1 & \text{if } i\text{th sausage is beef or pork,} \\ 0 & \text{if } i\text{th sausage is poultry.} \end{cases}$$

Such a variable is often called an *indicator variable* or a *dummy variable* and
allows the incorporation of a categorical variable such as meat type into a linear
regression model. We also call model (2.3) a *binary offset model* since the linear relationship between calories and sodium is offset by the binary variable
pork.beef.

As with the simple linear regression model, the coefficients can be estimated
using (2.2) with

$$\mathbf{X} = \begin{bmatrix} 1 & \text{sodium}_1 & \text{pork.beef}_1 \\ \vdots & \vdots & \vdots \\ 1 & \text{sodium}_{54} & \text{pork.beef}_{54} \end{bmatrix}, \quad \boldsymbol{\beta} = \begin{bmatrix} \beta_0 \\ \beta_1 \\ \beta_2 \end{bmatrix}, \quad \mathbf{y} = \begin{bmatrix} \text{calories}_1 \\ \vdots \\ \text{calories}_{54} \end{bmatrix}.$$

The fitted model is

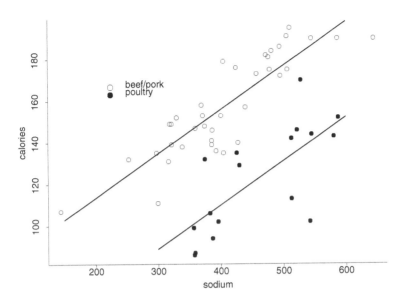

Figure 2.4 Fitted
model for the sausage
data. The model
corresponds to the
calories–sodium
relationship being
represented by two
parallel lines; one for
each meat type.

$$\widehat{\text{calories}} = 25.6 + 0.211\text{sodium} + 45.8\text{pork.beef}.$$

Because of the way we defined pork.beef, this is equivalent to

$$\widehat{\text{calories}} = \begin{cases} 71.4 + 0.211\text{sodium} & \text{for beef or pork sausages,} \\ 25.6 + 0.211\text{sodium} & \text{for poultry sausages.} \end{cases}$$

So the fitted model corresponds to two parallel lines with intercepts differing by the amount $\hat{\beta}_2$, as illustrated in Figure 2.4. The interpretations of the coefficients in the fitted model are:

- for both types of sausage, a one-unit increase in sodium corresponds to a mean increase of 0.211 calories;
- controlling for sodium content, beef and pork sausages have a mean calorie level that is 45.8 higher than poultry sausages.

However, before adopting these conclusions and perhaps going on to make inferential statements (e.g., confidence intervals or p-values), it is important to check the validity of the linear regression model. This is addressed in Section 2.3.

2.2.1 General Linear Model

The general multiple regression model with k predictor variables is

$$y_i = \beta_0 + \beta_1 x_{i1} + \cdots + \beta_k x_{ik} + \varepsilon_i. \tag{2.4}$$

Here, for $j = 1, \ldots, k$, we have x_{ij} as the value of the jth predictor variable for the ith case, $i = 1, \ldots, n$.

This model can be written in matrix notation as

$$\mathbf{y} = \mathbf{X}\boldsymbol{\beta} + \boldsymbol{\varepsilon}.$$

Here $\mathbf{y}$ is an $n \times 1$ vector of response variables, $\boldsymbol{\varepsilon}$ is an $n \times 1$ vector of "errors," and $\mathbf{X}$ is an $n \times (k+1)$ matrix of predictor variables. The ith row of $\mathbf{X}$, which we will denote by $\mathbf{x}_i^\mathsf{T}$, contains a 1 followed by the values of the predictor variables for the ith case, that is,

$$\mathbf{x}_i^\mathsf{T} = [1, x_{i1}, \dots, x_{ik}].$$

In the margin: *In the so-called no-intercept model, the intercept parameter is deleted and the 1 is omitted from $\mathbf{x}_i$ vector.*

The ordinary least-squares (OLS) estimator minimizes

$$\sum_{i=1}^{n}(y_i - \mathbf{x}_i^\mathsf{T}\boldsymbol{\beta})^2 = \|\mathbf{y} - \mathbf{X}\boldsymbol{\beta}\|^2,$$

where $\|\mathbf{v}\| = \sqrt{\mathbf{v}^\mathsf{T}\mathbf{v}}$ is the "length" of the vector $\mathbf{v}$. As seen in the preceding examples, the least-squares estimator of $\boldsymbol{\beta}$ is

$$\hat{\boldsymbol{\beta}} = (\mathbf{X}^\mathsf{T}\mathbf{X})^{-1}\mathbf{X}^\mathsf{T}\mathbf{y}.$$

The error vector $\boldsymbol{\varepsilon}$ has zero mean: $\mathsf{E}(\boldsymbol{\varepsilon}) = \mathbf{0}$. Additional assumptions about $\boldsymbol{\varepsilon}$ are often made. The first, often called *homoscedasticity,* is that

$$\text{var}(\varepsilon_i) = \sigma^2 \quad \text{for all } 1 \le i \le n. \tag{2.5}$$

It is also usual to assume that the errors are uncorrelated:

$$\text{cov}(\varepsilon_i, \varepsilon_j) = 0, \quad i \ne j. \tag{2.6}$$

Conditions (2.5) and (2.6) can be summarized by the expression

$$\text{Cov}(\boldsymbol{\varepsilon}) = \sigma^2 \mathbf{I}. \tag{2.7}$$

Finally, there is the *normality assumption*; given (2.7), this translates to

$$\boldsymbol{\varepsilon} \sim \mathrm{N}(\mathbf{0}, \sigma^2 \mathbf{I}). \tag{2.8}$$

2.3 Regression Diagnostics

The term *regression diagnostics* refers to a large collection of techniques used to check the quality of the data and the adequacy of a regression model. Often data are misrecorded or do not come from the population of interest. For example, a retailer analyzing weekly sales figures may be advised to exclude data from holiday periods since they are not part of the population of ordinary trading outcomes. If holiday sales were unintentionally included in the data set, then they would likely be outlying in some ways and might be detected by diagnostics. Outliers should not be removed simply because they are outlying, but if they are found to be erroneous then they should be removed or (if possible) corrected. In some circumstances outliers reveal an interesting feature of the data that had not been previously realized.

The two basic components of many diagnostics are the *fitted values* and the *residuals.* The ith fitted value is the estimate of $\mathsf{E}(y_i)$ from the model; that is, $\hat{y}_i = \mathbf{x}_i^\mathsf{T}\hat{\boldsymbol{\beta}}$, where as before $\mathbf{x}_i^\mathsf{T}$ is the ith row of $\mathbf{X}$. The vector of all n fitted values is

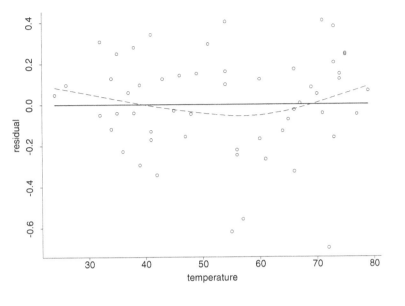

Figure 2.5 Residual plot for the least-squares fit to the simple linear regression model for the `logkwh` versus `temp` data from the electricity usage study. A scatterplot smoother (see Chapter 3) is included (dashed curve).

$$\hat{\mathbf{y}} = \mathbf{Hy}, \tag{2.9}$$

where

$$\mathbf{H} = \mathbf{X}(\mathbf{X}^\mathsf{T}\mathbf{X})^{-1}\mathbf{X}^\mathsf{T}.$$

The matrix $\mathbf{H}$ is called the *hat matrix* since multiplication by $\mathbf{H}$ converts $\mathbf{y}$ to $\hat{\mathbf{y}}$. As will be seen in the following chapters, the hat matrix plays an extremely important role in regression theory and practice.

The ith residual is defined to be

$$e_i = y_i - \hat{y}_i,$$

which is the difference between the ith observed response and its predicted value according to the fitted model. If the model provides an adequate fit to the data, then e_i predicts ε_i.

Most of the information for determining the adequacy of a linear regression model is contained in the residuals, since these estimate that part of the model that was assumed to be random. Therefore, any patterns in the residuals reflect extra structure that is not accommodated by the model. Residual analysis for diagnosis of linear regression models is a very large topic, with many more facets than we can go into here. We will just touch on some of the most basic principles. Fitted values and residuals are available for the semiparametric models we will be studying in later chapters, so the plots we discuss here are useful for semiparametric as well as linear regression.

Figure 2.5 shows the residuals for the fit to the `logkwh` and `temp` data plotted against the fitted values. The points are scattered about the zero line without any strong patterns, so this is an indication that the simple linear regression model is a reasonable one in this case. If there had been a curvilinear pattern to the residuals, then this would have been evidence of lack of fit of the simple linear model.

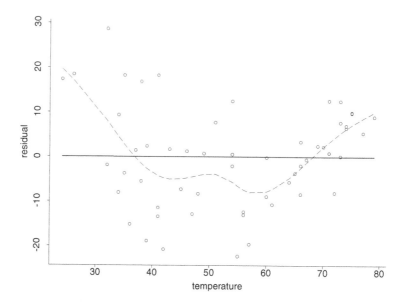

Figure 2.6 Residual
plot for the least-
squares fit to
the simple linear
regression model
for the kwh versus
temp data from the
electricity usage
study.

Sometimes a curvilinear pattern is difficult to detect, and it is helpful to include
a scatterplot smooth in the residual plot. This is shown in Figure 2.5 as a dashed
curve. A scatterplot smooth is a curve that shows the general trend of the data;
this will be discussed in depth in Chapter 3. Clearly, the smooth is close to the
horizontal line through zero and provides more evidence that there is no substantial pattern.

Figure 2.6 is the same plot as Figure 2.5 except that the response is kilowatt-
hours (kwh), not logkwh. In this figure, a curvilinear pattern is apparent, so the
straight line model appears not to fit. This lack of fit is the reason why logkwh,
rather than kwh, was used as the response variable.

In Figure 2.5 one sees that the amount of scatter in the residuals is roughly
constant across values of temp. This is good, since it supports the assumption
used later for making the inference (e.g., constructing confidence intervals) that
var(logkwh|temp) is constant. Another feature displayed by the plot is that none
of the residuals is unusually large in magnitude relative to the others, that is, there
are no outliers. This is also in keeping with the assumptions for inference.

The plot of the residuals against fitted values for the linear model fit to the
sausage data, shown in Figure 2.7, has the same patternless nature as Figure 2.5.
There is no lack of fit to the binary offset straight line model and the assump-
tion that var(calories|sodium) is constant. Also, there are no outliers in the
sausage data.

Noting that the vector of residuals is

$$\mathbf{e} = (\mathbf{I} - \mathbf{H})\mathbf{y}$$

and using

$$\mathbf{H} = \mathbf{H}^\mathsf{T} = \mathbf{H}^2, \tag{2.10}$$

we have

$$\mathrm{Cov}(\mathbf{e}) = (\mathbf{I} - \mathbf{H})\,\mathrm{Cov}(\mathbf{y})(\mathbf{I} - \mathbf{H})^\mathsf{T} = \sigma^2(\mathbf{I} - \mathbf{H}).$$

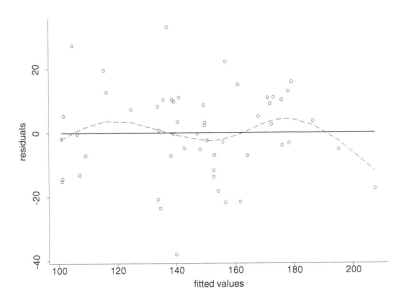

Figure 2.7 Residual plot for the sausage data.

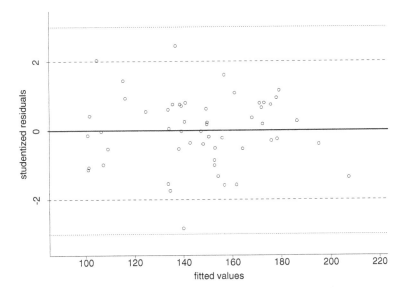

Figure 2.8 Studentized residual plot for the sausage data. The dashed lines are at ± 2 standard deviations and the dotted lines are at ± 3 standard deviations. The observations between the dashed lines are "mild" outliers. Observations outside the dotted lines, if they existed, would be strong outliers.

Therefore,

$$\widehat{\text{st.dev.}}(e_i) = \sigma\sqrt{1 - (i\text{th diagonal entry of } \mathbf{H})} = \sigma\sqrt{1 - H_{ii}}.$$

The ith *studentized residual* is then

$$e_i^* \equiv \frac{e_i}{\widehat{\text{st.dev.}}(e_i)} = \frac{e_i}{\hat{\sigma}\sqrt{1 - H_{ii}}}, \tag{2.11}$$

which, under normality and homoscedasticity, should behave like a N(0, 1) sample when the model is correct. Figure 2.8 shows the studentized residuals for the sausage example. Since all are between -3 and 3 and all but 3 out of 54 are between -2 and 2, we have no reason to doubt our assumption in this case.

The studentized residual is sometimes called the *internally studentized residual* to distinguish it from the externally studentized residual. The latter is defined by (2.11) but with $\hat{\sigma}$ replaced by an estimate of σ that does not use the ith case. In this book, studentized residual will always refer to internally studentized residual.

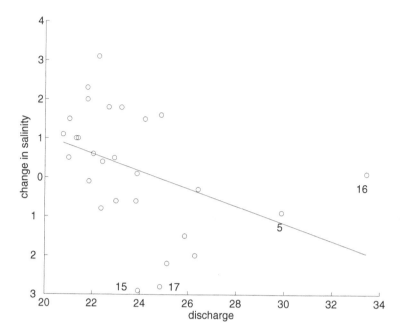

Figure 2.9 Salinity data. Scatterplot of discharge and delta.salinity with a least-squares line. Influential cases 5, 15, 16, and 17 are labeled.

2.3.1 *Influence Diagnostics*

Recall the salinity example of Chapter 1. The primary goal is to predict salinity from lagged.sal (i.e., the lagged value of salinity) and discharge, the discharge from rivers of fresh water into Pamlico Sound. As an example, we will regress the change in salinity, called delta.salinity, on discharge. Figure 2.9 is a scatter plot plus a least-squares fit. Figure 2.10 shows the studentized residuals. Cases 15, 16, and 17 have the large absolute studentized residuals. In Figure 2.9, cases 5 and 16 are seen to have outlying discharge values.

A natural question is: How much influence do these possible outliers have on the fitted line? Two diagnostics that are in common use to assessing influence are:

- the *hat diagonals* or *leverages,* which measure the potential of outliers in the predictors to influence the fit; and
- *Cook's distance* (also called *Cook's D*), which measures actual influence of an observation on the fit.

The *i*th *leverage* value is the *i*th diagonal of the hat matrix, H_{ii}. We know from (2.9) that the *i*th fitted value is

$$\hat{y}_i = \sum_{j=1}^{n} H_{ij} y_j = H_{i1} y_1 + \cdots + H_{ii} y_i + \cdots + H_{in} y_n, \qquad (2.12)$$

so that H_{ii} is the weight of y_i in the expression for $\hat{y}_i$, that is, the influence of y_i on its own fitted value. It should be appreciated that H_{ii} depends only on the predictors, not on the ys, so that H_{ii} measures only the *potential* for being influential and not actual influence. For example, in linear regression with a single

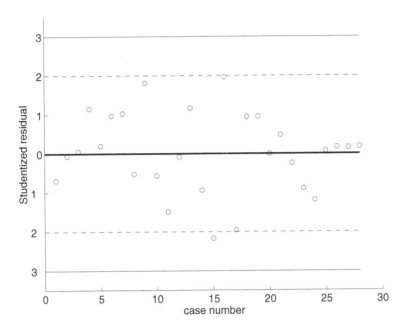

Figure 2.10 Salinity data with `discharge` as the predictor and `delta.salinity` as the response: studentized residuals.

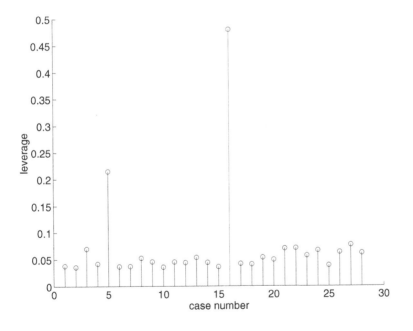

Figure 2.11 Salinity data with `discharge` as the predictor: leverage values.

predictor x, H_{ii} is a linear function of the squared distance from x_i to $\bar{x}$ and so is purely a measure of how "outlying" x_i is.

Figure 2.11 shows the leverages for the salinity data using `discharge` as the sole predictor. Cases 5 and especially 16 have high leverage.

Let $\hat{\mathbf{y}}$ and $\hat{\mathbf{y}}_{(i)}$ be, respectively, the vector of fitted values using all the data and with the ith case deleted. Let p be the number of regression parameters, including the intercept. Cook's D is defined as

Figure 2.12 Salinity data with `discharge` as the predictor and `delta.salinity` as the response: values of Cook's D.

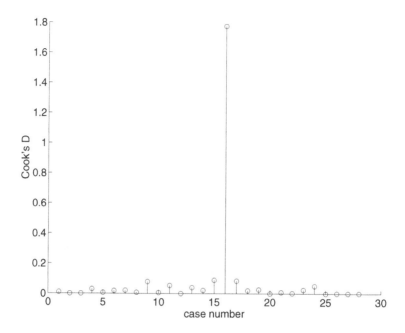

$$D_i = \frac{\|\hat{\mathbf{y}} - \hat{\mathbf{y}}_{(i)}\|^2}{p\hat{\sigma}^2}.$$

It can be shown that

$$D_i = \frac{(e_i^*)^2 H_{ii}}{p(1 - H_{ii})},$$

which expresses Cook's D in terms of the ith studentized residual (e_i^*) and the ith leverage.

Figure 2.12 shows the values of Cook's D for the salinity data using `discharge` as the predictor and `delta.salinity` as the response. Case 16 with high leverage and a large studentized residual has very large Cook's D, approximately ten times larger than the second largest value. We conclude that the fitted line would be changed substantially if case 16 were not included. This case might be an outlier or it might indicate that the linear model does not hold over a wide range of `discharge`. We will return to this example in Chapter 18.

2.3.2 *Autocorrelation*

The data in the electricity usage study were collected in time order. An important assumption in regression analysis is that the errors are independent or, at least, uncorrelated. If this assumption is false, then ordinary least-squares estimation could be inefficient. Perhaps more seriously, commonly used inferential procedures (as described in Section 2.4) are invalid when there is autocorrelation.

The main diagnostic is the autocorrelation function of the residuals. The kth lag sample autocorrelation, denoted $\hat{\rho}(k)$, is the sample correlation between e_i and e_{i-k}, where $k + 1 \leq i \leq n$. For example, the lag-1 autocorrelation measures the correlation between adjacent residuals. For general $k = 0, 1, \ldots, n - 1$,

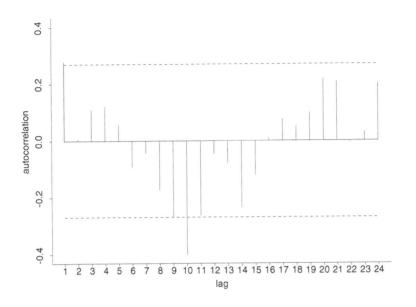

Figure 2.13
Autocorrelation plot
of the electricity
usage data. The
horizontal dashed
lines are at $\pm 2/\sqrt{n}$,
corresponding
to approximate
95% confidence
limits under the
independence
assumption.

$$\hat{\rho}(k) = \frac{\sum_{i=1}^{n-k} e_i e_{i-k}}{\sum_{i=1}^{n} e_i^2}$$

(see e.g. Brockwell and Davis 1996). If the errors are independent, then $\hat{\rho}(k)$ is approximately normally distributed with mean 0 and variance $1/n$. Thus, any value of $|\hat{\rho}(k)|$ that exceeds $2/\sqrt{n}$ is roughly significant. Of course, if one examines a large number of sample autocorrelations, then some should be significant by chance. For this reason – and because autocorrelation, if it exists, should manifest itself at short lags – one should look for significant autocorrelation at, say, $k = 1$ or 2. For monthly data, one should also look at autocorrelation at $k = 12$. If autocorrelation is not significant at these lags, then it is reasonable to assume that the errors are independent.

The electricity usage data are *longitudinal*. That is, the data were collected over time from a single residence. Longitudinal data are often correlated, and it is important to check for autocorrelation in this example.

Figure 2.13 is a plot of the first 24 autocorrelation values of the residuals from fitting (2.1). The horizontal dashed lines are at $\pm 2/\sqrt{n}$, corresponding to approximate 95% confidence limits under the independence assumption. The autocorrelation values at $k = 1$ and 10 are significant. The significant autocorrelation at lag 10 may be due to chance – as mentioned, we expect a few significant autocorrelations due solely to chance because we are examining 24 autocorrelations.

The significant lag-1 autocorrelation is likely to be real. Of course, statistical significance is not the same as practical significance. The lag-1 autocorrelation is 0.279 and the squared correlation is 0.0780. Squared correlation has an interpretation as the amount of squared variation in one variable that can be predicted by the second variable; see Section 2.4.7.1. Certainly, 7.8% is a somewhat small (though not negligible) percentage.

There are at least four ways one can deal with this autocorrelation. The first is to ignore the autocorrelation since it is small. We will take that approach in this chapter, since our intention is only to provide a simple illustrative example. Another option is to use additional predictor variables, in particular, lagged values of logkwh and temp. The model

$$\text{logkwh}_i = \beta_0 + \beta_1 \text{temp}_i + \beta_2 \text{logkwh}_{i-1} + \text{temp}_{i-1} + \text{error}_i$$

has lag-1 residual autocorrelation of only 0.0273. The fitted model is

$$\widehat{\text{logkwh}}_i = 3.837 - 0.034 \text{temp}_i + 0.278 \text{logkwh}_{i-1} + 0.011 \text{temp}_{i-1}.$$

A third approach is to use only temp_t as a predictor variable but to assume a more general form for $\text{Cov}(\mathbf{y})$.

The fourth approach is to add $f(\text{time})$, a random function of time, to the model. Since it is random, $f(\text{time})$ will induce correlation between the observations. We might model $f(\text{time})$ as a stationary stochastic process. This type of modeling strategy is used in the Milan mortality example in Chapter 8.

2.3.3   *The Building Blocks of Regression Diagnostics*

All of the diagnostics that we have discussed are constructed from two basic building blocks:

- hat diagonals, and
- residuals.

This fact has an important implication. In the later chapters of this book, we shall discuss the fitting of many types of semiparametric models by penalized least squares. These building blocks are available for such fits. For any fit there is, of course, a residual – the difference between the actual and the fitted response. For penalized least squares and other *linear* smoothers, there is also a hat matrix and therefore hat diagonals (see Section 3.10 for a definition of linear smoothers and their hat matrices). Thus, studentized residuals, leverages, and Cook's D are all easily generalized to the penalized least-squares fits that we will be discussing in the ensuing chapters. Eubank (1985) gives a nice discussion of influence diagnostics for smoothing splines, and his ideas are easily applied to penalized splines.

2.4   Inference

2.4.1   *Confidence and Prediction Intervals*

Although Figure 2.2 is a reasonable summary of the relationship between temp and logkwh, it does not give any indication of the uncertainty of the fitted line. If the study were re-run for a different time period, how much would this relationship between mean logkwh and temp level change? An enhancement that

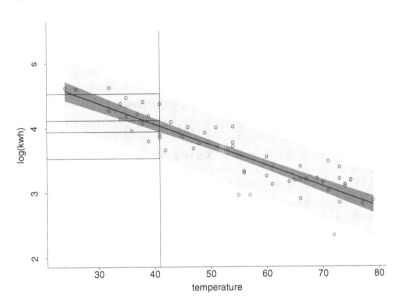

Figure 2.14 95% confidence (dark grey) and prediction (light grey) bands for the electricity usage example. The horizontal lines correspond to 95% confidence and prediction intervals at `temperature` = 41°F.

illustrates uncertainty is shown in Figure 2.14, which we will call a *confidence band plot*. The dark grey band gives an indication of the degree to which the mean of `logkwh`, for each level of `temp`, is unknown. It has the following interpretation:

> For a given level of `temp`, if a vertical line is drawn through that value then we can be 95% confident that the mean of `logkwh` given that `temp` is between the values cut off by the dark grey region. For example, there is a vertical line through `temp` = 41. The inner (i.e., second and third) horizontal lines show where this vertical line intersects the lower and upper boundaries of the confidence band and give a 95% confidence interval for $E(logkwh|temp = 41)$.

These confidence intervals for mean `logkwh` do not tell us where we can expect the actual values of `logkwh` to lie, since these will differ from their expected values by an amount equal to the value of `error` for that observation.

> The light grey band in Figure 2.14 is a 95% prediction band for `logkwh`. Suppose that we will observe `logkwh` at value of `temp` that is known now. We can be 95% confident that a new observation will lie in the interval above that value of `temp`. Thus, if `temp` = 41, then the 95% prediction interval for `logkwh` is the interval cut off by the outer (i.e., first and fourth) horizontal lines.

The dark grey region in Figure 2.14 was obtained by adding and subtracting twice the estimated standard deviation of the estimated line

$$\widehat{logkwh} = \hat{\beta}_0 + \hat{\beta}_1 temp.$$

Similarly, the light grey region is centered at $\widehat{logkwh}$, but the quantity that is added and subtracted is the estimated standard deviation of $y - \hat{y}$, the difference between a new observation and its predicted value.

In the remainder of this section we show how one obtains expected values and standard deviations of $\hat{\boldsymbol{\beta}}$, fitted values, and prediction errors, since these are the basis of confidence intervals.

2.4.2 *Inference about the Regression Coefficients*

Consider the general linear regression model

$$\mathbf{y} = \mathbf{X}\boldsymbol{\beta} + \boldsymbol{\varepsilon},$$

and let $\hat{\boldsymbol{\beta}}$ be the least-squares estimator of $\boldsymbol{\beta}$. First we will find the expectation vector of $\hat{\boldsymbol{\beta}}$, that is, the vector of its expected values.

The values of the predictor variables in the matrix $\mathbf{X}$ can be either fixed, as in a designed experiment, or random, as in observational data. However, in all calculations we will treat the matrix $\mathbf{X}$ as fixed; that is, all expectations, variances, and covariance are conditional on $\mathbf{X}$. For notational simplicity this convention will not be made explicit. We assume that $\mathbf{X}$ is of full rank – that its columns are linearly independent.

Recall the expression for the least-squares estimate of $\boldsymbol{\beta}$:

$$\hat{\boldsymbol{\beta}} = (\mathbf{X}^{\mathsf{T}}\mathbf{X})^{-1}\mathbf{X}^{\mathsf{T}}\mathbf{y}.$$

Since we are conditioning on $\mathbf{X}$, the only random variables in this $\hat{\boldsymbol{\beta}}$ expression are those in $\mathbf{y}$. Therefore, from (A.5) in Appendix A it follows that

$$\mathsf{E}(\hat{\boldsymbol{\beta}}) = (\mathbf{X}^{\mathsf{T}}\mathbf{X})^{-1}\mathbf{X}^{\mathsf{T}}\mathsf{E}(\mathbf{y}) = (\mathbf{X}^{\mathsf{T}}\mathbf{X})^{-1}\mathbf{X}^{\mathsf{T}}\mathbf{X}\boldsymbol{\beta} = \boldsymbol{\beta},$$

The covariance matrix of an m-dimensional random vector, $\mathrm{Cov}(\mathbf{U})$, is an $m \times m$ matrix whose (i, j)th entry is the covariance between the ith and jth components of the random vector $\mathbf{U}$ (see Appendix A).

showing that $\hat{\boldsymbol{\beta}}$ is unbiased. Next, we will find the covariance matrix of $\hat{\boldsymbol{\beta}}$, but first we need the covariance matrix of $\mathbf{y}$.

Using the homoscedasticity assumption and the assumption of no correlation (2.7), it follows that

$$\mathrm{Cov}(\mathbf{y}) = \sigma^2 \mathbf{I}.$$

Therefore, from equation (A.6) with $\mathbf{A} = (\mathbf{X}^{\mathsf{T}}\mathbf{X})^{-1}\mathbf{X}^{\mathsf{T}}$,

$$\mathrm{Cov}(\hat{\boldsymbol{\beta}}) = (\mathbf{X}^{\mathsf{T}}\mathbf{X})^{-1}\mathbf{X}^{\mathsf{T}}\,\mathrm{Cov}(\mathbf{y})\mathbf{X}(\mathbf{X}^{\mathsf{T}}\mathbf{X})^{-1}$$
$$= \sigma^2(\mathbf{X}^{\mathsf{T}}\mathbf{X})^{-1}. \tag{2.13}$$

Then, since

$$\mathrm{st.dev.}(\hat{\beta}_i) = \sqrt{i\text{th diagonal entry of } \mathrm{Cov}(\hat{\boldsymbol{\beta}})} = \sqrt{\{\mathrm{Cov}(\hat{\boldsymbol{\beta}})\}_{ii}},$$

we have

$$\widehat{\mathrm{st.dev.}}(\hat{\beta}_i) = \hat{\sigma} \times \sqrt{i\text{th diagonal entry of } (\mathbf{X}^{\mathsf{T}}\mathbf{X})^{-1}} = \hat{\sigma}\sqrt{\{(\mathbf{X}^{\mathsf{T}}\mathbf{X})^{-1}\}_{ii}},$$

where $\hat{\sigma}$ is an estimate of σ discussed in Section 2.4.6.

If the normality assumption (2.8) is made, then a $100(1 - \alpha)\%$ confidence interval for β_i is

$$\hat{\beta}_i \pm \widehat{\mathrm{st.dev.}}(\hat{\beta}_i)t\left(1 - \tfrac{\alpha}{2}; n - p\right), \tag{2.14}$$

Variable	Coeff.	St. dev.	95% CI
temp	−0.0319	0.00215	(−0.0362, −0.0276)
sodium	0.211	0.021	(0.170, 0.252)
pork.beef	45.8	4.22	(37.3, 54.2)

Table 2.1 Estimates, standard deviations, and 95% confidence intervals for the coefficients from the temp–logkwh example and from the sausage example.

where p is the number of regression parameters (including the intercept) in the model; $t(q; m)$ is the $100q$th percentile of the t-distribution with m degrees of freedom. For $n - p$ bigger than about 30 we can replace (2.14) by

$$\hat{\beta}_i \pm \widehat{\text{st.dev.}}(\hat{\beta}_i) z\left(1 - \tfrac{\alpha}{2}\right),$$

where $z(q)$ is the $100q$th percentile of the standard normal distribution.

These results can be used to construct Table 2.1, which shows coefficients, standard errors, and 95% confidence intervals (CIs) for the coefficients in electricity usage and sausage examples. Since none of the confidence intervals in Table 2.1 include zero, we see that all of the coefficients are significant at the 0.05 level.

The standard deviation of an estimator is usually called its *standard error.*

These confidence intervals have exact $100(1 - \alpha)\%$ coverage probability only if the errors are independent and normally distributed with a constant variance. However, the coverage probability is somewhat robust to nonnormality. Correlated errors, however, can cause the coverage probability to deviate wildly from $100(1 - \alpha)\%$. Since some residual autocorrelation was seen in the electricity usage data, the standard deviation and confidence interval for temp should be interpreted cautiously. In fact, in the expanded model that uses lagged values of temp and logkwh as additional predictor variables, the standard deviation of the coefficient of temp is 0.0039, rather larger than the value in Table 2.1.

2.4.3 *t-Statistics and p-Values*

One is often interested in testing the null hypothesis that β_i is zero versus the alternative that β_i is nonzero. The null hypothesis has the interpretation that there is no linear relationship between y and x_i such that one could drop x_i from the model. This test can be done using the t-statistic, defined as

$$t(\hat{\beta}_i) \equiv \frac{\hat{\beta}_i}{\widehat{\text{st.dev.}}(\hat{\beta}_i)}.$$

The p-value is

$$2[1 - F_t\{t(\hat{\beta}_i); n - p\}],$$

where $F_t(\cdot; k)$ is the cumulative distribution function of the t-distribution with k degrees of freedom.

An alternative way to test is to reject the null hypothesis if the $100(1 - \alpha)\%$ confidence interval does not include zero.

Table 2.2 Typical
regression package
output for the sausage
example.

Variable	Est.	St. dev.	t-Stat.	p-Value
intercept	25.5671	10.0543	2.5429	0.0141
sodium	0.2111	0.0206	10.2358	0.0000
pork.beef	45.7595	4.2173	10.8504	0.0000

Table 2.2 is a summary of the fit of (2.3) to the sausage data, which is typical
of that produced by regression packages.

2.4.4 *Inference about the Mean Response*

We are now in a position to describe how the grey regions in Figure 2.14 are computed. They rely on expressions for the expectation and standard deviation of
$\hat{y}_0 = \mathbf{x}_0^T \hat{\boldsymbol{\beta}}$, the estimator of $E(y|\mathbf{x}_0) = \mathbf{x}_0^T \boldsymbol{\beta}$, where $\mathbf{x}_0^T$ is a fixed and known row
vector of predictor variables. Of course, $\mathbf{x}_0^T$ could be $\mathbf{x}_i^T$, the ith row of $\mathbf{X}$, if we
were interested in $E(y_i)$.

Using (A.5) and the unbiasedness of $\hat{\boldsymbol{\beta}}$, we have

$$E(\hat{y}_0) = \mathbf{x}_0^T E(\hat{\boldsymbol{\beta}}) = \mathbf{x}_0^T \boldsymbol{\beta},$$

so that $\hat{y}_0$ inherits the unbiasedness of $\hat{\boldsymbol{\beta}}$. Also, using (A.6) and (2.13), we have

$$\text{var}(\hat{y}_0) = \mathbf{x}_0^T \text{Cov}(\hat{\boldsymbol{\beta}})\mathbf{x}_0 = \sigma^2 \mathbf{x}_0^T (\mathbf{X}^T \mathbf{X})^{-1} \mathbf{x}_0, \tag{2.15}$$

so that

$$\widehat{\text{st.dev.}}(\hat{y}_0) = \hat{\sigma} \sqrt{\mathbf{x}_0^T (\mathbf{X}^T \mathbf{X})^{-1} \mathbf{x}_0}.$$

A $100(1 - \alpha)\%$ confidence interval for $E(y)$ is

$$\hat{y}_0 \pm \widehat{\text{st.dev.}}(\hat{y}_0) t\left(1 - \tfrac{\alpha}{2}; n - p\right); \tag{2.16}$$

this result is the basis for the confidence band plot, the dark grey band in Figure 2.14.

For $n - p$ larger than about 30, we can replace $t\left(1 - \tfrac{\alpha}{2}; n - p\right)$ by

$$\hat{y}_0 \pm \widehat{\text{st.dev.}}(\hat{y}_0) z\left(1 - \tfrac{\alpha}{2}\right);$$

as before, $z(q)$ is the $100q$th percentile of the standard normal distribution.

2.4.5 *Inference about New Observations*

Suppose that we wish to predict the value of y at a new data point for which $\mathbf{x}_0$
is available; that is, we wish to predict $\mathbf{x}_0^T \boldsymbol{\beta} + \varepsilon$ when $\mathbf{x}_0$ is known. The predic-
tor is $\hat{y}_0 = \mathbf{x}_0^T \hat{\boldsymbol{\beta}}$, which predicts $\mathbf{x}_0^T \boldsymbol{\beta}$ by $\mathbf{x}_0^T \hat{\boldsymbol{\beta}}$ and ε by 0 – that is, both quantities
are predicted by estimates of their expected values. Uncertainty in the prediction
has two causes: $\boldsymbol{\beta}$ will differ from $\hat{\boldsymbol{\beta}}$, and ε will not equal 0.

Since the new ε is independent of $\hat{\boldsymbol{\beta}}$, it follows that

$$\text{var}(y - \hat{y}_0) = \text{var}(\varepsilon) + \text{var}(\hat{y}_0).$$

Hence, by (2.15), we have

$$\widehat{\text{st.dev.}}(y - \hat{y}_0) = \hat{\sigma}\sqrt{1 + \mathbf{x}_0^\mathsf{T}(\mathbf{X}^\mathsf{T}\mathbf{X})^{-1}\mathbf{x}_0}.$$

A $100(1 - \alpha)\%$ prediction interval for y is

$$\hat{y}_0 \pm \widehat{\text{st.dev.}}(y - \hat{y}_0)t\left(1 - \tfrac{\alpha}{2}; n - p\right); \tag{2.17}$$

this is the basis for the prediction band plot, the light band in Figure 2.14.

2.4.6 *Estimation of σ^2*

As we have seen, confidence and prediction intervals require an estimate of σ^2. The "natural" choice for this is the average of the squared residuals,

$$\frac{1}{n}\sum_{i=1}^{n} e_i^2.$$

Using (2.10), this can be expressed in matrix notation as

$$\begin{aligned}
\frac{1}{n}\|\mathbf{y} - \hat{\mathbf{y}}\|^2 &= \frac{1}{n}\|(\mathbf{I} - \mathbf{H})\mathbf{y}\|^2 \\
&= \frac{1}{n}\mathbf{y}^\mathsf{T}(\mathbf{I} - \mathbf{H})(\mathbf{I} - \mathbf{H})^\mathsf{T}\mathbf{y} \\
&= \frac{1}{n}\mathbf{y}^\mathsf{T}(\mathbf{I} - \mathbf{H})\mathbf{y}.
\end{aligned}$$

It is reasonable to ask that the estimate of σ^2 be unbiased when the model is correct. Note that

> The *trace* of a square matrix $\mathbf{A}$, denoted by $\text{tr}(\mathbf{A})$, is the sum of the diagonal elements of $\mathbf{A}$.

$$\begin{aligned}
\mathsf{E}\{\mathbf{y}^\mathsf{T}(\mathbf{I} - \mathbf{H})\mathbf{y}\} &= \mathsf{E}[\text{tr}\{\mathbf{y}\mathbf{y}^\mathsf{T}(\mathbf{I} - \mathbf{H})\}] \\
&= \text{tr}\{\mathsf{E}(\mathbf{y}\mathbf{y}^\mathsf{T})(\mathbf{I} - \mathbf{H})\} \\
&= \text{tr}[\{\text{Cov}(\mathbf{y}) + \mathsf{E}(\mathbf{y})\mathsf{E}(\mathbf{y})^\mathsf{T}\}(\mathbf{I} - \mathbf{H})] \\
&= \sigma^2\,\text{tr}(\mathbf{I} - \mathbf{H}) + (\mathbf{X}\boldsymbol{\beta})^\mathsf{T}(\mathbf{I} - \mathbf{H})\mathbf{X}\boldsymbol{\beta}.
\end{aligned}$$

Since $\mathbf{H} = \mathbf{X}(\mathbf{X}^\mathsf{T}\mathbf{X})^{-1}\mathbf{X}^\mathsf{T}$, it is easily shown that $(\mathbf{I} - \mathbf{H})\mathbf{X} = 0$. Therefore,

$$\begin{aligned}
\mathsf{E}\left(\frac{1}{n}\|\mathbf{y} - \hat{\mathbf{y}}\|^2\right) &= \frac{1}{n}\sigma^2\,\text{tr}\{\mathbf{I}_n - \mathbf{X}(\mathbf{X}^\mathsf{T}\mathbf{X})^{-1}\mathbf{X}^\mathsf{T}\} \\
&= \frac{1}{n}\sigma^2\{n - \text{tr}(\mathbf{I}_p)\} \\
&= \frac{n - p}{n}\sigma^2,
\end{aligned}$$

where $\mathbf{I}_p$ is the identity matrix of order p. This shows that our first proposal for estimation of σ^2 is biased by a factor of $\frac{n-p}{n}$. Therefore, it is common to use the unbiased estimator

$$\hat{\sigma}^2 \equiv \frac{1}{n - p}\|\mathbf{y} - \hat{\mathbf{y}}\|^2.$$

Figure 2.15 Linear and quadratic fits to the electricity usage data with kwh as the response variable.

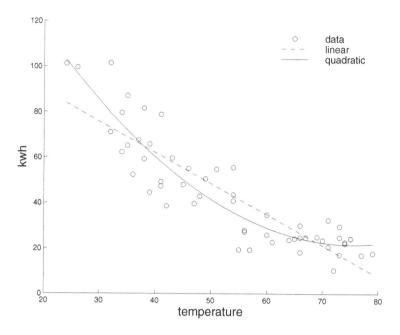

2.4.7 *Extra Sums of Squares and Hypothesis Testing*

In practice, we often need to compare two or more models. For example, in the electricity usage data, if one uses kwh as the response variable then there is evidence of lack of fit to a straight line model; see Figure 2.6. One might then consider a parabolic or quadratic model given by

$$\text{kwh}_i = \beta_0 + \beta_1 \text{temp}_i + \beta_2 \text{temp}_i^2 + \text{error}_i. \tag{2.18}$$

The fits to the linear and quadratic models are shown in Figure 2.15. The quadratic fit appears better, but is this just due to random variation or is there really curvature that requires an alternative to the straight line model?

We can address this question by starting with model (2.18) and testing

$$H_0 : \beta_2 = 0 \quad \text{versus} \quad H_1 : \beta_2 \neq 0.$$

The least-squares estimator minimizes

$$\sum_{i=1}^{n} (y_i - \beta_0 - \beta_1 \text{temp}_i - \beta_2 \text{temp}_i^2)^2 = \sum_{i=1}^{n} e_i^2,$$

which is often called the *residual sum of squares* (RSS). The RSS provides a cursory measure of the quality of the fit. If one model fits better than another, then this difference in fit should be evident in the RSS values for the two models. Therefore, a test can be performed by comparing the sum of squared residuals from the quadratic model to the same quantity under the null hypothesis – that is, from the straight line model. The former is always smaller than the later. The difference between the two, called the *extra sum of squares*, measures the improvement in fit gained by adopting the quadratic model instead of the linear. The extra sum of squares is

$$\text{ExtraSS} \equiv \text{RSS}_{\text{linear}} - \text{RSS}_{\text{quadratic}},$$

where $\text{RSS}_{\text{linear}}$ and $\text{RSS}_{\text{quadratic}}$ are the RSS values for the respective models.

This idea can be generalized. Suppose that two models are being considered, where one is smaller than the other (e.g., the smaller model can be obtained by dropping predictor variables from the larger model). We will refer to these as the *larger* and *smaller* models. Then

$$\text{ExtraSS} \equiv \text{RSS}_{\text{smaller}} - \text{RSS}_{\text{larger}}.$$

The extra sum of squares can be used to test the larger versus smaller models. The null hypothesis is that the smaller model fits the data, while the alternative is that the smaller does not fit but the larger does. Let p_{smaller} and p_{larger} be the number of parameters in the respective models and let $\hat{\sigma}^2_{\text{larger}}$ be the estimate of σ^2 from the larger model. The F-test statistic is

$$F \equiv \frac{\text{ExtraSS}/(p_{\text{larger}} - p_{\text{smaller}})}{\hat{\sigma}^2_{\text{larger}}}. \tag{2.19}$$

It can be shown that, under the null hypothesis and the normality assumption (2.8), F has an F-distribution with $p_{\text{larger}} - p_{\text{smaller}}$ and $n - p_{\text{larger}}$ degrees of freedom.

The F-statistic has an intuitive interpretation. The numerator is the ratio of the improvement in fit of the larger model over the smaller one to the increase in number of parameters; the denominator is an estimate of the variance of the response. The F-statistic compares the improvement per parameter to the data variability.

Given this interpretation, one sees that the null hypothesis should be rejected when F is large. An α-level test is obtained by rejecting the null hypothesis when F exceeds the $(1 - \alpha)$ quantile of the F-distribution with $p_{\text{larger}} - p_{\text{smaller}}$ and $n - p_{\text{larger}}$ degrees of freedom. Therefore, the p-value is $F_F(F; p_{\text{larger}} - p_{\text{smaller}}, n - p_{\text{larger}})$, where $F_F(\cdot; m_1, m_2)$ is the cumulative distribution function of the F-distribution with m_1 and m_2 degrees of freedom.

These F-tests have exact α-level only if the errors are independent and normally distributed with a constant variance. Like the t-intervals in Section 2.4, they are somewhat robust to nonnormality and nonconstant variance but not to correlated errors.

In the electricity usage example, $\text{RSS}_{\text{linear}} = 6833$, $\text{RSS}_{\text{quadratic}} = 4766$, and $\hat{\sigma}^2 = 91.66$. Therefore,

$$F = \frac{(6833 - 4766)/(2 - 1)}{91.66} = 22.55.$$

This gives a p-value of 0.000016, which is strong evidence that the straight line model does not fit well compared to the quadratic one.

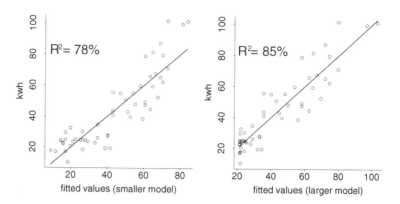

2.4.7.1 R^2 Values and F-Tests

An alternative (and perhaps more intuitive) way to think about an F-test is in terms of the increase in the R^2 *value* of the fit. Assuming the model contains an intercept, the R^2 value is given by

$$R^2 = \text{square of correlation coefficient between } \mathbf{y} \text{ and } \hat{\mathbf{y}}.$$

The idea is that a model with a "better" fit exhibits a stronger correlation between the data and fitted values, and thus it has a higher R^2 value. The caveat is that the R^2 value is *always* increased by adding more terms to a linear regression, so the pertinent issue is whether the addition of one or more terms leads to a *significant* increase in R^2. This is illustrated in Figure 2.16 for the linear and quadratic models for the electricity usage data described in Section 2.2.

As shown in these plots, the R^2 values are 78% for the smaller (linear) model and 85% for the larger (quadratic) model. Does this represent a significant enough increase in R^2 to adopt the latter model?

With a bit of algebra, one can show that the F-statistic is

$$F = \frac{R^2_{\text{larger}} - R^2_{\text{smaller}}}{(1 - R^2_{\text{smaller}})(p_{\text{larger}} - p_{\text{smaller}})/(n - p_{\text{larger}})},$$

which shows how the F-test can be thought of as deciding whether or not an increase in the R^2 value is significant. Because the F-statistic is highly significant in this case, we can conclude that the second plot in Figure 2.16 represents a significantly higher R^2 value.

Another interpretation of R^2 is as the proportion of the variation in the response that can be predicted or "explained" by the predictor variables. An R^2 value of 0.25 means that only 25% of the variability in the response is predictable using the regression model.

2.5 Parametric Additive Models

Perhaps the most widely used semiparametric regression model is the *additive model* (see e.g. Hastie and Tibshirani 1990). It assumes that the effect of the predictor variables on the response variable is *additive*, which essentially means that

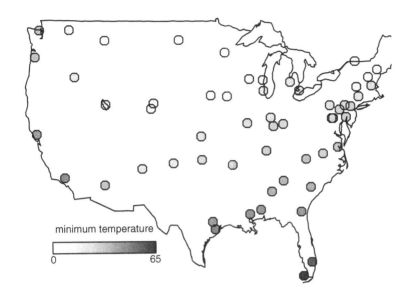

Figure 2.17 Location of 56 U.S. cities with points shaded with respect to min.temp. A map of the United States is superimposed.

minimum temperature

0   65

the predictor variables affect the response variable independently of one another. We will give a thorough treatment of additive models in Chapter 8. As we will see there, the term is usually taken to mean that the predictors are *nonparametric* functions. But many parametric regression models are also additive models, so it is expedient to introduce the notion of an additive model – and the mechanics of its fitting and interpretation – in this more familiar setting.

Figure 2.17 displays a "shaded" scatterplot of data corresponding to the average minimum January temperature of 56 cities in the United States and the geographical location of the cites (Peixoto 1990). We will refer to these data as the *U.S. temperature* data. In Peixoto (1990) it is demonstrated that an accurate prediction model for minimum temperature is

$$\text{min.temp}_i = \beta_0 + \beta_1 \text{lat}_i + \beta_{21} \text{lon}_i + \beta_{22} \text{lon}_i^2 + \beta_{23} \text{lon}_i^3 + \varepsilon_i, \quad (2.20)$$

where lat_i is the number of degrees latitude of the ith city and lon_i is the number of degrees longitude. We work with the negative version of degrees longitude so that the longitude axis matches geographical maps.

Equation (2.20) is an example of an additive model. It assumes that lat and lon act additively on the mean minimum temperature value. There is no *interaction* (sometimes called *effect modification*). This means that the effect of lat on min.temp is the same for all values of lon, and vice versa.

2.5.1   *Displaying Additive Fits*

Figure 2.18 shows the fits corresponding to (2.20). The main attraction of an additive model is that, assuming the model is correct, the two panels show the estimated effects in a simple, interpretable way. The additive model shows that lat has a negative linear effect on min.temp while lon has a striking nonlinear effect due to, for example, altitudinal variation. This result suggests that one

Figure 2.18
Components of fit
of (2.20) to the U.S.
temperature data.

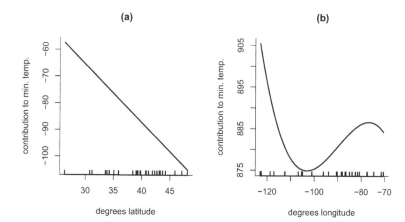

might add altitude as another predictor variable, though altitude is not included in the data set we are using.

Notice that Figure 2.18 also shows the values of the predictor variables through the use of a *rug plot* at the base of each panel. Apart from conveying the distribution of the predictor, this plot can be useful for identifying observations with high influence. However, there is still room for improvement. For example, the vertical axes are somewhat arbitrary and there is no indication of the amount of variability in the estimates.

The next few subsections describe a number of embellishments to Figure 2.18. For easier digestion we will use the following notation:

$$y_i = \text{min.temp}_i, \quad s_i = \text{lat}_i, \quad t_i = \text{lon}_i.$$

2.5.1.1 *Vertical Alignment*

The first aspect of Figure 2.18 that warrants attention is that the vertical positions of the curves are arbitrary. If the line in Figure 2.18(a) is moved up or down 10 units then the interpretation is unchanged: a 1-degree increase in latitude results in a 2.36-degree decrease in mean minimum temperature. The scales on the vertical axes are only meaningful in a *relative* sense; they have no *absolute* interpretation. Since we have the freedom to choose the vertical positionings, we should try to make them meaningful in the absolute sense. A reasonable solution

The *response surface* of a regression fit is the multivariate function of the predictors with all parameters replaced by their estimates.

is to plot, for each predictor, the profile of the response surface with each of the other predictors set at their average. For the current example this would involve plotting

$$\hat{\beta}_0 + \hat{\beta}_1 s + \hat{\beta}_{21}\bar{t} + \hat{\beta}_{22}\bar{t}^2 + \hat{\beta}_{23}\bar{t}^3 \quad \text{against } s$$

and

$$\hat{\beta}_0 + \hat{\beta}_1\bar{s} + \hat{\beta}_{21}t + \hat{\beta}_{22}t^2 + \hat{\beta}_{23}t^3 \quad \text{against } t.$$

Operationally, this can be achieved by setting up grids of size M,

$$\mathbf{g}_s \equiv [g_{s1}, \ldots, g_{sM}]^\mathsf{T} \quad \text{and} \quad \mathbf{g}_t \equiv [g_{t1}, \ldots, g_{tM}]^\mathsf{T},$$

in the s and t directions (respectively) and then defining the matrices

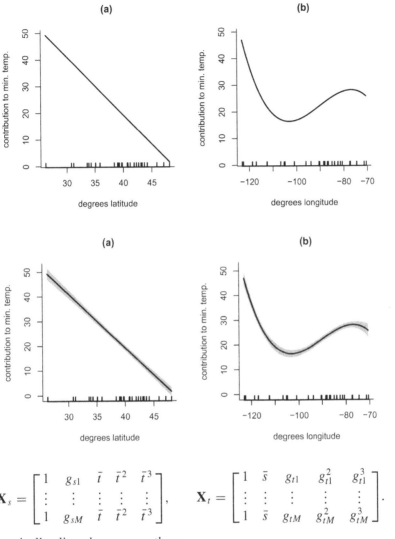

Figure 2.19
Components of parametric additive model fit to the U.S. temperature data with vertical alignment as described in the text.

Figure 2.20
Components of fit to the U.S. temperature data as effects about the mean, with approximate pointwise 95% confidence bands indicated by the shaded regions.

$$\mathbf{X}_s = \begin{bmatrix} 1 & g_{s1} & \bar{t} & \bar{t}^2 & \bar{t}^3 \\ \vdots & \vdots & \vdots & \vdots & \vdots \\ 1 & g_{sM} & \bar{t} & \bar{t}^2 & \bar{t}^3 \end{bmatrix}, \qquad \mathbf{X}_t = \begin{bmatrix} 1 & \bar{s} & g_{t1} & g_{t1}^2 & g_{t1}^3 \\ \vdots & \vdots & \vdots & \vdots & \vdots \\ 1 & \bar{s} & g_{tM} & g_{tM}^2 & g_{tM}^3 \end{bmatrix}.$$

The vertically aligned curves are then

$$\mathbf{X}_s \hat{\boldsymbol{\beta}} \text{ versus } \mathbf{g}_s \quad \text{and} \quad \mathbf{X}_t \hat{\boldsymbol{\beta}} \text{ versus } \mathbf{g}_t,$$

where $\hat{\boldsymbol{\beta}} = [\hat{\beta}_0\ \hat{\beta}_1\ \hat{\beta}_{21}\ \hat{\beta}_{22}\ \hat{\beta}_{22}]^\mathsf{T}$. These are shown in Figure 2.19. The vertical scales are the same and correspond to the minimum temperature values.

2.5.1.2   *Variability Bands*

Just as we argued in the single predictor case, it is useful to show some estimate of the variability in the estimated curves. Such an embellishment is shown in Figure 2.20. The shaded regions correspond to approximate 95% pointwise confidence intervals for mean minimum temperature for a given value of the predictor on the horizontal axis, and the contribution of the other predictor set at its average.

Figure 2.21
Components to the
U.S. temperature
data as effects about
the mean, with 95%
confidence bands and
partial residuals.

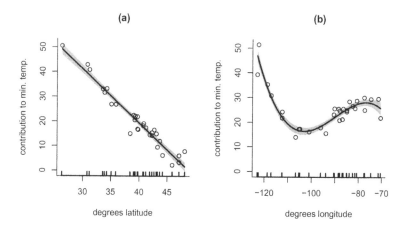

How are these variability bands computed? If, for a general random vector $\mathbf{v} = [v_1, \ldots, v_M]$, we define

$$\text{st.dev.}(\mathbf{v}) \equiv \left[\sqrt{\text{var}(v_1)}, \ldots, \sqrt{\text{var}(v_M)}\right]^{\mathsf{T}},$$

then it is straightforward to show that

$$\text{st.dev.}(\mathbf{X}_s\hat{\boldsymbol{\beta}}) = \sqrt{\text{diagonal}\{\mathbf{X}_s \, \text{Cov}(\hat{\boldsymbol{\beta}})\mathbf{X}_s^{\mathsf{T}}\}}$$

$$= \sigma\sqrt{\text{diagonal}\{\mathbf{X}_s(\mathbf{X}^{\mathsf{T}}\mathbf{X})^{-1}\mathbf{X}_s^{\mathsf{T}}\}}.$$

The vertical boundaries of the band in Figure 2.20(b) are then

$$\mathbf{X}_s\hat{\boldsymbol{\beta}} \pm 2 \times \widehat{\text{st.dev.}}(\mathbf{X}_s\hat{\boldsymbol{\beta}}),$$

where $\widehat{\text{st.dev.}}(\mathbf{X}_s\hat{\boldsymbol{\beta}})$ involves replacement of σ by its usual estimate (Section 2.4.6). As usual, the multiplicative factor of 2 is an approximation to the 0.975 quantile of the normal and t-distributions, and it leads to approximate 95% pointwise confidence.

2.5.1.3 *Partial Residuals*
The ith residual from the fit to the U.S. temperature data is

$$e_i = y_i - \hat{\beta}_1 s_i - \hat{\beta}_{21} t_i - \hat{\beta}_{22} t_i^2 - \hat{\beta}_{23} t_i^3.$$

If the model provides an adequate fit to the data then the e_i should behave like a random sample with zero mean (Section 2.3). If the e_i are added to each of the curves in Figure 2.20 then, for an adequate fit, the residuals should be randomly scattered about the curve. The result is shown in Figure 2.21. In this case the scatter about the curves does not show any systematic patterns, so we have a positive diagnosis of model adequacy.

In general, the sum of the componentwise fitted values and residuals are called *partial residuals* and provide a useful diagnostic for adequacy of an additive model fit.

2.5.1.4 *Derivative Plots*

In the situation where both predictors enter the model linearly, that is, where the response surface is

$$\hat{\beta}_0 + \hat{\beta}_1 s + \hat{\beta}_2 t, \tag{2.21}$$

it is usual to report the coefficient estimates $\hat{\beta}_1$ and $\hat{\beta}_2$ and their estimated standard deviations (see e.g. Table 2.1). This is because they correspond to the *rates of change* or *slopes* of the mean response for each of the predictors. Another way of expressing this is in terms of partial derivatives: $\hat{\beta}_1$ and $\hat{\beta}_2$ are the partial derivatives of (2.21) with regard to s and t, respectively.

Partial derivatives are obtained via differentiation with respect to one variable while the others are held fixed.

In the U.S. temperature example the second component is not linear, but we can extend the notion of predictor slope by using partial differentiation. From

$$\hat{y}(s, t) \equiv \hat{\beta}_0 + \hat{\beta}_1 s + \hat{\beta}_{21} t + \hat{\beta}_{22} t^2 + \hat{\beta}_{23} t^3$$

it follows that

$$\frac{\partial}{\partial s} \hat{y}(s, t) = \hat{\beta}_1,$$

$$\frac{\partial}{\partial t} \hat{y}(s, t) = \hat{\beta}_{21} + 2\hat{\beta}_{22} t + 3\hat{\beta}_{23} t^2.$$

Curve estimates and variability bands can be computed in the same fashion as for Figure 2.20 by replacing the columns of $\mathbf{X}_s$ and $\mathbf{X}_t$ by their respective partial derivatives,

$$\mathbf{X}'_s = \begin{bmatrix} 0 & 1 & 0 & 0 & 0 \\ \vdots & \vdots & \vdots & \vdots & \vdots \\ 0 & 1 & 0 & 0 & 0 \end{bmatrix}, \qquad \mathbf{X}'_t = \begin{bmatrix} 0 & 0 & 1 & 2g_{t1} & 3g_{t1}^2 \\ \vdots & \vdots & \vdots & \vdots & \vdots \\ 0 & 0 & 1 & 2g_{tM} & 3g_{tM}^2 \end{bmatrix},$$

and then plotting

$$\mathbf{X}'_s \hat{\boldsymbol{\beta}} \text{ versus } \mathbf{g}_s \quad \text{and} \quad \mathbf{X}'_t \hat{\boldsymbol{\beta}} \text{ versus } \mathbf{g}_t.$$

Variability bands can be obtained by adding plus and minus twice the estimated standard deviations:

$$\widehat{\text{st.dev.}}(\mathbf{X}'_s \hat{\boldsymbol{\beta}}) = \hat{\sigma} \sqrt{\text{diagonal}\{\mathbf{X}'_s (\mathbf{X}^\mathsf{T} \mathbf{X})^{-1} (\mathbf{X}'_s)^\mathsf{T}\}},$$

$$\widehat{\text{st.dev.}}(\mathbf{X}'_t \hat{\boldsymbol{\beta}}) = \hat{\sigma} \sqrt{\text{diagonal}\{\mathbf{X}'_t (\mathbf{X}^\mathsf{T} \mathbf{X})^{-1} (\mathbf{X}'_t)^\mathsf{T}\}}.$$

Figure 2.22 is the resulting graphic for the U.S. temperature data. The left-hand panel is none other than a graphical representation of the regression coefficient for latitude, along with its approximate 95% confidence interval. The right-hand side shows that the slope for longitude changes depending on location. For example, for a given latitude, the longitudinal effect is significantly negative on the U.S. west coast but is significantly positive at around -90 degrees longitude.

A summary of the fitted model that relates the additive fits to the geography is shown as Figure 2.23.

Figure 2.22
Derivatives of
components of
fit to the U.S.
temperature data,
with approximate
pointwise 95%
confidence bands.

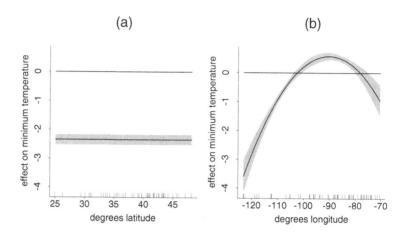

Figure 2.23
Summary of fitted
additive model in
relation to the United
States map.

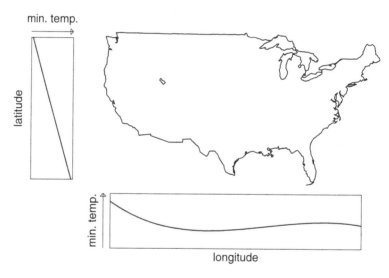

2.5.2 *Degrees of Freedom*

The fit to the U.S. temperature data involves five parameters. Alternatively, we can say that the fit uses 5 *degrees of freedom*. The latter terminology has the advantage that it generalizes more naturally to semiparametric regression fits, so we will mainly use this alternative in the remainder of the book.

In general parametric regression models such as

$$\mathbf{y} = \mathbf{X}\boldsymbol{\beta} + \boldsymbol{\varepsilon},$$

the number of degrees of freedom is simply p, the length of the $\boldsymbol{\beta}$-vector or (equivalently) the number of columns in the $\mathbf{X}$-matrix. However, these are not ideal definitions of "degrees of freedom" because they do not generalize to semiparametric models.

Recall from Section 2.3 that the "hat matrix" $\mathbf{H} = \mathbf{X}(\mathbf{X}^{\mathsf{T}}\mathbf{X})^{-1}\mathbf{X}^{\mathsf{T}}$ is such that

$$\hat{\mathbf{y}} = \mathbf{H}\mathbf{y},$$

Table 2.3
Breakdown of degrees
of freedom (df) for
fit to U.S. temperature
data.

Term	df
intercept	1
latitude	1
longitude	3
Fit	5

so that H_{ii} is the multiplying factor for the contribution of y_i to its corresponding fitted value. The total of these factors is

$$\sum_{i=1}^{n} H_{ii} = \mathrm{tr}(\mathbf{H}) = \mathrm{tr}\{\mathbf{X}(\mathbf{X}^\mathsf{T}\mathbf{X})^{-1}\mathbf{X}^\mathsf{T}\}$$
$$= \mathrm{tr}\{\mathbf{X}^\mathsf{T}\mathbf{X}(\mathbf{X}^\mathsf{T}\mathbf{X})^{-1}\} = \mathrm{tr}(\mathbf{I}_p) = p,$$

so this also coincides with the number of parameters being fit. Moreover, its derivation can be applied to any regression fit of the form $\hat{\mathbf{y}} = \mathbf{L}\mathbf{y}$, where $\mathbf{L}$ is an $n \times n$ matrix. This flexibility will prove to be useful in the extension to semiparametric models.

Here is an intuitive interpretation of $\sum_{i=1}^{n} H_{ii}$. First note that H_{ii} is the coefficient of y_i in equation (2.12) for $\hat{y}_i$. A large value of H_{ii} indicates that the ith data point has a high influence on its own fitted value. For this reason, H_{ii} is called the *leverage* of the ith data point. Thus, the degrees of freedom p is the total influence of all the observations – the greater the number of parameters, the greater the aggregate influence of the data.

As mentioned earlier, the fit to the U.S. temperature data uses 5 degrees of freedom. But it is also useful to break this down according to the contribution from each term in the model, as shown in Table 2.3. The degrees-of-freedom values correspond to the number of parameters used for each component of the additive model. But we will now describe an alternative definition that will be useful when we extend additive models to nonparametric fits. Note that the vector of fitted values is

$$\hat{\mathbf{y}} = \hat{\beta}_0 \mathbf{1} + \hat{\beta}_{11}\mathbf{s} + \hat{\beta}_{21}\mathbf{t} + \hat{\beta}_{22}\mathbf{t}^2 + \hat{\beta}_{23}\mathbf{t}^3$$
$$= \mathbf{H}_0 \mathbf{y} + \mathbf{H}_1 \mathbf{y} + \mathbf{H}_2 \mathbf{y}, \tag{2.22}$$

where, for example,

$$\mathbf{H}_1 = \mathbf{s}\mathbf{e}_2^\mathsf{T}(\mathbf{X}^\mathsf{T}\mathbf{X})^{-1}\mathbf{X}^\mathsf{T}, \qquad \mathbf{X} = [\mathbf{1}\ \mathbf{s}\ \mathbf{t}\ \mathbf{t}^2\ \mathbf{t}^3],$$

and $\mathbf{e}_2 = [0\ 1\ 0\ 0\ 0]^\mathsf{T}$.

Specifically, let $\mathbf{E}_0 = \mathrm{diag}(1, 0, 0, 0, 0)$, $\mathbf{E}_1 = \mathrm{diag}(0, 1, 0, 0, 0)$, and $\mathbf{E}_2 = \mathrm{diag}(0, 0, 1, 1, 1)$. Then $\mathbf{E}_0 + \mathbf{E}_1 + \mathbf{E}_2 = \mathbf{I}$ and $\mathbf{H}_i = \mathbf{X}\mathbf{E}_i(\mathbf{X}^\mathsf{T}\mathbf{X})^{-1}\mathbf{X}^\mathsf{T}$. Using the identity $\mathrm{tr}(\mathbf{AB}) = \mathrm{tr}(\mathbf{BA})$, it follows that $\mathrm{tr}(\mathbf{H}_i) = \mathrm{tr}(\mathbf{E}_i)$ and so

$$\mathrm{tr}(\mathbf{H}_0) = \mathrm{tr}(\mathbf{H}_1) = 1 \quad \text{and} \quad \mathrm{tr}(\mathbf{H}_2) = 3.$$

This suggests adoption of the degrees of freedom being the trace of the *componentwise* hat matrices for additive models in general.

2.6   Model Selection

In applied multiple regression there are usually several possible models for the data set at hand. *Model selection* concerns the problem of choosing among them. Linhart and Zucchini (1986) provide a summary of the many approaches to this problem.

In a nutshell, good model selection methods aim to achieve a balance between *goodness of fit* and *parsimony*. Better fits to the data can always be achieved by adding more parameters. But parsimonious models (i.e., those with few parameters) are attractive because of their simplicity and interpretability; they are also less subject to estimation variability and so can yield more accurate predictions.

One approach to model selection is hypothesis testing. For example, if the models can be ordered by complexity – linear, quadratic, and cubic polynomial regression models, say – then we can test the first model versus the second. If the first model is rejected, then we test the second versus the third. One continues in this fashion until one finds a null model that is not rejected. Hypothesis testing, however, is not entirely satisfactory as a model selection method. The asymmetry between the null and alternative hypothesis is problematic in the model selection context. Hypothesis testing is "biased" toward the null hypothesis in that we retain the null unless there is strong evidence to the contrary. When we want to select between two models, we often want to treat them on an equal footing rather than accepting one unless there is strong evidence against it. Moreover, the error probabilities of a sequence of tests are difficult to determine. For example, suppose we test linear versus quadratic regression, then quadratic versus cubic, then cubic versus quartic, and so forth – until we accept the null hypothesis of the smaller of the two models being tested. Since each individual test statistic has an null F-distribution, we could keep the type I error rate fixed at (say) 0.05 for each individual test; but determining the probability that a correct model is selected by the entire sequence of tests is an unsolved problem.

A more satisfactory approach to model selection is to estimate the predictive ability of the various models and then select the model with the best estimated predictive power. This method of model selection has the advantage that it extends easily to semiparametric modeling.

The residual sum of squares (RSS) of a model is a measure of predictive ability, since a residual is the difference between an observation of a response and its fitted or predicted value:

$$e_i = y_i - \hat{y}_i.$$

However, RSS is not satisfactory as a model selector. The problem is that $\hat{y}_i$ uses y_i as well as the other observations to predict y_i. The result is that the most complex model – that is, the model with the largest degrees of freedom and containing the other models as special cases – always has the smallest RSS.

There is a simple remedy to this problem: when predicting y_i, use all the observations except the ith one. Thus, define $\hat{y}_{(i)}$ to be the predicted value for the ith case when the ith case is not used to estimate $\boldsymbol{\beta}$. Then, let $e_{(i)} = y_i - \hat{y}_{(i)}$

be the ith *deleted residual*. The *predicted residual sum of squares* (PRESS) is defined by

$$\text{PRESS} = \sum_{i=1}^{n} e_{(i)}^2.$$

PRESS is sometimes called the *cross-validation* statistic. Cross-validation is the technique of model validation that splits the data into two disjoint sets, fits the model to one set, predicts the data in the second set using only the fit to the first set, and then compares these predictions to the actual observations in the second set. This process can be repeated using different partitionings of the data into two sets. PRESS uses *leave-one-out* cross-validation, where in each partition the second set contains a single observation and the first set contains the remaining observations.

The cross-validation statistic is not as difficult to calculate as it might first appear. One does *not* need to fit the model n times, thanks to an important identity. Let H_{ii} be the ith diagonal element of the hat matrix $\mathbf{H}$. Then, the ith deleted residual is related to the ith ordinary residual by

$$e_{(i)} = \frac{e_i}{1 - H_{ii}}.$$

Cross-validation selects the model with the smallest value of the cross-validation statistic or (equivalently) of the PRESS statistic.

A popular model selection criterion related to cross-validation is the Mallows C_p statistic (Mallows 1973). It can be written in various forms, but one is

$$C_p = \text{RSS}(p) + 2\hat{\sigma}^2 p, \tag{2.23}$$

where p is the number of terms in the candidate model, $\text{RSS}(p)$ is the residual sum of squares for the same candidate model, and

$$\hat{\sigma}^2 = \frac{1}{n - p_{\text{largest}}} \text{RSS}(p_{\text{largest}})$$

is the estimate of the residual variance based on the largest model being considered. The idea is to choose the model with the smallest C_p value. The justification for the C_p expression (2.23) is given in Section 5.3.3.

Consider once again the electricity usage data with kwh as the response variable. We saw that the quadratic model improves upon the straight line model. Might a cubic polynomial model offer further improvement? The PRESS statistics for the linear, quadratic, and cubic regression models are 7414, 5248, and 5356, respectively. Therefore, cross-validation selects the quadratic model. The linear polynomial model suffers from large bias and thus has a large PRESS value. The cubic model includes the quadratic model as a special case, so bias is not a problem for the cubic model. However, compared to the quadratic model, the cubic model must estimate an additional parameter. The result is that cubic model predictions have slightly more random variation and are therefore somewhat less accurate compared to predictions from the quadratic model.

Figure 2.24 Example
of quadratic regression
improving on linear
regression: (a) linear
fit to Janka hardness
data, (b) residual plot;
(c) quadratic fit to
Janka hardness data,
(d) residual plot.

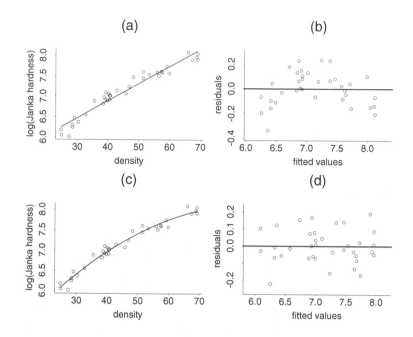

2.7   Polynomial Regression Models

Most of the remainder of this book is concerned with the situation where linearity between mean response and predictor cannot be reasonably assumed. Perhaps the most common means of dealing with nonlinearity in regression is to use higher-degree polynomials. A pth-degree regression model for the scatterplot (x_i, y_i), $1 \le i \le n$, is

$$y_i = \beta_0 + \beta_1 x_i + \cdots + \beta_p x_i^p + \varepsilon_i.$$

Figure 2.24 illustrates polynomial regression via some forestry data. The response is the logarithm of *Janka hardness,* a structural property of timber that is difficult to measure. The predictor is the timber's density. In Figures 2.24(a) and (b), the linear model is seen to be inadequate. However, the *quadratic* regression model ($p = 2$) fits the data quite well, as shown in panels (c) and (d).

> The Janka hardness data are from *Regression Analysis* by E. J. Williams (1959).

We now turn to an example in which polynomial regression does not seem to help. The technique known as LIDAR (light detection and ranging) uses the reflection of laser-emitted light to detect chemical compounds in the atmosphere. The LIDAR technique has proven to be an efficient tool for monitoring the distribution of several atmospheric pollutants of importance; see Sigrist (1994).

> The LIDAR example was taken from Holst et al. (1996).

A typical LIDAR data set is shown in Figure 2.25. The horizontal variable, range, is the distance traveled before the light is reflected back to its source. The vertical variable, logratio, is the logarithm of the ratio of received light from two laser sources. One source had a frequency equal to the resonance frequency of the compound of interest, which was mercury in this study. The other source had a frequency off this resonance frequency. For details see Ruppert et al. (1997).

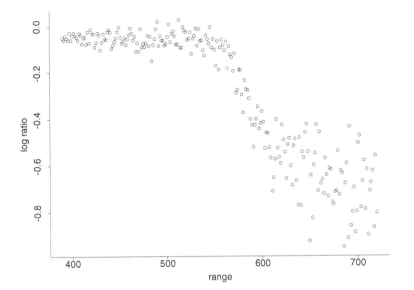

Figure 2.25
Scatterplot of LIDAR
data.

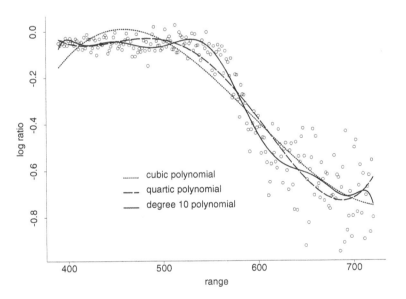

Figure 2.26 Higher-
degree polynomial fits
to the LIDAR data.

Important features of this scatterplot are the nonlinear pattern and the evidence
of nonconstant variance (heteroscedasticity) – there is more vertical scatter on
the right compared to the left. Apparently, `logratio` is more variable at larger
values of `range`. In nonparametric regression we estimate

$$f(\text{range}) \equiv \text{E}(\text{logratio}|\text{range}).$$

A noteworthy feature of this example is that there is scientific interest in the first
derivative of f as well as in f itself.

The mean response is neither linear nor quadratic in the predictor. Can higher-
degree polynomial models provide an adequate fit to these data? Figure 2.26

shows fits of degree 3, 4, and 10 to the LIDAR data. The lower-degree fits do not go through the data very well. In particular, the cubic fit turns downward at the left boundary, though the data do not. There is a similar problem for the quartic fit at the right boundary. Neither the cubic nor the quartic fits can follow the sudden downturn in the data around range equal to 550–580. The degree-10 fit goes through the data reasonably well but has wiggles that are representative not of any features in the data but rather of high-degree polynomials generally. These wiggles are a serious problem in this example, where the derivative of the regression function is the focus of study. Small wiggles perturb the derivative of the fitted function far more than they perturb the fitted function itself. The derivative is proportional to the negative concentration and should therefore be nonpositive. Disturbingly, the derivatives of the polynomial fits in Figure 2.26 can be quite large and positive. This is not true of the nonparametric fits in Chapter 6.

We might use high-degree polynomial models if nothing better were available, but fortunately much better fitting methods are available. This is the focus of Chapter 3.

2.8   Nonlinear Regression

All of the parametric models that have been discussed so far are *linear,* meaning that the model is linear in the parameters. For example, a polynomial model is linear in the coefficients even though it is a nonlinear function of the predictor variable, so polynomial regression is a linear model. In many subject matter disciplines there are theoretical models relating the predictors and the response, and often these models are nonlinear in their parameters. Such nonlinear theoretical models lead us to nonlinear regression. The nonlinear regression model is

$$y_i = f(\mathbf{z}_i; \boldsymbol{\beta}) + \varepsilon_i,$$

where $\mathbf{z}_i$ is a vector of predictors, $\boldsymbol{\beta}$ is a vector of unknown regression coefficients, and $f(\mathbf{z}; \boldsymbol{\beta})$ is a known function.

For example, the term structure model in Section 1.5,

$$P(0) = P(T) \exp\left(-\int_0^T f(x)\,dx\right),$$

leads to a nonlinear regression model if we have a parametric model for the forward rate f. In our AT&T bond example, the response is the time-0 price expressed as a percentage of par:

$$y_i = 100\frac{P(0)}{P(T_i)},$$

where T_i is the maturity of the ith bond and $P(T_i)$ is its par value. Suppose that we model f as linear:

$$f(t) = \beta_0 + \beta_1 t.$$

Then the regression model is

$$y_i = 100 \exp(-\beta_0 T_i - \beta_1 T_i^2/2) + \varepsilon_i.$$

Here ε_i represents noise in the price $P(0)$, mostly due to prices being some-what "stale" (that is, based on the last trade). Also, $\mathbf{z}_i = T_i$ and $f(T; \boldsymbol{\beta}) = 100 \exp(-\beta_0 T - \beta_1 T^2)$.

Nonlinear regression models are usually estimated by least squares, so that $\hat{\boldsymbol{\beta}}$ minimizes

$$\sum_{i=1}^{n} \{y_i - f(\mathbf{z}_i; \boldsymbol{\beta})\}^2.$$

The least-squares estimate cannot be found explicitly as for linear regression, but there are iterative algorithms such as the Gauss–Newton and Levenberg–Marquardt algorithms that can be used. These are readily available in, for exam-ple, PROC NLIN of SAS, nls of S-Plus, and lsqnonlin of MATLAB.

Define

$$\mathbf{Z} = \begin{bmatrix} \mathbf{z}_1^{\mathsf{T}} \\ \vdots \\ \mathbf{z}_n^{\mathsf{T}} \end{bmatrix} \quad \text{and} \quad \mathbf{f}(\mathbf{Z}; \boldsymbol{\beta}) = \begin{bmatrix} f(z_1; \boldsymbol{\beta}) \\ \vdots \\ f(z_n; \boldsymbol{\beta}) \end{bmatrix}.$$

The analog of the matrix $\mathbf{X}$ in linear regression is

$$\mathbf{X} = \frac{\partial}{\partial \boldsymbol{\beta}} f(\mathbf{Z}; \boldsymbol{\beta}),$$

so that the (i, j)th element of $\mathbf{X}$ is

$$X_{ij} = \frac{\partial}{\partial \beta_j} f(\mathbf{z}_i; \hat{\boldsymbol{\beta}}).$$

There is no exact formula for the covariance matrix of $\hat{\boldsymbol{\beta}}$, but for large samples we have the approximation

$$\text{Cov}(\hat{\boldsymbol{\beta}}) \simeq \hat{\sigma}^2 (\mathbf{X}^{\mathsf{T}} \mathbf{X})^{-1}.$$

In the term structure example,

$$[X_{i1} \ X_{i2}] = -[100T_i \ 50T_i^2] \exp(-\hat{\beta}_0 T_i - \hat{\beta}_1 T_i^2/2) \qquad (2.24)$$

and

$$\mathbf{X}^{\mathsf{T}} \mathbf{X} = \sum_{i=1}^{n} \exp(-2\hat{\beta}_0 T_i - \hat{\beta}_1 T_i^2) \begin{bmatrix} (100)^2 T_i^2 & (50)(100) T_i^3 \\ (50)(100) T_i^3 & (50)^2 T_i^4 \end{bmatrix}.$$

Using (2.8), inference for $\boldsymbol{\beta}$, and confidence intervals for the mean response, prediction intervals for new observations can be obtained using the method intro-duced in Section 2.4 for linear regression.

We fit the linear forward rate curve model to the U.S. STRIPS data. We also tried the models where the forward curve was constant and where it was qua-dratic. The estimated forward curves are shown in Figure 2.27.

Residuals can be defined for a nonlinear regression model in the obvious way:

$$e_i = y_i - f(\mathbf{z}_i; \hat{\boldsymbol{\beta}}).$$

Figure 2.27 Term structure example: estimated forward rate curves assuming a constant, linear, and quadratic model.

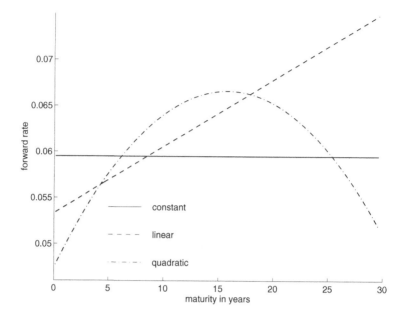

Figure 2.28 Term structure example: residuals from constant, linear, and quadratic model.

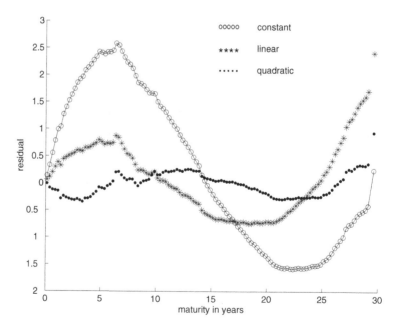

Residuals from the constant, linear, and quadratic models are plotted in Figure 2.28. For the constant curve model, the residuals are shown as circles connected by solid lines. For the other two models, only lines are used. The residuals do not have the "random" appearance one expects from a nearly independent independent series but instead are a nearly smooth curve. This is especially true of the constant forward rate model and, for T large, the linear forward rate model – these show poor fit. There is a problem either with the polynomial models or with the assumption that the errors are independent, or (more likely) with both. We will return to this example in Chapter 18. There we will see that a nonparametric model fits better than the parametric polynomial models considered here, but the residuals do not appear independent even with a more sophisticated model for the forward rate. We were puzzled by the observation with the largest maturity, which appears as an outlier owing to an extremely large residual in all three models. Eventually, while this book was in production, we learned that this observation is a contaminant and in future work we will delete it. This STRIPS did not trade recently and its price was imputed, not determined by an actual trade as was the case for the other STRIPS in the data set.

We do not have space to discuss nonlinear regression in much detail, but more information is available via the references given in Section 2.10.

2.9 Transformations in Regression

As we discussed in Section 2.6, simple models, if they fit the data well, generally predict with more accuracy than complex models. They are also in keeping with the statistician's role of summarizing data in a form that is easy to explain and interpret. If a simple model does not fit the data very well, it is often possible to *transform* some of the variables so that the model fits the transformed data. Transforming a variable means replacing the variable by some function of that variable – for example, replacing kwh by its logarithm logkwh, as was done in Section 2.1 so that a straight line model was appropriate.

As we mentioned in Section 2.2.1, several assumptions are made in linear regression:

(1) $E(y|\mathbf{x}) = \mathbf{x}^T\boldsymbol{\beta}$;
(2) $\text{var}(y|\mathbf{x})$ is constant;
(3) $\varepsilon = y - E(y|\mathbf{x})$ is normally distributed; and
(4) ε_i and ε_j are uncorrelated if $i \neq j$.

Transformations can be used to meet one or more of assumptions (1)–(3). Unfortunately, data transformation cannot help if the data are correlated. Atkinson (1985) and Carroll and Ruppert (1988) are good sources for information about data transformation. These references discuss advanced methods for selecting a data transformation. Here we will consider only the simplest method: trial and error.

In regression we can transform either the response, the predictor variables, or both. Transforming either the predictor variables or the response can help meet assumption (1), but only transformation of the response affects the distribution of the response and can help meet assumptions (2)–(3).

The most commonly used transformations are the square root, log, and reciprocal transformations, although other power transformations are sometimes used. Consider a power transformation of a variable u, which replaces u by

$$g_\lambda(u) \equiv u^\lambda, \quad \lambda \neq 0. \tag{2.25}$$

The "strength" of a transformation depends on how much its first derivative differs from a constant function. Since

$$\frac{\partial g_\lambda(u)}{\partial u} = \text{constant} \times u^{\lambda-1}, \quad \lambda \neq 0,$$

and

$$\frac{\partial \log(u)}{\partial u} = u^{-1},$$

it is natural to consider the log transformation to correspond to $\lambda = 0$; hence we define

$$g_0(u) = \log(u). \tag{2.26}$$

However, definitions (2.25) and (2.26) have the drawback that

$$g_\lambda(u) \not\to g_0(u) \text{ as } \lambda \to 0.$$

This makes choice of λ via, say, maximum likelihood messy. It is thus more common to work with the *Box–Cox* (Box and Cox 1964) family of transformations:

$$g_\lambda(u) = \begin{cases} (u^\lambda - 1)/\lambda, & \lambda \neq 0, \\ \log(u), & \lambda = 0. \end{cases}$$

Observe that, with this definition, $g_\lambda(u)$ is continuous and in fact infinitely differentiable at $\lambda = 0$.

We have seen that the electricity usage data can be fit adequately by either (a) a straight line model to logkwh or (b) a quadratic polynomial model to kwh. Model (a) might be considered simpler than (b), though we feel that they are of comparable complexity since (a) involves a choice of transformation, which is similar to using an additional regression parameter. However, there is one good reason for using model (a) instead of (b). Assumption (2) concerning constant variance is much closer to being fulfilled when the log transformation in (a) is used. We can see this by using residual plots. In these plots, we use the studentized residuals as defined in Section 2.3.

When checking for a constant variance, it is better to plot absolute studentized residuals rather than the studentized residuals themselves. Using absolute studentized residuals increases the data density by combining the negative and positive residuals. Also, a scatterplot smooth to the absolute residuals can be of considerable help when deciding whether the variation in y is constant or not. In contrast, a scatterplot smooth of the residuals can reveal lack of fit but tells us nothing about possible heteroscedasticity.

Figure 2.29 is a plot of the absolute studentized residuals from the quadratic fit of kwh to temp. Superimposed is a scatterplot smoother. One can see the general pattern that at low temperatures, where average kwh is highest, variability in kwh is also highest.

The scatterplot smoother dips down at the far left in Figure 2.29. This seems not to be a "real" effect but due rather to random variation caused by data sparsity

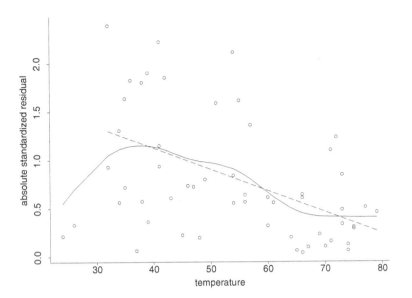

Figure 2.29 Absolute residuals from the quadratic polynomial fit of kwh to temp. The solid curve is a smooth to all the data and the dashed line is a smooth to the data with the two leftmost observations removed.

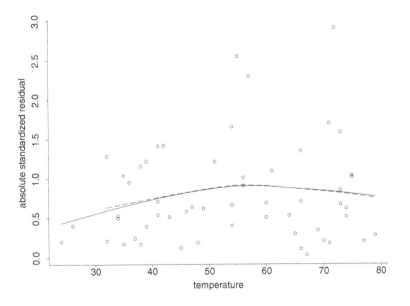

Figure 2.30 Absolute residuals from the fit of logkwh to temp. The solid curve is a smooth to all the data and the dashed line is a smooth to the data with the two leftmost observations removed.

in that region. For comparison, we have included a scatterplot smooth that does not use the two leftmost data points. Thus, the quadratic model fits kwh well, but the residual variation is heteroscedastic.

Figure 2.30 is a similar plot of the absolute studentized residuals from the straight line fit to logkwh. One can see that the residual variation is nearly constant, especially if one ignores the two data points at the extreme left. When a response transformation induces a nearly constant variance, one says that it *stabilizes* the variance and calls it the *variance-stabilizing* transformation.

Figure 2.31
Scatterplots of all
nine combinations
of the identity,
square root, and
log transformations
($\lambda = 1$, 1/2, and 0)
on range and on
logratio for the
LIDAR data.

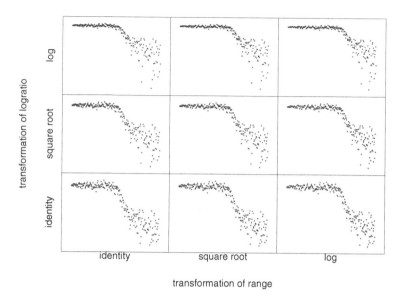

Despite the many examples that show the usefulness of data transformation, there is a limit to what transformations can accomplish. The next example illustrates this unfortunate fact.

2.9.1 *LIDAR Monitoring of Air Pollutants*

The LIDAR data set is an example where transformations seem to be unhelpful. We tried all nine combinations of the identity, square root, and log transformations ($\lambda = 1$, 1/2, and 0) on range and on logratio. Since logratio can assume negative values, a constant was first added to logratio to ensure that its smallest value was 0.1 before transformation. The resulting scatterplots are shown in Figure 2.31.

None of the nine combinations of transformations could remove the severe heteroscedasticity or induce a relationship between range and logratio that appeared linear or even parabolic.

Usually a power transformation can straighten only a convex or concave $x-y$ relationship and not a more complex relationship, such as between range and logratio in the LIDAR example. The next chapter is concerned with remedies to this shortcoming of parametric regression models and transformations.

As explained, for example, in Carroll and Ruppert (1988), transformations can stabilize a response variance only when that variance is a function of the response mean. In the LIDAR example, the response variance is a function of range but not of the response mean. This fact can be appreciated by noticing that, as range increases from 400 to 550, the mean response is nearly constant but the variability of the response increases noticeably. Thus, it should come as no surprise that transformations cannot stabilize the variance in the LIDAR example.

Transformations are wonderful tools, but they are not a panacea. Not every example can be transformed into fitting a standard parametric model. For this reason, it is essential to include semiparametric models in one's statistical toolkit.

2.10 Bibliographic Notes

Several general textbooks on parametric regression were mentioned in the introduction to this chapter. Here we will mention some more specialized textbooks.

Good books on regression diagnostics include Belsley, Kuh, and Welsch (1980), Cook and Weisberg (1982), and Chatterjee and Hadi (1988). Diagnostics for smoothing splines are discussed in Eubank (1984, 1985) and Eubank and Gunst (1986) – see also Eubank (1988). Diagnostics are useful for finding outliers but do not address the question of what to do about them. The robust regression methods discussed in Rousseeuw and Leroy (1987) are designed to provide stable and reliable estimates in the presence of possible outliers. Atkinson (1985) is an excellent textbook of transformations in regression. We have only briefly discussed parametric nonlinear regression, where the conditional expectation has a known parametric form that is nonlinear in the parameters, but that topic is treated in depth in Bates and Watts (1988) and Seber and Wild (1989).

Ryan's (1997) textbook covers many of the newer topics in regression analysis, including diagnostics, transformations, robust estimation, smoothing, and non-linear regression. Also covered by Ryan are ridge regression, which is related to the penalized least-squares estimators used in this book, and logistic regression, which is a special case of generalized regression that we discuss in Chapter 10.

2.11 Summary of Formulas

General linear model

$$\mathbf{y} = \mathbf{X}\boldsymbol{\beta} + \boldsymbol{\varepsilon}, \quad \mathrm{E}(\boldsymbol{\varepsilon}) = \mathbf{0}$$

Homoscedasticity and noncorrelation assumptions

$$\mathrm{Cov}(\boldsymbol{\varepsilon}) = \sigma^2 \mathbf{I}$$

Normality assumption

$$\boldsymbol{\varepsilon} \sim \mathrm{N}(\mathbf{0}, \sigma^2 \mathbf{I})$$

Least-squares estimates

$$\hat{\boldsymbol{\beta}} = (\mathbf{X}^\mathsf{T}\mathbf{X})^{-1}\mathbf{X}^\mathsf{T}\mathbf{y}$$

Fitted values and hat matrix

$$\hat{\mathbf{y}} = \mathbf{H}\mathbf{y}, \quad \mathbf{H} = \mathbf{X}(\mathbf{X}^\mathsf{T}\mathbf{X})^{-1}\mathbf{X}^\mathsf{T}$$

Residuals

$$e_i = y_i - \hat{y}_i, \quad 1 \le i \le n$$

$$\mathbf{e} = (\mathbf{I} - \mathbf{H})\mathbf{y}$$

Residual sum of squares

$$\text{RSS} = \sum_{i=1}^{n} e_i^2 = \|\mathbf{y} - \hat{\mathbf{y}}\|^2 = \mathbf{y}^{\mathsf{T}}(\mathbf{I} - \mathbf{H})\mathbf{y}$$

Variability estimation

$$\text{Cov}(\hat{\boldsymbol{\beta}}) = \sigma^2 (\mathbf{X}^{\mathsf{T}}\mathbf{X})^{-1}$$

$$\text{st.dev.}(\hat{\boldsymbol{\beta}}_i) = \sigma \sqrt{i\text{th diagonal entry of } (\mathbf{X}^{\mathsf{T}}\mathbf{X})^{-1}}$$

$$\hat{\sigma}^2 = \|\mathbf{y} - \hat{\mathbf{y}}\|^2/(n - p), \quad p = \text{number of terms}$$

Confidence intervals

Under a normality assumption, a $100(1 - \alpha)\%$ confidence interval for β_i is

$$\hat{\beta}_i \pm \widehat{\text{st.dev.}}(\hat{\beta}_i) t\left(1 - \tfrac{\alpha}{2}; n - p\right)$$

where p is the number of terms in the model and $t(q, m)$ is the $100q$th percentile of the t distribution with m degrees of freedom.

Under a normality assumption, a $100(1 - \alpha)\%$ confidence interval for $\text{E}(y)$ at x_0 is

$$\hat{y}_0 \pm \widehat{\text{st.dev.}}(\hat{y}_0) t\left(1 - \tfrac{\alpha}{2}; n - p\right)$$

Prediction interval for new observation

A $100(1 - \alpha)\%$ prediction interval for y at x_0 is

$$\hat{y}_0 \pm \widehat{\text{st.dev.}}(y - \hat{y}_0) t\left(1 - \tfrac{\alpha}{2}; n - p\right)$$

R^2 value

$$R^2 = \text{square of correlation coefficient between } \mathbf{y} \text{ and } \hat{\mathbf{y}}$$

F-test

$$F = \frac{R_{\text{larger}}^2 - R_{\text{smaller}}^2}{(1 - R_{\text{smaller}}^2)(p_{\text{larger}} - p_{\text{smaller}})/(n - p_{\text{larger}})}$$

Under the smaller model and normality assumptions, F has an F-distribution with $p_{\text{larger}} - p_{\text{smaller}}$ and $n - p_{\text{larger}}$ degrees of freedom.

Model selection

$$\text{PRESS} = \sum_{i=1}^{n} e_{(i)}^2, \quad e_{(i)} = \frac{e_i}{1 - H_{ii}}$$

where $e_{(i)}$ is the ith residual from fit with ith observation omitted;

$$C_p = \text{RSS} + 2\hat{\sigma}^2 p, \quad \hat{\sigma}^2 = \frac{1}{n - p_{\text{largest}}} \text{RSS}(p_{\text{largest}})$$

where p is the number of terms in the candidate model.

3

Scatterplot Smoothing

3.1 Introduction

As we saw at the end of the previous chapter, the LIDAR data, shown again in Figure 3.1, are virtually impossible to model using traditional parametric techniques. Many of the problems described in Chapter 1 also involve nonlinear effects that are difficult to model parametrically. There is a clear imperative to be able to handle such nonlinear relationships effectively through more flexible techniques. Even in circumstances where transformations and/or quadratic terms can be used to handle nonlinearities, it should be kept in mind that their use can require a good deal of expertise and time. In some applications, particularly those where many regression fits are required, it is not feasible for such delicate modeling to be done. Rather, one requires the nonlinear components to be handled automatically.

In this chapter we will look at some ways of freeing oneself of the restrictions of parametric regression models. The title of this chapter, *scatterplot smoothing*, has become commonplace in data-analytic contexts where one is interested in highlighting the "underlying trend" in the scatterplot. Here the scatterplot points are simply treated as a collection of points on a plane, without any regard to an

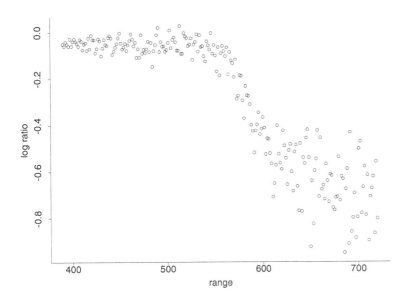

Figure 3.1
Scatterplot of LIDAR data.

underlying probabilistic model. Alternatively, we could think of the vertical position of each point as a realization of a random variable y conditional on the variable x with value corresponding to the horizontal position of the point. In this chapter, x will be univariate. Nonparametric regression with multiple predictor variables will be discussed in Chapters 7–13. The underlying trend would then be a function such as

$$f(x) = E(y|x).$$

This can also be written as

$$y_i = f(x_i) + \varepsilon_i, \quad E(\varepsilon_i) = 0,$$

in which case the problem is often referred to as *nonparametric regression,* where the function f is some unspecified "smooth" function that needs to be estimated from the (x_i, y_i).

There are several available methods for smoothing a scatterplot. The method that we focus on in this chapter, *penalized splines,* has the attractiveness of being a relatively straightforward extension of linear regression modeling (see e.g. O'Sullivan 1986; Kelly and Rice 1990; Gray 1992, 1994; Eilers and Marx 1996; Hastie 1996). Some alternatives are discussed at the end of the chapter.

3.2 Preliminary Ideas

We will start with the *straight line regression model*:

$$y_i = \beta_0 + \beta_1 x_i + \text{error}_i. \tag{3.1}$$

Figure 3.2 provides some graphical representations of this model. Figure 3.2(a) shows an example of (3.1), with the line representing the underlying regression function and the points representing a typical data set for such a model.

Figure 3.2(b) displays the *corresponding basis* for the model. These are the functions:

$$1 \text{ and } x. \tag{3.2}$$

From linear algebra, a *basis* of a vector space is a set, $\mathcal{V}$, of elements of that space such that any element of the space can be expressed uniquely as a linear combination of elements of $\mathcal{V}$. For example, $\beta_0 + \beta_1 x$ is a *linear combination* of the basis functions 1 and x. Thus, $\{1, x\}$ is a basis for the vector space of all linear polynomials in x. Further details are given in Appendix A.

Note that the right-hand side of (3.1) is a linear combination of these functions, which is the reason for use of the word *basis.* The basis functions correspond to the columns of the **X**-matrix for fitting the regression:

$$\mathbf{X} = \begin{bmatrix} 1 & x_1 \\ \vdots & \vdots \\ 1 & x_n \end{bmatrix}.$$

As we described in Chapter 2, the vector of fitted values $\hat{\mathbf{y}}$ can be obtained from this matrix and **y** through the formula

$$\hat{\mathbf{y}} = \mathbf{X}(\mathbf{X}^\mathsf{T}\mathbf{X})^{-1}\mathbf{X}^\mathsf{T}\mathbf{y} = \mathbf{H}\mathbf{y}. \tag{3.3}$$

A simple extension of the simple linear model is the *quadratic model*:

$$y_i = \beta_0 + \beta_1 x_i + \beta_2 x_i^2 + \text{error}_i. \tag{3.4}$$

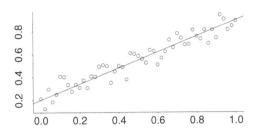

(a) Straight Line Model

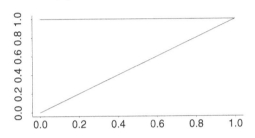

(b) Corresponding Basis

Figure 3.2 The simple linear regression model.

It is illustrated in Figure 3.3. Notice that there is an extra basis function: x^2, which corresponds to the addition of the $\beta_2 x_i^2$ term to the model. The **X**-matrix for the quadratic model is

$$\mathbf{X} = \begin{bmatrix} 1 & x_1 & x_1^2 \\ \vdots & \vdots & \vdots \\ 1 & x_n & x_n^2 \end{bmatrix},$$

and fitted values can be obtained using (3.3) with this particular **X**.

The quadratic model is an example of how the simple linear model might be extended to handle nonlinear structure. We will now look at how it can be extended to accommodate a different type of nonlinear structure.

Consider the model depicted in Figure 3.4(a). We call this the *broken stick* model because it consists of two differently sloped lines that join together at $x = 0.6$. How might one choose the set of basis functions to handle this type of structure? One possible answer (there are several others) is to introduce a basis function that is zero to the left of 0.6 and then is a positively sloped function from 0.6 onward. One should be able to see from Figure 3.4 that the broken line in the top panel can be obtained as a linear combination of the three basis functions in the bottom panel. A compact mathematical way of expressing the new basis function is

$$(x - 0.6)_+,$$

where, for any number u, u_+ is equal to u if u is positive and is equal to 0 otherwise. The broken stick model (with a break at $x = 0.6$) is therefore

$$y_i = \beta_0 + \beta_1 x_i + \beta_{11}(x_i - 0.6)_+ + \texttt{error}_i, \tag{3.5}$$

We often call $(x - 0.6)_+$ the *positive part* of the function $x - 0.6$ because the "+" sets it to zero for those values of x where $x - 0.6$ is negative (i.e. $x < 0.6$). A function such as $(x - 0.6)_+$ is also sometimes referred to as a *truncated line*. The reason for this name is illustrated in Figure 3.4(b).

Figure 3.3 The
quadratic regression
model.

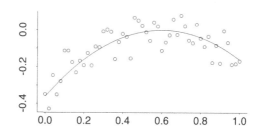

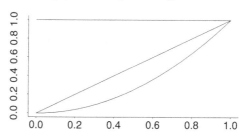

Figure 3.4 The
broken stick
regression model.

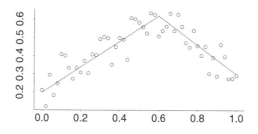

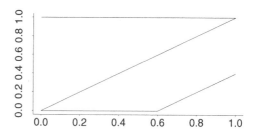

which can be fit using (3.3) with

$$\mathbf{X} = \begin{bmatrix} 1 & x_1 & (x_1 - 0.6)_+ \\ \vdots & \vdots & \vdots \\ 1 & x_n & (x_n - 0.6)_+ \end{bmatrix}.$$

Figure 3.5 The whip regression model.

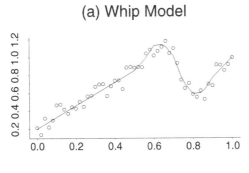

(a) Whip Model

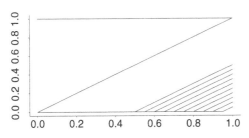

(b) Corresponding Basis

Now suppose that we have structure that is more complicated than the broken stick model. Figure 3.5(a) gives an example for which there is a straight line structure in the left-hand half, but the right-hand half is prone to a high amount of detailed structure. We will descriptively label this the *whip model* since the right-hand half is free to move around like the lash of a whip, while the left-hand side corresponds to the whip's stiff handle and is linear. If we have good reason to believe that our underlying structure is of this basic form, then how might one change the basis? One possible answer is provided by Figure 3.5(b), where the functions

$$(x - 0.5)_+, \ (x - 0.55)_+, \ \ldots, \ (x - 0.95)_+$$

are included. It is not too hard to see that this basis will do a reasonable job of modeling any "whiplike" structure with a handle between $x = 0$ and $x = 0.5$. Once again we can appeal to ordinary least squares to fit such a model with the **X**-matrix:

$$\mathbf{X} = \begin{bmatrix} 1 & x_1 & (x_1 - 0.5)_+ & (x_1 - 0.55)_+ & \cdots & (x_1 - 0.95)_+ \\ \vdots & \vdots & \vdots & \vdots & \ddots & \vdots \\ 1 & x_n & (x_n - 0.5)_+ & (x_n - 0.55)_+ & \cdots & (x_n - 0.95)_+ \end{bmatrix}.$$

From this example you should now appreciate that it is possible to handle any complex type of structure you like by simply adding more functions of the form $(x - \kappa)_+$ to the basis or, equivalently, by adding a column of $(x_i - \kappa)_+$ values to the **X**-matrix. The value of κ corresponding to the function $(x - \kappa)_+$ is usually referred to as a *knot*. This is because the function is made up of two lines that are "tied together" at $x = \kappa$. Figure 3.5 displays the function $(x - \kappa)_+$ for $\kappa = 0.5, 0.55, \ldots, 0.95$.

Figure 3.6 Linear spline regression fit to LIDAR data with knots at range $= 575$ and range $= 600$. The bar at the base of the plot shows the position of the knots.

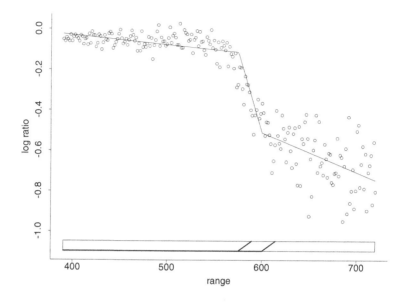

A function such as $(x - 0.6)_+$ is called a *linear spline basis function* and a set of such functions is called a *linear spline basis*. Note that any linear combination of linear spline basis functions $1, x, (x - \kappa_1)_+, \ldots, (x - \kappa_K)_+$ is a piecewise linear function with knots at $\kappa_1, \ldots, \kappa_K$. Such a function is called a *spline*.

The traditional definition of the word *spline* is a thin strip of flexible timber. A mathematical spline is so named because of the analogy of a flexible function able to adapt to the data.

Rather than referring to the spline basis function $(x - \kappa)_+$, it is common to simply refer to its knot κ. Thus, we say that a model has a knot at 0.35 if the function $(x - 0.35)_+$ is in the basis. The spline model for f is

$$f(x) = \beta_0 + \beta_1 x + \sum_{k=1}^{K} b_k (x - \kappa_k)_+. \tag{3.6}$$

3.3 Practical Implementation

In the previous section we knew the form of the underlying function, so selection of the appropriate basis was relatively easy. In the real world we are only given the scatterplot, so selection of a good basis is usually more challenging. Consider the scatterplot for the LIDAR data shown in Figure 3.1. One could start by trying to choose appropriate knots by trial and error. Figure 3.6 shows the result of a fit with knots placed at range $= 575$ and range $= 600$. The bar at the base of the plot shows the position of the knots. While this fit displays the essential qualitative features of the data, it is somewhat lacking in quality. For instance, for low values of range the points follow a much flatter trend than the fitted line over this region.

An obvious remedy is to use four knots instead of two; say, at range $= 500$, 550, 600, and 650. The resulting fit is shown in Figure 3.7. This fit is clearly much more pleasing than the one shown in Figure 3.6 and would suffice for most practical purposes. However, the piecewise linear nature of the fit is an artifact of the

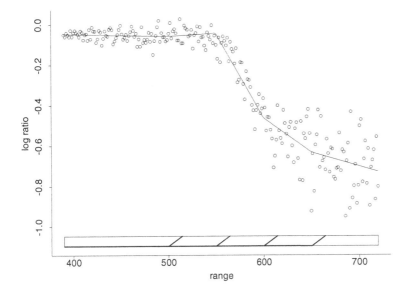

Figure 3.7 Linear spline regression fit to LIDAR data with knots at `range` = 500, 550, 600, and 650.

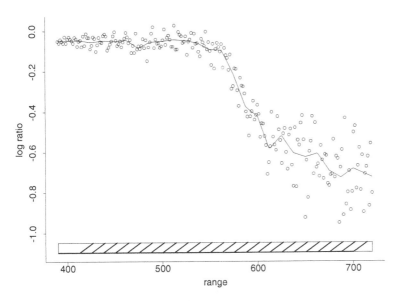

Figure 3.8 Linear spline regression fit to LIDAR data with knots at `range` = 400, 412.5, 425, ..., 700.

fitting method rather than the underlying mean structure in these data. In other words, we do not really believe that the underlying mean is piecewise linear with knots at `range` = 500, 550, 600, and 650 – this answer has arisen because of our model specification. One might try to alleviate this problem by adding even more knots, say at every 12.5 meters of the `range`. This results in Figure 3.8. Because the fit in Figure 3.8 is based on a larger set of knots, the long piecewise linear segments of Figure 3.7 are not present and the fitting procedure has much more flexibility. However, for these data, it appears that there is now *too much* flexibility: the plot is heavily "overfitted", meaning that the fitted function is following small, apparently random, fluctuations in the data as well as the main features.

Figure 3.9 Linear
spline regression fit
to LIDAR data with
knots at range =
612.5, 650, 662.5, and
687.5 deleted from the
basis that was used
to construct the fit in
Figure 3.8.

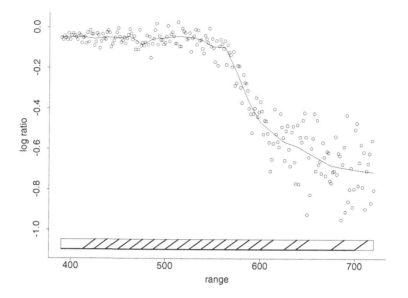

One could try pruning the knots to overcome this problem. For example, if the
knots at range = 612.5, 650, 662.5, and 687.5 are deleted then we obtain the fit
in Figure 3.9.

This fit is quite pleasing because it fits the data well without overfitting it.
However, it was arrived at after a lot of time-consuming trial and error. Clearly
it would be better to have the computer choose the best knots for us. The next
section discusses some ideas for achieving this objective.

3.4 Automatic Knot Selection

A natural first attempt at automatic selection of the knots is to use a *model selection
criterion*. There are several such criteria available in the multiple linear regression framework, for example, cross-validation and Mallows's C_p as described in
Section 2.6. The idea is to choose that combination of knots that optimizes the
chosen criterion. However, before contemplating such an approach it is worth
considering the number of possible models. If there are K candidate knots then
there are 2^K possible models, assuming the overall intercept and linear term are
always present. For the example depicted in Figure 3.8, $K = 24$ and so the
number of models that could be generated from this set of knots is 2^{26}, which
is approximately equal to 67 million! Already we can see that application of the
usual model selection ideas becomes either highly computationally intensive or
impossible for this approach to automatic knot selection.

The recent literature has seen the proposal of several approaches to automatic
knot selection that circumvent the need to fit all possible models. Many of them
are based on stepwise regression ideas. See, for example, Smith (1982), Friedman and Silverman (1989), Stone et al. (1997), Smith and Kohn (1996), Denison,

Stepwise regression
means building
regression models by
adding or deleting
predictors one at
a time based on
some goodness-of-fit
criterion.

Mallick, and Smith (1997), and DiMatteo, Genovese, and Kass (2001). The approach of the latter three references is to couch the problem in a Bayesian framework and use Monte Carlo Markov chain procedures to select the knots. A review and comparison of some of these approaches is given by Wand (2000).

Although most of the automatic knot selection procedures mentioned in the previous paragraph have exhibited good performance, they are each quite complicated and computationally intensive. In particular, their extension to the semiparametric models that are required to handle the problems given in Chapter 1 can, in some cases, be quite difficult. We therefore seek a simpler method for flexible spline-based regression.

3.5 Penalized Spline Regression

As we have already discussed, the roughness of the fit in Figure 3.8 is due to there being too many knots in the model. Another way to overcome this problem is to retain all of the knots but to *constrain* their influence. The hope is that this will result in a less variable fit.

Consider a general spline model with K knots, where K is large ($K = 24$ in Figure 3.8). The ordinary least-squares fit can be written as

$$\hat{\mathbf{y}} = \mathbf{X}\hat{\boldsymbol{\beta}}, \quad \text{where } \hat{\boldsymbol{\beta}} \text{ minimizes } \|\mathbf{y} - \mathbf{X}\boldsymbol{\beta}\|^2$$

and $\boldsymbol{\beta} = [\beta_0, \beta_1, \beta_{11}, \ldots, \beta_{1K}]^\mathsf{T}$, with β_{1k} the coefficient of the kth knot. As discussed previously, unconstrained estimation of the β_{1k} leads to a wiggly fit. Constraints on the β_{1k} that might rectify this situation are

(1) $\max|\beta_{1k}| < C$,
(2) $\sum|\beta_{1k}| < C$, and
(3) $\sum \beta_{1k}^2 < C$.

For judicious choice of C, each of these will lead to a smoother fit to the scatterplot. However, the third constraint is much easier to implement than the first two. If we define the $(K + 2) \times (K + 2)$ matrix

$$\mathbf{D} = \begin{bmatrix} 0 & 0 & 0 & 0 & 0 & \cdots & 0 \\ 0 & 0 & 0 & 0 & 0 & \cdots & 0 \\ 0 & 0 & 1 & 0 & 0 & \cdots & 0 \\ 0 & 0 & 0 & 1 & 0 & \cdots & 0 \\ \vdots & \vdots & \vdots & \vdots & \vdots & \ddots & \vdots \\ 0 & 0 & 0 & 0 & 0 & \cdots & 1 \end{bmatrix} = \begin{bmatrix} \mathbf{0}_{2\times 2} & \mathbf{0}_{2\times K} \\ \mathbf{0}_{K\times 2} & \mathbf{I}_{K\times K} \end{bmatrix},$$

then our minimization problem can be written as

$$\text{minimize } \|\mathbf{y} - \mathbf{X}\boldsymbol{\beta}\|^2 \quad \text{subject to } \boldsymbol{\beta}^\mathsf{T}\mathbf{D}\boldsymbol{\beta} \le C.$$

It can be shown, using a Lagrange multiplier argument, that this is equivalent to choosing $\boldsymbol{\beta}$ to minimize

Lagrange multipliers are a mathematical tool for solving optimization problems when the solution is subject to constraints.

Figure 3.10
Penalized linear
spline regression fit
to LIDAR data with
$\lambda = 30$.

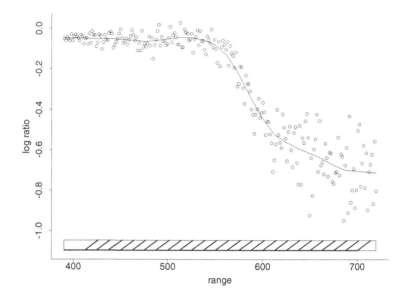

Equation (3.7) can be
minimized by using
exactly the same
calculus techniques
that are used to derive
the solution to the
OLS problem. See
Section A.2.12 for
an introduction to
calculus of functions
of a vector.

$$\|\mathbf{y} - \mathbf{X}\boldsymbol{\beta}\|^2 + \lambda^2 \boldsymbol{\beta}^\mathsf{T} \mathbf{D} \boldsymbol{\beta} \qquad (3.7)$$

for some number $\lambda \geq 0$.

This has the solution

$$\hat{\boldsymbol{\beta}}_\lambda = (\mathbf{X}^\mathsf{T}\mathbf{X} + \lambda^2 \mathbf{D})^{-1}\mathbf{X}^\mathsf{T}\mathbf{y}.$$

The term $\lambda^2 \boldsymbol{\beta}^\mathsf{T}\mathbf{D}\boldsymbol{\beta}$ is called a *roughness penalty* because it penalizes fits that are too rough, thus yielding a smoother result. The amount of smoothing is controlled by λ, which is therefore usually referred to as a *smoothing parameter*. The fitted values for a penalized spline regression are then given by

$$\hat{\mathbf{y}} = \mathbf{X}(\mathbf{X}^\mathsf{T}\mathbf{X} + \lambda^2 \mathbf{D})^{-1}\mathbf{X}^\mathsf{T}\mathbf{y}. \qquad (3.8)$$

The reason for using λ^2 rather than λ is given in Section 3.7. Penalized estimation shrinks all coefficients of spline basis functions toward zero and can be contrasted with knot selection, which shrinks some coefficients all the way to zero while leaving the remaining coefficients unshrunk. Note that (3.8) is a type of *ridge regression*. Ridge regression is sometimes used in parametric multiple regression modeling to reduce the variability of estimated coefficients (see e.g. Draper and Smith 1998).

Figure 3.10 shows a fit to the LIDAR data obtained by applying (3.8) with $\lambda = 30$. This fit is quite pleasing. It depends on the set of knots and the smoothing parameter λ. In Chapter 5 we will show that, provided the knots cover the range of x_i values reasonably well, their number and positioning does not make much difference to the result. However, λ has quite a big effect. This is illustrated in Figure 3.11, where fits to the LIDAR data are shown for various values of λ with the knot sequence used in Figure 3.8. The case $\lambda = 0$ corresponds to the unconstrained case, so the fit is identical to that shown in Figure 3.8. For $\lambda = 10$ we have downweighted the influence of the knots, so the fit is a little less rough. Increasing λ threefold leads to a very pleasing fit, as shown in Figure 3.11(c). If we

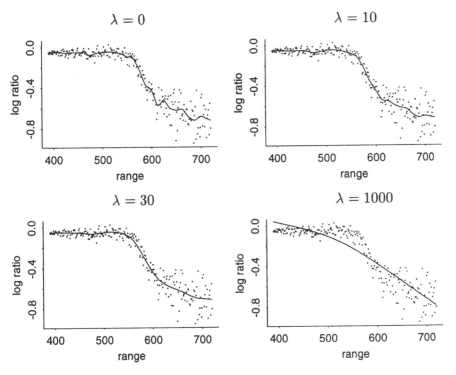

Figure 3.11 Linear penalized spline regression fits to LIDAR data for λ values of 0, 10, 30, and 1000 (24 knots are used).

take λ to be very large, as in Figure 3.11(d), then the effect of the knots diminishes and the least-squares line is approached.

Methods for choosing λ and the knot locations from the data are described in Chapter 5.

3.6   Quadratic Spline Bases

Each of the regression models that we have fit so far are linear splines – that is, continuous, piecewise linear functions. They are conceptually very simple and intuitive and will suffice in many applications.

As we discussed in Section 3.2, the reason for the piecewise linear nature of the functions is that they are a linear combination of piecewise linear functions of the form $(x - \kappa)_+$. A simple way of escaping from piecewise linearity is to add x^2 to the basis and also to replace each $(x - \kappa)_+$ by its square, $(x - \kappa)_+^2$. The function $(x - 0.6)_+^2$ is illustrated in Figure 3.12.

Notice that this function does not have a sharp corner like $(x - 0.6)_+$ does. In other words, $(x - 0.6)_+^2$ has a *continuous first derivative*. It follows that any linear combination of the functions

$$1, x, x^2, (x - \kappa_1)_+^2, \ldots, (x - \kappa_K)_+^2 \qquad (3.9)$$

The potentially ambiguous notation $(x - \kappa)_+^p$ will always be taken to mean $\{(x - \kappa)_+\}^p$ rather than $\{(x - \kappa)^p\}_+$.

will also have a continuous first derivative and not have any sharp corners. This will usually result in an aesthetically more appealing fit. We call (3.9) a *quadratic spline basis* with knots at $\kappa_1, \ldots, \kappa_K$.

Figure 3.12 A
quadratic basis
function with a knot
at 0.6.

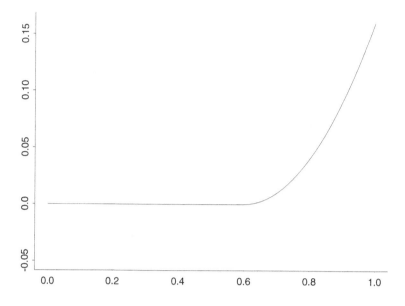

Figure 3.13
Penalized spline
regression fits to the
fossil data based
on (a) linear spline
basis functions and
(b) quadratic spline
basis functions. In
each case, eleven
equally space knots
are used.

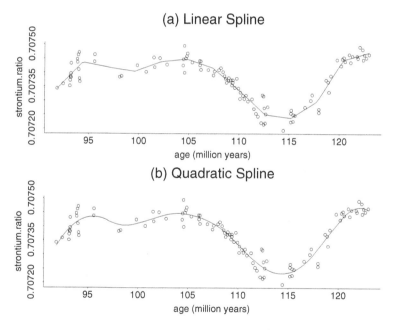

The fossil data
were collected by
T. Bralower of
the University of
North Carolina
and are analyzed
by Chaudhuri and
Marron (1999).
We are grateful to
Professor J. S. Marron
for providing us with
these data.

Another advantage of using quadratic spline basis functions is that quadratics tend to do a better job of fitting peaks and valleys in a scatterplot. This is illustrated in Figure 3.13 for data on ratios of strontium isotopes found in fossil shells and their age. Both estimates are based on eleven equally spaced knots. The quadratic spline fit smooths out the valley much more effectively than the linear spline fit; and there are no unsightly corners. However, if one uses enough knots and penalized least squares, then the difference between a linear and quadratic spline fit is usually negligible – see Section 5.5, in particular, Figure 5.12.

When fitting quadratic splines by penalized least squares, the coefficients of 1, x, and x^2 are unpenalized. If the ith row of $\mathbf{X}$ contains the basis functions (3.9) in that order, then $\mathbf{D} = \mathrm{diag}(0, 0, 0, 1, \ldots, 1)$.

3.7   Other Spline Models and Bases

All of the previous examples in this chapter use the linear spline or the quadratic spline model with the truncated power basis. Two types of generalizations are possible:

(1) to other spline models; and
(2) to other bases for a given spline model.

One reason for considering other models is to achieve smoother fits. Smoother fits are especially important if one plans to differentiate the fit to estimate a derivative of the regression function; see Section 6.8.

In principle, a change of basis does not change the fit – though some bases are more numerically stable and allow computation of a fit with greater accuracy. Besides numerical stability, reasons for selecting one basis over another are ease of implementation (especially of penalties) and interpretability. The latter consideration is usually not too important since one is generally interested only in the fit, not the estimated coefficients.

An obvious generalization is to a spline model of general degree. Using the truncated power functions, the basis is:

$$1, x, \ldots, x^p, (x - \kappa_1)_+^p, \ldots, (x - \kappa_k)_+^p, \tag{3.10}$$

which is known as the *truncated power basis of degree p*. Since the function $(x - \kappa)_+^p$ has $p - 1$ continuous derivatives, higher values of p lead to smoother spline functions. The pth-degree spline model is

$$f(x) = \beta_0 + \beta_1 x + \cdots + \beta_p x^p + \sum_{k=1}^{p} \beta_{pk} (x - \kappa_k)_+^p. \tag{3.11}$$

When fitting a pth-degree spline by penalized least squares, none of the polynomial coefficients is penalized. For $p > 0$, the formula for the fitted values becomes

$$\hat{\mathbf{y}} = \mathbf{X}(\mathbf{X}^\mathsf{T}\mathbf{X} + \lambda^{2p}\mathbf{D})^{-1}\mathbf{X}^\mathsf{T}\mathbf{y}, \tag{3.12}$$

where

$$\mathbf{D} = \mathrm{diag}(\mathbf{0}_{p+1}, \mathbf{1}_K). \tag{3.13}$$

The power of $2p$ on the λ is justified as follows. Suppose that the x-variable undergoes a transformation of the form

$$x \mapsto \alpha x$$

for some $\alpha > 0$. This will happen if different units of measurement are used for x. Then it is natural to ask that application of the same transformation on the smoothing parameter,

Figure 3.14 *B*-spline
bases of degrees
(a) one, (b) two,
and (c) three. The
positions of the knots
are indicated by the
solid diamonds.

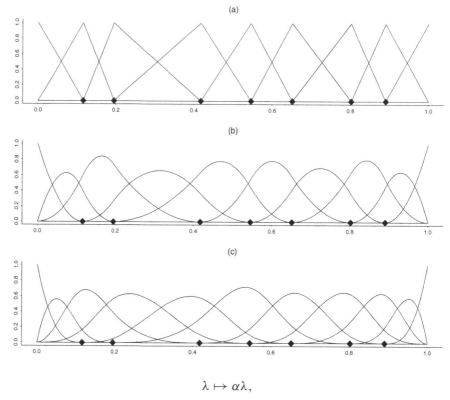

$$\lambda \mapsto \alpha\lambda,$$

results in the same fit $\hat{\mathbf{y}}$. It can be shown that the parameterization used in (3.12)
is necessary and sufficient for λ to have this property.

3.7.1 *B-Splines*

Truncated power bases are useful for understanding the mechanics of spline-based
regression, and they can be used in practice if the knots are selected carefully or a
penalized fit is used. However, the truncated power bases have the practical disadvantage that they are far from orthogonal. This can sometimes lead to numerical
instability when there is a large number of knots and the penalty parameter λ is
small (or zero in the case of ordinary least squares). Therefore, in practice, especially for OLS fitting, it is advisable to work with equivalent bases with more
stable numerical properties. The most common choice is the *B-spline* basis.

    Figure 3.14 shows the *B*-spline bases of degrees 1, 2, and 3 for the case of seven
irregularly spaced knots. Each of these are equivalent to the truncated power basis
of the same degree. In the regression context, this means that if one used an **X**-
matrix with columns corresponding to the functions in Figure 3.14, then the fits
would be identical to those obtained using the truncated polynomial of the same
degree with knots in the same locations.

    Mathematically, we can quantify the equivalence as follows. If $\mathbf{X}_T$ is an **X**-
matrix with columns corresponding to (3.10) and if $\mathbf{X}_B$ is the **X**-matrix corresponding to the the *B*-spline basis of the same degree and same knot locations,

Two bases are
equivalent if they
span the same set of
functions. By *span*
we mean the set of
all possible linear
combinations of the
basis functions. Some
details are given in
Appendix A.

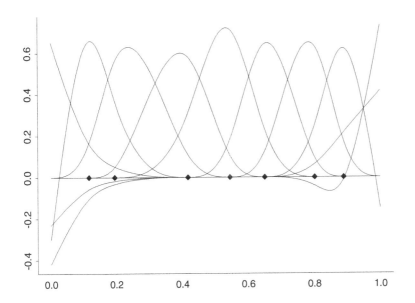

Figure 3.15 Natural spline basis for the same set of knots used in Figure 3.14.

then

$$\mathbf{X}_B = \mathbf{X}_T \mathbf{L}_p, \tag{3.14}$$

where $\mathbf{L}_p$ is a square invertible matrix. Substitution into (3.8) allows us to express a penalized spline fit of degree p in terms of the B-spline basis as

$$\hat{\mathbf{y}} = \mathbf{X}_B (\mathbf{X}_B^\mathsf{T} \mathbf{X}_B + \lambda^{2p} \mathbf{L}_p^\mathsf{T} \mathbf{DL}_p)^{-1} \mathbf{X}_B^\mathsf{T} \mathbf{y}. \tag{3.15}$$

This form of the estimator is used by Eilers and Marx (1996).

Regression packages routinely transform the columns of the $\mathbf{X}$-matrix to a version that is more numerically stable (see Appendix B). Therefore, the basis functions used in the formulation of a model will not correspond to those that are used during its fitting. For this reason, we will not concern ourselves about numerical issues when formulating spline-based smoothers.

3.7.2   *Natural Cubic Splines*

A commonly used modification of the cubic spline model is the *natural cubic spline basis*. Natural cubic splines are cubic splines with the constraint that they are linear in their tails beyond the boundary knots. For example, in Figure 3.15 the basis functions would be linear to the left of 0 and to the right of 1; here 0 and 1 are called the *boundary knots* and the other knots are *interior knots*. The linearity is enforced through the constraints that the spline f satisfy $f'' = f''' = 0$ at the boundary knots. The natural spline basis shown in Figure 3.15 has the knot locations used in Figure 3.14.

A *cubic smoothing spline*, $\hat{f}$, minimizes the residual sum of squares plus a penalty on the integral of the squared second derivative, $(\hat{f}'')^2$. The resulting penalized RSS is

$$\sum_{i=1}^{n} \{y_i - \hat{f}(x_i)\}^2 + \lambda^3 \int \{\hat{f}^{(2)}(x)\}^2 \, dx, \qquad (3.16)$$

where the $\lambda > 0$ is the smoothing parameter and the cubic power is based on the scale arguments given earlier in this section. Although there is no prior constraint imposed on $\hat{f}$ that it even be a spline, it has been proved that the minimizer of this penalized sum of squares is a natural cubic spline with knots (boundary and interior) at the x_i. Smoothing splines have enjoyed widespread use in nonparametric regression and have a vast literature; see, for example, the monographs of Eubank (1988), Wahba (1990), and Green and Silverman (1994). In Chapter 13 we will describe their connection with spatial statistical methodology.

Natural cubic splines are called "natural" because they arise as the solution of an optimization problem. Also, the draftsman's spline – a thin flexible strip of wood once used to draw curves and the prototype of mathematical splines – assumes a linear shape beyond the pegs used to constrain it. However, the boundary constraints have no natural statistical interpretation of which we are aware. Nor are the boundary constraints natural in any application we have seen. We see little reason for preferring the natural cubic spline model to the cubic spline model without these constraints.

The smoothing spline penalty (3.16) can be implemented with penalized splines as well. In that case, penalized splines are much like smoothing splines. However, penalized splines are more general than smoothing splines in that one can use as many or as few knots as desired and natural cubic spline boundary constraints can be either imposed or not.

3.7.3   *Radial Basis Functions*

The truncated polynomial basis functions (3.10) and the B-spline basis functions defined via (3.14) span the space of pth-degree polynomials with knots at $\kappa_1, \ldots, \kappa_K$. When p is odd, yet another set of basis functions with this property is

$$1, x, \ldots, x^p, |x - \kappa_1|^p, \ldots, |x - \kappa_k|^p.$$

Figure 3.16 shows the basis functions for $p = 1$ when the knots are the same as those used in Figure 3.14.

Note that

$$|x - \kappa_k|^p = r(|x - \kappa_k|), \quad \text{where } r(u) = u^p.$$

This shows that the basis functions $|x - \kappa_k|^p$ $(1 \le k \le K)$ depend only on the distance $|x - \kappa_k|$ and the univariate function r. An advantage of this property is that the extension to higher-dimensional predictor variables is straightforward. That is, if $\mathbf{x} \in \mathbb{R}^d$ and $\kappa_1, \ldots, \kappa_K$ are knots in $\mathbb{R}^d$, then plausible basis functions are those of the form

$$r(\|\mathbf{x} - \kappa_k\|); \qquad (3.17)$$

as before, $\|\mathbf{v}\| = \sqrt{\mathbf{v}^{\mathsf{T}}\mathbf{v}}$ is the length of the vector $\mathbf{v}$. Functions of general form (3.17) are radially symmetric about κ_k. They are sometimes called *radial* basis functions.

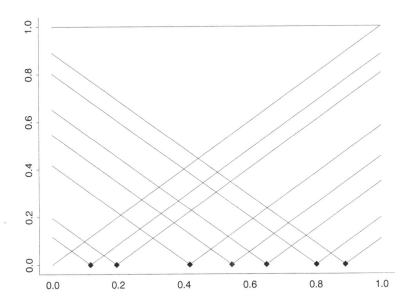

Figure 3.16 Linear radial basis functions for the same set of knots used in Figure 3.14.

Smoothing splines have a natural representation in terms of radial basis functions. For a given value of $\lambda > 0$, the cubic smoothing spline defined in the previous section can be written as

$$\hat{f}(x) = \hat{\beta}_0 + \hat{\beta}_1 x + \sum_{j=1}^{n} \hat{\beta}_{1j} |x - x_j|^3, \tag{3.18}$$

where $\hat{\beta}_0$, $\hat{\beta}_1$ and $\hat{\beta}_{11}, \ldots, \hat{\beta}_{1n}$ minimize

$$\|\mathbf{y} - \mathbf{X}_0 \boldsymbol{\beta}_0 - \mathbf{X}_1 \boldsymbol{\beta}_1\|^2 + \lambda^3 \boldsymbol{\beta}_1^{\mathsf{T}} \mathbf{K} \boldsymbol{\beta}_1$$

subject to

$$\mathbf{X}_0^{\mathsf{T}} \boldsymbol{\beta}_1 = \mathbf{0}; \tag{3.19}$$

see Green and Silverman(1994). Here $\boldsymbol{\beta}_0 \equiv [\beta_0, \beta_1]^{\mathsf{T}}$, $\boldsymbol{\beta}_1 = [\beta_{11}, \ldots, \beta_{1n}]^{\mathsf{T}}$, $\mathbf{X}_0 = [1, x_i]_{1 \le i \le n}$, and

$$\mathbf{X}_1 = \mathbf{K} \equiv \left[|x_i - x_j|^3 \right]_{1 \le i, j \le n}. \tag{3.20}$$

Note that constraint (3.19) means that smoothing splines use n basis functions rather than the $n + 2$ implied by the combined number of columns in $\mathbf{X}_0$ and $\mathbf{X}_1$.

Approximations to smoothing splines requiring considerably less computation can be obtained by specifying a knot sequence $\kappa_1, \ldots, \kappa_K$, using the basis functions $1, x, |x - \kappa_1|^3, \ldots, |x - \kappa_K|^3$, and replacing (3.20) by

$$\mathbf{X}_1 = \left[|x_i - \kappa_k|^3 \right]_{\substack{1 \le k \le K \\ 1 \le i \le n}} \quad \text{and} \quad \mathbf{K} = \left[|\kappa_k - \kappa_{k'}|^3 \right]_{1 \le k, k' \le K}.$$

This family of smoothers can be extended to those of arbitrary odd degree by replacing the cubic power on the radial basis functions by $2m - 1$ ($m = 1, 2, \ldots$) and adding polynomial terms up to degree $m - 1$ as in Section 13.2.1 (see also Nychka 2000). Section 13.5 describes some implementational details of this approach.

Smoothers with radial basis functions also arise in the field of *geostatistics,* where the term *kriging* is used in place of smoothing. Chapter 13 provides details.

3.8   Other Penalties

As the penalty parameter λ converges to infinity, the smooth fit will converge toward some limit. One good way to understand a penalty is to identify the limit.

If we use a pth-degree spline with the truncated power function basis and if the penalty is

$$\lambda^{2p} \sum_{k=1}^{K} \beta_{pk}^2, \tag{3.21}$$

then the fit is shrunk toward the least-squares fit to a pth-degree polynomial. Thus, the degree of the spline and the shrinkage target are tied together. With penalty (3.21), for example, to shrink toward a straight line one must use linear splines.

In some instances, one wants the flexibility of choosing the spline model and the shrinkage limit independently. To achieve this flexibility, one can introduce other penalties. Eilers and Marx (1996) use B-splines with *equally spaced knots* and penalize the $(q+1)$th-order difference in the B-spline coefficients. Thus, by using $q = 1$, for example, shrinkage is toward a straight line regardless of the degree of the spline; whereas, for $q = 2$, the shrinkage is toward a parabola. The Eilers and Marx penalty is not tied to the use of B-splines, though they do not mention this point. We can replace one basis by an *equivalent* basis if the penalty matrix is also transformed as discussed in Section 3.7.1; see equations (3.14) and (3.15). Thus, the Eilers and Marx penalty could be implemented using truncated power functions as the basis, provided the knots were equally spaced.

The Eilers and Marx penalty is certainly useful when equally spaced knots are appropriate. In practice, however, we often find irregularities in the spacings of the data that require the knots to be unequally spaced. We prefer using knots that are at equally spaced quantiles. An alternative to (3.21) that allows shrinkage to a wide range of limits and that is suitable for any spacing of the knots is a quadratic integral penalty of the form

$$\lambda^{2p} \int_a^b \{f^{(q+1)}(x)\}^2 \, dx, \tag{3.22}$$

where $q+1 \le p$ and where a and b are the smallest and largest x-values. Penalty (3.22) shrinks toward a qth-degree polynomial. Using $p = 3$ and $q = 1$ gives a fit very similar to a cubic smoothing spline. In fact, using natural cubic splines with knots at every x gives precisely a cubic smoothing spline.

Let $\mathbf{B}(x) = [B_1(x), \dots, B_N(x)]^\mathsf{T}$ be the vector of spline basis functions, so that the ith row of $\mathbf{X}$ is $\mathbf{B}(x_i)^\mathsf{T}$ and

$$\hat{f}(x) = \hat{\boldsymbol{\beta}}^\mathsf{T} \mathbf{B}(x).$$

Then (3.22) is equal to

$$\boldsymbol{\beta}^{\mathsf{T}} \mathbf{D} \boldsymbol{\beta},$$

where $\mathbf{D}$ is no longer the diagonal matrix of zeros and ones given by (3.13) but rather the second moment matrix of $\mathbf{B}$:

$$\mathbf{D} = \int_a^b \mathbf{B}^{(q+1)}(x) \{\mathbf{B}^{(q+1)}(x)\}^{\mathsf{T}} \, dx.$$

The fit is still given by (3.12) with only the definition of $\mathbf{D}$ in that expression changed.

3.9   General Definition of a Penalized Spline

The various types of penalized splines that have been mentioned can be tied together with a broader concept. Our general definition of a penalized spline is $\hat{\boldsymbol{\beta}}^{\mathsf{T}} \mathbf{B}(x)$, where $\hat{\boldsymbol{\beta}}$ is the minimizer of

$$\sum_{i=1}^n \{y_i - \boldsymbol{\beta}^{\mathsf{T}} \mathbf{B}(x_i)\}^2 + \alpha \boldsymbol{\beta}^{\mathsf{T}} \mathbf{D} \boldsymbol{\beta}$$

for some symmetric positive semidefinite matrix $\mathbf{D}$ and scalar $\alpha > 0$.

The explicit form of $\hat{\boldsymbol{\beta}}$ is (3.12) with λ^{2p} replaced by α. Throughout this book, we will ordinarily use spline basis (3.10) and $\mathbf{D}$ given by (3.13). However, as shown in the previous section, other choices of $\mathbf{D}$ have appeal, especially to those familiar with smoothing splines.

When applying penalized splines, there are two basic choices to be made:

(1) the spline model – that is, the degree and knot locations and whether to impose constraints such as natural spline boundary constraints;
(2) the penalty – or, more explicitly, the form of the penalty up to a nonnegative smoothing parameter.

Once these two choices have been made, there follow two secondary choices:

(3) the basis functions – for example, truncated power functions or B-splines – used to represent the model (here we are concerned mostly with interpretability of the regression coefficients, not numerical stability);
(4) the basis functions used in the computations (here we are mostly concerned with numerical stability).

Choices (3) and (4) do not affect the fitted curve, except for the effects of numerical error. Choice (4) is discussed in Appendix B. If one uses mixed model software, then choice (4) is made automatically by the software.

Constraints such as natural spline constraints are part of the model in (1) and are subsumed in the basis functions in (3) and (4). For example, the natural cubic spline basis functions are linear combinations of cubic spline basis functions that satisfy the natural spline constraints, with the number of natural cubic spline basis functions equal to the number of cubic spline basis functions minus the number of independent constraints.

Once the penalty and the basis functions have been determined, then a fifth "choice" is automatically determined:

(5) the penalty matrix $\mathbf{D}$.

As discussed in Section 3.7.1, if in (3) or (4) we change the basis to an equivalent basis, then the penalty matrix can be changed so that the penalty itself is unchanged.

3.10   Linear Smoothers

Although penalized spline fits are fundamentally different from ordinary least-squares fits, they do share some common ground. In particular they are both *linear* functions of the data vector $\mathbf{y}$. As noted in Chapter 2, for a linear regression model

$$\mathbf{y} = \mathbf{X}\boldsymbol{\beta} + \boldsymbol{\varepsilon},$$

the ordinary least-squares fit to the data is

$$\hat{\mathbf{y}} = \mathbf{Hy}, \quad \text{where} \quad \mathbf{H} = \mathbf{X}(\mathbf{X}^{\mathsf{T}}\mathbf{X})^{-1}\mathbf{X}^{\mathsf{T}} \tag{3.23}$$

is the hat matrix. The penalized spline model generalizes this to

$$\hat{\mathbf{y}} = \mathbf{S}_{\lambda}\mathbf{y}, \tag{3.24}$$

where

$$\mathbf{S}_{\lambda} = \mathbf{X}(\mathbf{X}^{\mathsf{T}}\mathbf{X} + \lambda^{2p}\mathbf{D})^{-1}\mathbf{X}^{\mathsf{T}}$$

and $\mathbf{X}$ corresponds to the pth-degree truncated polynomial basis. In nonparametric regression contexts, $\mathbf{S}_{\lambda}$ is usually called the *smoother matrix,* though the term *hat matrix* is also used.

From (3.23) and (3.24) it is immediate that each can be written in the form

$$\hat{\mathbf{y}} = \mathbf{Ly}$$

for some $n \times n$ matrix $\mathbf{L}$ that does not depend on $\mathbf{y}$. We call this the class of *linear smoothers.*

Often, $\mathbf{L}$ depends on $\mathbf{y}$ through a smoothing parameter. In this case, the smoother is not really linear. However, it is common practice to pretend the data-based smoothing parameter is fixed and, as an approximation, to treat the smoother as linear.

3.11   Error of a Smoother

Let $\hat{f}$ be an estimator of f in the general scatterplot smoothing model

$$y_i = f(x_i) + \varepsilon_i,$$

and let $\mathcal{X} \subseteq \mathbb{R}$ be a set of values of x for which estimation of $f(x)$ is of interest. Then, for each $x \in \mathcal{X}$,

$$\hat{f}(x) \text{ is an estimate of } f(x).$$

A cornerstone of statistical estimation theory is the *error* incurred by an estimator with respect to a given target. The most common measure of error is the *mean squared error* (MSE), which in this case is

$$\text{MSE}\{\hat{f}(x)\} \equiv \text{E}[\{\hat{f}(x) - f(x)\}^2].$$

The MSE has the advantage of admitting the following decomposition:

$$\text{MSE}\{\hat{f}(x)\} = [\text{E}\{\hat{f}(x)\} - f(x)]^2 + \text{var}\{\hat{f}(x)\},$$

which represents squared bias and variance contributions to the overall error.

Usually the entire curve fit is of interest (rather than individual points), so it is common to measure the error *globally* across several values of x. One possibility is the *mean integrated squared error* (MISE):

$$\text{MISE}\{\hat{f}(\cdot)\} \equiv \int_{\mathcal{X}} \text{MSE}\{\hat{f}(x)\}\, dx.$$

A simpler alternative – and one that avoids dependence on $\mathcal{X}$ – is the *mean summed squared error* (MSSE):

$$\text{MSSE}\{\hat{f}(\cdot)\} \equiv \text{E} \sum_{i=1}^{n} \{\hat{f}(x_i) - f(x_i)\}^2,$$

where only error at the observations are considered. An advantage of MSSE is that it has matrix algebraic representations that simplify. Let $\hat{\mathbf{f}} = [\hat{f}(x_1), \ldots, \hat{f}(x_n)]^\mathsf{T}$ denote the vector of fitted values and let $\mathbf{f}$ similarly define the vector of $f(x_i)$ values. Then

$$\text{MSSE}(\hat{\mathbf{f}}) = \text{E}\|\hat{\mathbf{f}} - \mathbf{f}\|^2.$$

In the case of linear smoothers (Section 3.10),

$$\hat{\mathbf{f}} = \mathbf{L}\mathbf{y}.$$

Then the mean summed squared error can be written as

$$\text{MSSE}(\hat{\mathbf{f}}) = \sum_{i=1}^{n} \{\text{E}\hat{f}(x_i) - f(x_i)\}^2 + \text{Var}\{\hat{f}(x_i)\}$$

$$= \sum_{i=1}^{n} \{\text{E}(\mathbf{L}\mathbf{y})_i - f_i\}^2 + \text{Var}\{(\mathbf{L}\mathbf{y})_i\}$$

$$= \sum_{i=1}^{n} \{\text{E}(\mathbf{L}\mathbf{y})_i - f_i\}^2 + \{\text{Cov}(\mathbf{L}\mathbf{y})\}_{ii}$$

$$= \|(\mathbf{L} - \mathbf{I})\mathbf{f}\|^2 + \text{tr}\{\text{Cov}(\mathbf{L}\mathbf{y})\}$$

$$= \|(\mathbf{L} - \mathbf{I})\mathbf{f}\|^2 + \text{tr}\{\mathbf{L}\,\text{Cov}(\mathbf{y})\mathbf{L}^\mathsf{T}\}.$$

Assuming homoscedasticity via $\text{Cov}(\mathbf{y}) = \sigma_\varepsilon^2 \mathbf{I}$ yields

$$\text{MSSE}(\hat{\mathbf{f}}) = \|(\mathbf{L} - \mathbf{I})\mathbf{f}\|^2 + \sigma_\varepsilon^2 \, \text{tr}(\mathbf{L}\mathbf{L}^\mathsf{T}). \tag{3.25}$$

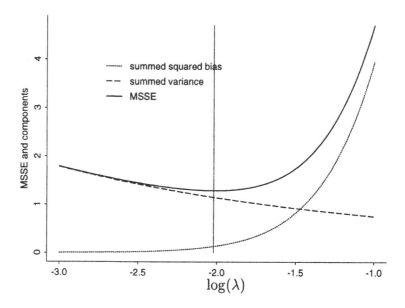

The first term of (3.25) represents the squared bias contribution, while the second corresponds to the sum of the variances. In smoothing, bias and variance work against one another in what is known as the *bias–variance trade-off*. For penalized spline smoothers, where $\mathbf{L} = \mathbf{S}_\lambda$, larger values of λ lead to an increase in bias but to a decrease in variance. Smaller values of λ lead to the opposite results. Figure 3.17 illustrates the variance–bias trade-off for data simulated from $f(x) = \sin(3\pi x)$, where the x_i ($1 \leq i \leq 200$) are equally spaced between 0 and 1 and where $\sigma_\varepsilon = 0.4$ (see Figure 4.6 for a realization of these data).

From Figure 3.17 we see that the MSSE-optimal amount of smoothing corresponds to $\log(\lambda) \simeq -2$ and represents the optimal trade-off between summed squared bias and summed variance. However, note that this is only obtainable in the unrealistic situation where f and σ_ε are known. Chapter 5 deals with estimation of the optimal amount of smoothing.

3.12   Rank of a Smoother

Classical smoothing splines use n basis functions, whereas linear penalized splines or cubic radial smoothers (3.18) with K knots use $K + 2$ basis functions. For large n there are considerable computational savings available from the latter approach. But what is the cost? We will now show that "reduced knot" smoothers tend to discard components of "full knot" smoothers that are unimportant to the final smooth.

Consider a general linear smoother with $n \times n$ smoother matrix $\mathbf{L}$, where $\mathbf{L}$ is symmetric and positive semidefinite (as is true for all of the spline smoothers we consider). Let $\lambda_1 \geq \cdots \geq \lambda_n$ be the ordered *eigenvalues* of $\mathbf{L}$ with corresponding *eigenvectors* $\mathbf{v}_1, \ldots, \mathbf{v}_n$. Figure 3.18 shows the eigenvalues of (a) a full knot smoothing spline and (b) a cubic radial smoother with $K = 9$ knots, both for

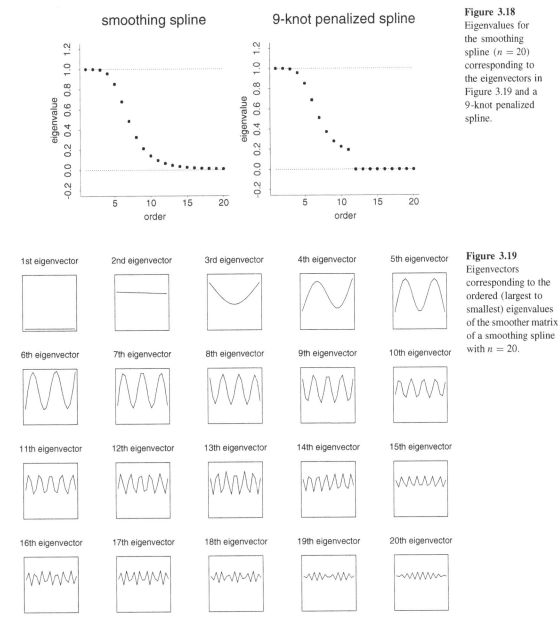

Figure 3.18
Eigenvalues for the smoothing spline ($n = 20$) corresponding to the eigenvectors in Figure 3.19 and a 9-knot penalized spline.

Figure 3.19
Eigenvectors corresponding to the ordered (largest to smallest) eigenvalues of the smoother matrix of a smoothing spline with $n = 20$.

$n = 20$ equally spaced observations. In Figure 3.19, the eigenvectors corresponding to the full knot smoothing spline are plotted against the predictor variable. Notice that the eigenvectors corresponding to the smaller eigenvalues become more oscillatory.

Eigenvalues and eigenvectors are defined through the relationship

$$\mathbf{L}\mathbf{v}_i = \lambda_i \mathbf{v}_i, \quad i = 1, \dots, n. \tag{3.26}$$

The eigenvectors form a basis for $\mathbb{R}^n$, so for a general response vector $\mathbf{y}$ one has the representation

$$\mathbf{y} = \sum_{i=1}^{n} \alpha_i \mathbf{v}_i.$$

Using (3.26), fitted values can then be represented as

$$\hat{\mathbf{y}} = \sum_{i=1}^{n} \alpha_i \lambda_i \mathbf{v}_i.$$

With this representation, Figure 3.18(a) and Figure 3.19 show that the main part of the smoothing spline fit is in the first nine or ten eigenvectors. The higher-order eigenvectors contribute relatively little to the fit.

In Figure 3.18(b) one sees that the eigenvalues are exactly zero for all but the first eleven eigenvectors. These correspond to a linear transformation of the eleven basis functions in the fit. The eigenvectors of this fit are not plotted here but are similar in nature to those plotted in Figure 3.19. It is apparent that the reduced knot smoother has a built-in omission of the least important component of the smoother. Hastie (1996) gives a fuller mathematical treatment of this basic idea.

Using the matrix algebraic result

A set of vectors are *linearly independent* if none of the vectors can be expressed as linear combinations of the others; see Appendix A.

rank of a matrix $\mathbf{L} \equiv$ number of linearly independent columns in $\mathbf{L}$

= number of eigenvalues of $\mathbf{L}$ that are nonzero,

one can use the rank of the smoother matrix to quantify its computational complexity. Smoothers that use considerably less than n basis functions will be called *low-rank,* while those with basis functions approximately the same as the sample size will be called *full-rank*.

Low-rank smoothers have risen to prominence in recent years owing to articles such as Eilers and Marx (1996) and Hastie (1996). However, the idea of using fewer than n basis functions in spline smoothing had been discussed in several earlier articles, including Parker and Rice (1985), O'Sullivan (1986, 1988), Kelly and Rice (1990), and Gray (1994), The S-PLUS function `smooth.spline()` uses low-rank approximations for $n > 50$.

For scatterplot smoothing with sample sizes in the hundreds, there is not a great deal to be gained from the use of low-rank smoothers. But for larger sample sizes and more complicated models involving several smooths, as treated in the later chapters of this book, the reduction in computational overhead enabled by low-rank smoothers can be enormous.

3.13   Degrees of Freedom of a Smoother

As seen in Figure 3.11, a small value of the smoothing parameter λ leads to a wiggly scatterplot smooth, somewhat close to interpolation of the data, whereas a very large λ results in a parametric fit that depends on the degree of the spline basis functions. What is not so clear is the relationship between intermediate values of λ and the resulting amount of smoothing. This problem is illustrated in Figure 3.20, which shows 24-knot linear penalized spline smooths of the LIDAR

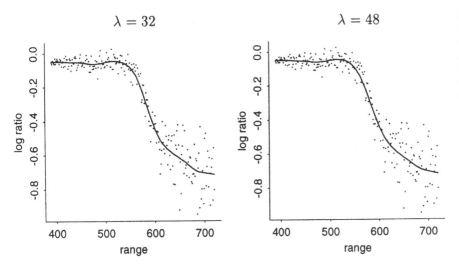

Figure 3.20 24-knot linear spline fits to the LIDAR data with $\lambda = 32$ and $\lambda = 48$ but with respective degrees of freedom equal to 9.86 and 8.34.

data with $\lambda = 32$ and $\lambda = 48$. Even though the second smoothing parameter is 50% larger than the first, the two fits look almost identical. So the value of the smoothing parameter does not have a direct interpretation as to how much structure is being imposed in the fit.

A transformation $t(\lambda)$ of λ that provides a reasonable solution to this problem is

$$t(\lambda) = \mathrm{tr}(\mathbf{S}_\lambda),$$

which we call the *degrees of freedom* of the fit corresponding to the smoothing parameter λ. We saw in Section 2.5.2 that, in parametric regression, the trace of the hat matrix $\mathbf{H}$ is

$$\mathrm{tr}(\mathbf{H}) = \text{number of fitted parameters} = \text{degrees of freedom},$$

so $\mathrm{tr}(\mathbf{S}_\lambda)$ is a generalization of this definition to scatterplot smoothers. It has the rough interpretation as the *equivalent number of parameters* and can be calibrated with polynomial fits: a scatterplot smooth with v degrees of freedom summarizes the data to about the same extent as a $(v-1)$-degree polynomial.

We will use the notation

$$df_{\mathrm{fit}} \equiv \mathrm{tr}(\mathbf{S}_\lambda) \tag{3.27}$$

as an abbreviation. For the penalized spline we have, by the matrix result $\mathrm{tr}(\mathbf{AB}) = \mathrm{tr}(\mathbf{BA})$,

$$df_{\mathrm{fit}} = \mathrm{tr}\{(\mathbf{X}^\mathsf{T}\mathbf{X} + \lambda^2\mathbf{D})^{-1}\mathbf{X}^\mathsf{T}\mathbf{X}\}.$$

For a penalized spline scatterplot smooth with K knots and degree p, it is easily shown that

$$\mathrm{tr}(\mathbf{S}_0) = p + 1 + K.$$

At the other extreme,

$$\mathrm{tr}(\mathbf{S}_\lambda) \to p + 1 \quad \text{as } \lambda \to \infty,$$

so positive values of λ correspond to

$$p + 1 < df_{\mathrm{fit}} < p + 1 + K.$$

Figure 3.21 Plot of
$df_{\text{fit}}(\lambda)$ against λ for
24-knot linear spline
fits to the LIDAR
data.

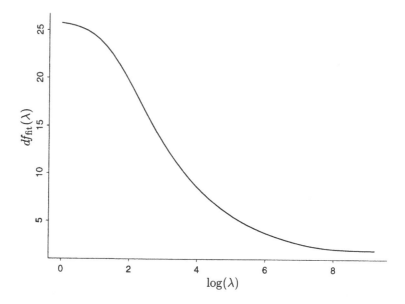

Figure 3.21 shows df_{fit} plotted against λ for 24-knot linear penalized spline smooths of the LIDAR data. From this one sees that there is a monotonically decreasing relationship between λ and df_{fit}. Thus, there is a unique df_{fit} corresponding to each value of $\lambda > 0$. For the values of λ depicted in Figure 3.20, the values of df_{fit} are 9.86 and 8.34. This explains their similar appearance: they have roughly the same number of equivalent parameters.

3.14 Residual Degrees of Freedom

As seen in Section 2.4.6, $\mathsf{E}(\text{RSS}) = (n - p)\sigma^2$ in parametric regression models. Therefore, the *degrees of freedom for residuals*, $n - p$, is used to correct for bias when using RSS to estimate σ^2.

In nonparametric regression, we define the *residual degrees of freedom* to be

$$df_{\text{res}} \equiv n - 2\,\text{tr}(\mathbf{S}_\lambda) + \text{tr}(\mathbf{S}_\lambda \mathbf{S}_\lambda^\mathsf{T}). \tag{3.28}$$

As in parametric estimation, the residual degrees of freedom is used in estimation of σ^2. To see the motivation for the definition, suppose that the true model is

$$\mathbf{y} = \mathbf{f} + \boldsymbol{\varepsilon}, \quad \text{where } \text{Cov}(\boldsymbol{\varepsilon}) = \sigma^2 \mathbf{I}. \tag{3.29}$$

Let

$$\hat{\mathbf{f}}_\lambda \equiv \mathbf{S}_\lambda \mathbf{y}. \tag{3.30}$$

Using the result

$$\mathsf{E}(\mathbf{v}^\mathsf{T}\mathbf{A}\mathbf{v}) = \mathsf{E}(\mathbf{v})^\mathsf{T}\mathbf{A}\mathsf{E}(\mathbf{v}) + \text{tr}\{\mathbf{A}\,\text{Cov}(\mathbf{v})\}$$

for a general random vector $\mathbf{v}$ (see Appendix A), we have

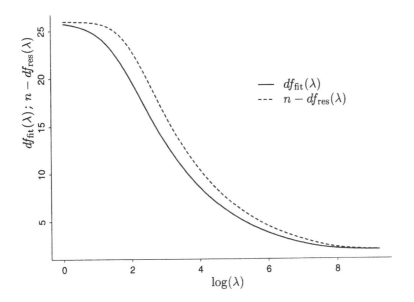

Figure 3.22
Comparison of
$n - df_{res}(\lambda)$ and
$df_{fit}(\lambda)$ for 24-knot
linear spline fits to the
LIDAR data.

$$
\begin{aligned}
E(RSS) &= E\|\hat{\mathbf{f}}_\lambda - \mathbf{y}\|^2 \\
&= E\|(\mathbf{S}_\lambda - \mathbf{I})\mathbf{y}\|^2 \\
&= E\{\mathbf{y}^\mathsf{T}(\mathbf{S}_\lambda - \mathbf{I})^\mathsf{T}(\mathbf{S}_\lambda - \mathbf{I})\mathbf{y}\} \\
&= \mathbf{f}^\mathsf{T}(\mathbf{S}_\lambda - \mathbf{I})^\mathsf{T}(\mathbf{S}_\lambda - \mathbf{I})\mathbf{f} + \sigma^2 \operatorname{tr}\{(\mathbf{S}_\lambda - \mathbf{I})^\mathsf{T}(\mathbf{S}_\lambda - \mathbf{I})\} \\
&= \|(\mathbf{S}_\lambda - \mathbf{I})\mathbf{f}\|^2 + \sigma^2\{\operatorname{tr}(\mathbf{S}_\lambda\mathbf{S}_\lambda^\mathsf{T}) - 2\operatorname{tr}(\mathbf{S}_\lambda) + n\} \\
&= \|(\mathbf{S}_\lambda - \mathbf{I})\mathbf{f}\|^2 + \sigma^2 df_{res}. \quad\quad (3.31)
\end{aligned}
$$

Assuming that the bias term $\|(\mathbf{S}_\lambda - \mathbf{I})\mathbf{f}\|^2$ is negligible, it follows that

$$RSS/df_{res}$$

is an unbiased estimate of σ^2.

Note that

$$n - df_{res} = 2\operatorname{tr}(\mathbf{S}_\lambda) - \operatorname{tr}(\mathbf{S}_\lambda\mathbf{S}_\lambda^\mathsf{T}) \quad\quad (3.32)$$

is an alternative measure to

$$df_{fit} = \operatorname{tr}(\mathbf{S}_\lambda) \quad\quad (3.33)$$

for the effective number of parameters being fit by $\hat{f}_\lambda$. For parametric regression models fitted by ordinary least squares, (3.32) and (3.33) coincide because $\mathbf{S}_\lambda\mathbf{S}_\lambda^\mathsf{T} = \mathbf{S}_\lambda$. But for nonparametric models they can differ substantially. For example, in the 24-knot linear spline fit to the LIDAR data,

$$df_{fit} = 18 \text{ corresponds to } n - df_{res} = 21.3.$$

Figure 3.22 is an embellishment of Figure 3.21 with $n - df_{res}$ added. The difference is greater for "mid-range" amounts of smoothing, whereas for low and high amounts of smoothing the measures tend to coincide. This is because zero smoothing and "infinite" smoothing correspond to parametric regression fits.

For regression methods that are based on residual sums of squares (such as F-tests), use of df_{res} is more appropriate than df_{fit}. Details are given in Section 6.6.2.

3.15   Other Approaches to Scatterplot Smoothing

Spline-based smoothers form just one class of the large collection of scatterplot smoothers developed over the years. In this section we briefly describe some of the other main classes.

3.15.1   *Local Polynomial Fitting*

One of the most popular methods for smoothing a scatterplot is local polynomial fitting. One of its advantages compared with spline-based smoothers is simpler theoretical analysis. This has allowed greater insight into the smoothing process. Summaries of this theory are given in Wand and Jones (1995), Fan and Gijbels (1996), and Loader (1999).

Figure 3.23 provides an illustration of the basic idea. The smooth at $x = u$ is obtained by fitting a weighted least-squares line where the weights correspond to the height of the *kernel function,* which is shown at the base of the plot. The estimate at $x = v$ is obtained similarly and also illustrated in Figure 3.23. If this procedure is applied over a grid of x-values then the solid curve results.

In Figure 3.23, local lines are being fitted. However, polynomials of any degree could be used. Let p be the degree of the polynomial being fit. At a point x, the smooth is obtained by fitting the pth-degree polynomial model

$$E(y_i) = \beta_0 + \beta_1(x_i - x) + \cdots + \beta_p(x_i - x)^p$$

using weighted least squares with *kernel weights* $K\{b^{-1}(x_i - x)\}$. The kernel function K is usually taken to be a symmetric positive function with $K(x)$ decreasing as $|x|$ increases. For example, Figure 3.23 uses the standard normal density function. The parameter $b > 0$ is the smoothing parameter for local polynomial smoothers and is usually referred to as the *bandwidth*. The value of the curve estimate is the height of the fit $\hat{\beta}_0$, where $\hat{\boldsymbol{\beta}} = [\hat{\beta}_0, \ldots, \hat{\beta}_p]^\mathsf{T}$ minimizes

$$\sum_{i=1}^{n} \{y_i - \beta_0 - \cdots - \beta_p(x_i - x)^p\}^2 K\left(\frac{x_i - x}{b}\right).$$

Assuming the invertibility of $\mathbf{X}_x^\mathsf{T}\mathbf{W}_x\mathbf{X}_x$, standard weighted least-squares theory leads to the solution

$$\hat{\boldsymbol{\beta}} = (\mathbf{X}_x^\mathsf{T}\mathbf{W}_x\mathbf{X}_x)^{-1}\mathbf{X}_x^\mathsf{T}\mathbf{W}_x\mathbf{y},$$

where

$$\mathbf{X}_x = \begin{bmatrix} 1 & x_1 - x & \cdots & (x_1 - x)^p \\ \vdots & \vdots & \ddots & \vdots \\ 1 & x_n - x & \cdots & (x_n - x)^p \end{bmatrix}$$

is an $n \times (p + 1)$ design matrix and

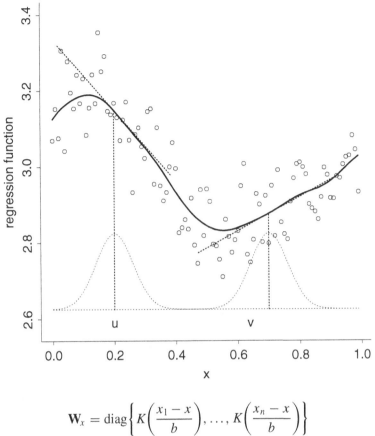

Figure 3.23 Local linear scatterplot smooth (solid curve) based on 100 simulated observations (represented by circles). The dotted curves are the kernel weights and cubic fits at the points u and v.

$$\mathbf{W}_x = \text{diag}\left\{ K\left(\frac{x_1 - x}{b}\right), \ldots, K\left(\frac{x_n - x}{b}\right) \right\}$$

is an $n \times n$ diagonal matrix of weights. Since the estimator of $f(x) = \mathsf{E}(y|x)$ is the intercept coefficient, we obtain

$$\hat{f}(x; p, b) = \mathbf{e}_1^\mathsf{T}(\mathbf{X}_x^\mathsf{T}\mathbf{W}_x\mathbf{X}_x)^{-1}\mathbf{X}_x^\mathsf{T}\mathbf{W}_x\mathbf{y},$$

where $\mathbf{e}_1$ is the $(p+1) \times 1$ vector having 1 in the first entry and 0 elsewhere.

The case $p = 0$ results in the *Nadaraya–Watson* (Nadaraya 1964; Watson 1964) estimator:

$$\hat{m}(x; 0, b) = \frac{\displaystyle\sum_{i=1}^{n} K\left(\frac{x_i - x}{b}\right) y_i}{\displaystyle\sum_{i=1}^{n} K\left(\frac{x_i - x}{b}\right)}.$$

The data analyst must choose p and b. Our experience is that $p = 1$ works well if f appears to be monotonically increasing; otherwise, $p = 2$ is satisfactory. The bandwidth b can be chosen by trial and error with visual inspection, but it can also be chosen from the data using one of the automatic smoothing parameter selection approaches discussed in Chapter 5.

The Nadaraya–Watson (or "local constant") estimator has long been studied by theoreticians, but the local linear ($p = 1$) estimator seems to have been more widely used in practice after the seminal paper of Cleveland (1979). The reasons for the superior practical performance of local linear over local constant estimation became clearer with the papers of Fan (1992, 1993). Near the boundaries of the data – and also in the interior, if the x are unequally spaced – local linear estimation is less biased than local constant estimation. Fan (1992, 1993) showed that, as $n \to \infty$ and $b \to 0$, the bias of $\hat{f}(x; b, p)$ is $O(b^2)$ for all x but the bias of $\hat{m}(x; b, p)$ is $O(b)$ at the boundaries and $O(b^2)$ at the interior. Ruppert and Wand (1994) showed that this effect of greater asymptotic bias near the boundary than in the interior holds for all even values of p. However, experience with data and simulation studies is required when interpreting this asymptotic result. The effect of this "boundary bias" is most severe for $p = 0$. In practice, $p = 2$ is an excellent choice for the degree of the local polynomials and is much less variable near the boundaries as compared to $p = 3$. In simulation studies, $p = 2$ often outperforms $p = 1$ and $p = 3$.

> A sequence a_n is $O(c_n)$ if there exists a constant M such that $|a_n| \leq M|c_n|$ for n. In other words, a_n is bounded by a multiple of c_n. Thus, saying that the bias is $O(b^2)$ means that there is a constant M such that the bias when using bandwidth b is bounded in absolute value by Mb^2 for any b.

There are several variations on the basic local polynomial fitting idea depicted in Figure 3.23. Mostly they involve changing the value of the bandwidth across the estimation region. For example, the method of Cleveland (1979) sets the bandwidth so that the number of points used to estimate $f(x)$ is fixed, regardless of the estimation location x. The resulting scatterplot smooth is named LOESS (short for "local regression").

Relative to penalized splines, local polynomial regression is slow to compute if programmed directly. However, there are several strategies for speeding up the calculations (see e.g. Cleveland and Grosse 1991; Härdle and Scott 1992; Fan and Marron 1994; Seifert et al. 1994).

3.15.2 Series-Based Smoothers

Without loss of generality, assume that the regression function f is defined on the unit interval $[0, 1]$. Under certain regularity conditions, f admits the *Fourier series* representation

$$f(x) = \beta_0 + \sum_{j=1}^{\infty} \{\beta_j^s \sin(j\pi x) + \beta_j^c \cos(j\pi x)\}.$$

For higher values of j, the functions $\sin(j\pi x)$ and $\cos(j\pi x)$ become more oscillatory, as shown in Figure 3.24. The more oscillatory functions account for the finer structure in f. For smoother f, the corresponding coefficients will be small. This suggests the model

$$\hat{f}(x) = \hat{\beta}_0 + \sum_{j=1}^{J} \{\hat{\beta}_j^s \sin(j\pi x) + \hat{\beta}_j^c \cos(j\pi x)\},$$

where $\hat{\beta}_j^s$, $\hat{\beta}_j^c$ ($1 \leq j \leq J$) and $\hat{\beta}_0$ are all estimated by least squares. The cut-off value J is the smoothing parameter in this case.

Other basis functions that are ordered by amount of oscillation may be used instead of the trigonometric basis functions. An example is the *Demmler–Reinsch*

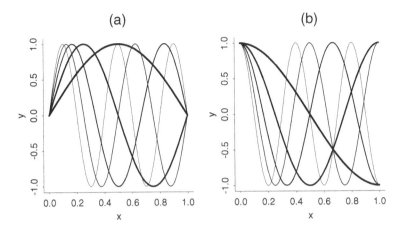

Figure 3.24 Fourier series basis functions: (a) $\sin(j\pi x)$ and (b) $\cos(j\pi x)$, both for $1 \le j \le 5$.

basis, which (like the B-spline basis) is a transformation of the truncated polynomial basis used throughout most of this chapter (see e.g. Nychka and Cummins 1996). Another possibility is one of the *wavelet* bases. The use of wavelets in nonparametric regression has been the focus of a great deal of research since the early 1990s, in part because of the ability of wavelets to handle discontinuities, kinks, and other sharp features in a systematic fashion. We have no hope of doing justice to this burgeoning research area here. Instead we refer to the texts of Ogden (1996) and Härdle et al. (1998).

3.16   Choosing a Scatterplot Smoother

In this chapter we have described several ways of smoothing a scatterplot. Numerous embellishments (see e.g. Chapter 17) are possible. How should one decide between these various choices?

First, it should be noted that there are approximate mathematical equivalences between the various linear smoothers described in this chapter (e.g., Silverman 1984; Eubank 1994). This means that, for fixed degrees of freedom, the fits from two different scatterplot smoothing methods are approximately the same. This is illustrated in Figure 3.25 for low-rank and full-rank spline, local linear, and Fourier series smoothers. Each smooth uses 11 degrees of freedom on the LIDAR data and are difficult to tell apart.

Some criteria for evaluating and deciding between scatterplot smoothers may be listed as follows.

The list of criteria for evaluating and deciding between scatterplot smoothers is an adaptation of one given in Marron (1996).

(1) *Convenience.* Is it available on the analyst's favorite computer package?
(2) *Implementability.* If not immediately available, how easy is it to implement in the analyst's favorite programming language?
(3) *Flexibility.* Is the smoother able to handle a wide range of types of relationships that may exist among the variables of interest?
(4) *Simplicity.* Is it easy to understand how the technique processes the data to obtain answers?

Figure 3.25
Scatterplot smooths
of the LIDAR data
using four different
approaches: 24-knot
penalized spline (low-
rank); smoothing
spline (full-rank);
local linear estimator;
Fourier series
estimator. Each
smooth uses 11
degrees of freedom.

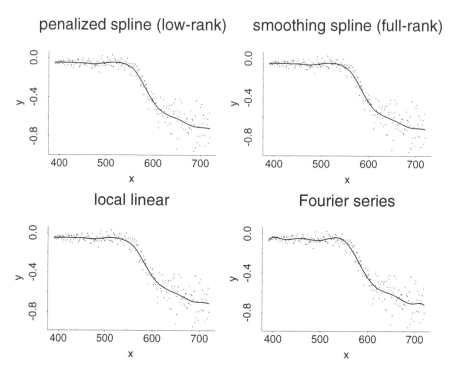

(5) *Tractability*. Is it easy to analyze the mathematical properties of the technique?

(6) *Reliability*. Can the answers be trusted?

(7) *Efficiency*. Does the technique use the data in the most efficient way?

(8) *Extendibility*. Is the technique easily extended to more complicated settings?

This book is geared toward analysis of real data sets, some of which are quite
complex (see Chapter 1). Therefore, we place a high premium on (2) and (8).
Criteria (4), (6), and (7) are also quite important to us, and each of the techniques
described in this chapter do quite well in this regard.

A trade-off exists regarding (3). Most of the smoothers described in this chapter require the specification of a single smoothing parameter. However, there is a
cost in flexibility and – for functions with differing amounts of curvature across
their domain – improvements are possible by using, for example, the ideas described in Section 3.4 and Chapter 17. On the other hand, these methods require a
bigger implementational effort and can be difficult to extend to more complicated
settings.

3.17   Bibliographic Notes

Scatterplot smoothing has a very large literature both inside and outside of statistics. In the past decade or so, several books have summarized various aspects of

the problem from a statistical viewpoint, so we will use these as pointers to the wider literature.

Books mainly on spline approaches to scatterplot smoothing are Eubank (1988), Wahba (1990), Green and Silverman (1994), Dierckx (1995), Gu (2002), and Hansen et al. (2003). The latter book concentrates on the automatic basis function approach described in Section 3.4.

Books mainly on local polynomial and kernel approaches to scatterplot smoothing are Müller (1988), Nadaraya (1989), Härdle (1990, 1991), Wand and Jones (1995), Fan and Gijbels (1996), Simonoff (1996), Bowman and Azzalini (1997), Hart (1997), Loader (1999), Pagan and Ullah (1999), and Fox (2000). Loader's book contains an interesting early history on local regression.

Books on classical series approaches are Thompson and Tapia (1990), Tarter and Lock (1993), and Efromovich (1999). The wavelet approaches are treated in Ogden (1996), Louis, Maass, and Rieder (1997), Härdle et al. (1998), Müller and Vidakovic (1999), Vidakovic (1999), Nason and Silverman (2000), and Walter and Shen (2001).

3.18   Summary of Formulas

Nonparametric regression model
Given scatterplot data (x_i, y_i), $1 \le i \le n$,

$$y_i = f(x_i) + \varepsilon_i$$

where $f(x) = \mathsf{E}(y|x)$.

Penalized spline with truncated polynomial basis
Model is

$$f(x) = \beta_0 + \beta_1 x + \cdots + \beta_p x^p + \sum_{k=1}^{K} \beta_{pk}(x - \kappa_k)_+^p$$

for $p = 1, 2, \ldots$.
   Fitting criterion is

$$\text{minimize } \|\mathbf{y} - \mathbf{X}\boldsymbol{\beta}\|^2 + \lambda^{2p} \boldsymbol{\beta}^\mathsf{T} \mathbf{D}\boldsymbol{\beta}$$

where

$$\mathbf{X} = \begin{bmatrix} 1 & x_1 & \cdots & x_1^p & (x_1 - \kappa_1)_+^p & \cdots & (x_1 - \kappa_K)_+^p \\ \vdots & \vdots & \ddots & \vdots & \vdots & \ddots & \vdots \\ 1 & x_n & \cdots & x_n^p & (x_n - \kappa_1)_+^p & \cdots & (x_n - \kappa_K)_+^p \end{bmatrix}$$

and $\mathbf{D} = \mathrm{diag}(\mathbf{0}_{p+1}, \mathbf{1}_K)$.
   Fitted values are

$$\hat{\mathbf{y}} = \mathbf{X}(\mathbf{X}^\mathsf{T}\mathbf{X} + \lambda^{2p}\mathbf{D})^{-1}\mathbf{X}^\mathsf{T}\mathbf{y}$$

Smoother matrix is

$$\mathbf{S}_\lambda = \mathbf{X}(\mathbf{X}^\mathsf{T}\mathbf{X} + \lambda^{2p}\mathbf{D})^{-1}\mathbf{X}^\mathsf{T}$$

Penalized spline with radial basis
Model is

$$f(x) = \beta_0 + \beta_1 x + \cdots + \beta_{m-1} x^{m-1} + \sum_{k=1}^{K} \beta_{mk} |x - \kappa_k|^{2m-1}$$

for $m = 1, 2, \ldots$.

Fitting criterion is

$$\text{minimize } \|\mathbf{y} - \mathbf{X}\boldsymbol{\beta}\|^2 + \lambda^{2m-1} \boldsymbol{\beta}^{\mathsf{T}} \mathbf{K} \boldsymbol{\beta}$$

where

$$\mathbf{X} = \begin{bmatrix} 1 & x_1 & \cdots & x_1^{m-1} & |x_1 - \kappa_1|^{2m-1} & \cdots & |x_1 - \kappa_K|^{2m-1} \\ \vdots & \vdots & \ddots & \vdots & \vdots & \ddots & \vdots \\ 1 & x_n & \cdots & x_n^{m-1} & |x_n - \kappa_1|^{2m-1} & \cdots & |x_n - \kappa_K|^{2m-1} \end{bmatrix}$$

and

$$\mathbf{K} = \left[|\kappa_k - \kappa_{k'}|^{2m-1} \right]_{1 \le k, k' \le K}$$

Fitted values are

$$\hat{\mathbf{y}} = \mathbf{X}(\mathbf{X}^{\mathsf{T}} \mathbf{X} + \lambda^{2m-1} \mathbf{K})^{-1} \mathbf{X}^{\mathsf{T}} \mathbf{y}$$

Smoother matrix is

$$\mathbf{S}_\lambda = \mathbf{X}(\mathbf{X}^{\mathsf{T}} \mathbf{X} + \lambda^{2m-1} \mathbf{K})^{-1} \mathbf{X}^{\mathsf{T}}$$

Linear smoothers
Fitted values $\hat{\mathbf{y}}$ of the form

$$\hat{\mathbf{y}} = \mathbf{L}\mathbf{y}$$

for some $n \times n$ matrix $\mathbf{L}$.

Rank of a symmetric linear smoother
For symmetric $\mathbf{L}$:

rank of a linear smoother = rank of $\mathbf{L}$

= number of eigenvalues of $\mathbf{L}$ that are nonzero

Degrees of freedom of a linear smoother

$$df_{\text{fit}} = \text{tr}(\mathbf{L})$$

Residual degrees of freedom of a linear smoother

$$df_{\text{res}} = n - 2\,\text{tr}(\mathbf{L}) + \text{tr}(\mathbf{L}\mathbf{L}^{\mathsf{T}})$$

4

Mixed Models

4.1 Introduction

Mixed models are an extension of regression models that allow for the incorporation of *random effects*. However, they also turn out to be closely related to smoothing. In fact, we will show in Section 4.9 that the penalized spline smoother exactly corresponds to the optimal predictor in a mixed model framework. This link allows for mixed model methodology and software to be used in semiparametric regression analysis, as we will demonstrate in subsequent chapters.

This chapter begins with a brief review of mixed models. Readers with detailed knowledge of mixed models could skip these sections and proceed directly to Section 4.9.

4.2 Mixed Models

Much of the early work on mixed models – in particular, the special case of *variance component* models – was motivated by the analysis of data from animal breeding experiments and driven by the need to incorporate heritabilities and genetic correlations in a parsimonious fashion. They have also played an important role in establishing quality control procedures and determination of sampling designs, among other applications. Overviews of this vast topic are provided by Searle, Casella, and McCulloch (1992), Vonesh and Chinchilli (1997), Pinheiro and Bates (2000), Verbeke and Molenberghs (2000), and McCulloch and Searle (2001).

A more contemporary application of mixed models is the analysis of longitudinal data sets (see e.g. Laird and Ware 1982; Diggle et al. 2002). We will use this setting to illustrate the essence of mixed modeling.

Figure 4.1 shows two representations of data pertaining to weight measurements of 48 pigs for nine successive weeks. Figure 4.1(a) is simply a scatterplot of the weights against their corresponding week number; in Figure 4.1(b), lines are drawn connecting those measurements that belong to the same pig. This second panel shows the longitudinal aspect of the data: there are nine repeated measurements for each pig.

Let weight$_{ij}$ denote the weight of pig i on week j, and let week$_j = j$ be the corresponding week number. If the data are treated *cross-sectionally* (i.e.,

Figure 4.1
Representations of pig
weight data. Panel (a)
is a scatterplot of
weight against week
number. In (b), lines
are used to connect
those points pertaining
to the same pig.

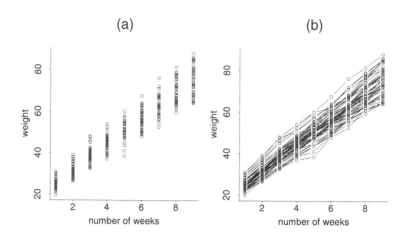

without taking the repeated measure aspect into account), then the ordinary least-squares model

$$\texttt{weight}_{ij} = \beta_0 + \beta_1 \texttt{week}_j + \varepsilon_{ij}, \quad 1 \le i \le 48, \ 1 \le j \le 9, \quad (4.1)$$

with ε_{ij} i.i.d. N$(0, \sigma_\varepsilon^2)$, leads to a slope estimate (with estimated standard deviation) of

$$\hat{\beta}_1 = 6.21, \qquad \widehat{\text{st.dev.}}(\hat{\beta}_1) = 0.0818.$$

But there are some problems with (4.1). First of all, inspection of Figure 4.1(b) shows that the scatterplot *for each individual pig* is less variable, so one would expect that utilization of *within-pig* information would be beneficial. Related to this is the fact that (4.1) ignores the correlation of measurements pertaining to the same pig. An analysis of the residuals shows patterns arising from this correlation, so the assumption that the ε_{ij} are independent does not hold.

An initial remedy would be to extend (4.1) to allow for a different intercept for each pig. This models the data shown in Figure 4.1(b) as 48 parallel lines and would be written

$$\texttt{weight}_{ij} = \alpha_i + \beta_1 \texttt{week}_j + \varepsilon_{ij}, \quad (4.2)$$

where α_i represents the intercept for the ith pig. This leads to a noticeably more precise estimate of β_1, but model (4.2) is unappealing on a few counts. First, it contains 49 parameters, 48 intercepts and one slope, which is somewhat large for such a simple data set. Secondly, it gives too much credence to the pigs used in the study. If 48 different pigs were sampled, then the estimated α_i would be completely different. Normally we think of parameters as being population dependent rather than sample dependent. Other longitudinal data sets have many more subjects and even fewer repeated measurements. So a model such as (4.2) is not very satisfactory when the data are longitudinal.

Longitudinal data
sets consist of
measurements made
on a set of individuals
repeatedly over time.

A remedy is to replace α_i by a *random intercept*:

$$\texttt{weight}_{ij} = \beta_0 + U_i + \beta_1 \texttt{week}_j + \varepsilon_{ij}. \quad (4.3)$$

Here

$$U_1, \ldots, U_{48}$$

are treated as a random sample from, say, a $N(0, \sigma_U^2)$ distribution for some $\sigma_U^2 > 0$. The U_i-term is an example of a *random effect* and has the advantage of requiring just a single parameter, σ_U^2, which is commonly referred to as a *variance component*. Moreover, it takes into account the randomness due to other samples of pigs. For these data, one may want to consider a random slope model in which β_1 is replaced by $\beta_1 + V_i$, where V_i is a random effect that accounts for possible variability in the slopes of the growth curves. However, since we are using this example for illustrative purposes, we will assume that (4.3) is adequate.

Model (4.3) is an example of a *mixed model*. It has a *fixed component*,

$$\beta_0 + \beta_1 \texttt{week}_j,$$

as well as a *random component*,

$$U_i \sim N(0, \sigma_U^2).$$

The next two sections describe estimation techniques for fitting (4.3). For these data they result in

$$\hat{\beta}_1 = 6.21, \qquad \widehat{\text{st.dev.}}(\hat{\beta}_1) = 0.0391,$$

which is somewhat more precise than an ordinary regression model. Moreover, the random intercept U_i allows for the within-pig correlation. To appreciate this, note that the covariance between the weights of pig i, measured at two different times ($j \neq j'$), is

$$\text{cov}(\texttt{weight}_{ij}, \texttt{weight}_{ij'}) = \text{var}(U_i) = \sigma_U^2.$$

The correlation coefficient is then

$$\text{corr}(\texttt{weight}_{ij}, \texttt{weight}_{ij'}) = \frac{\sigma_U^2}{\sigma_\varepsilon^2 + \sigma_U^2}.$$

This is estimated to be 0.775, indicating considerable within-pig correlation in this case.

4.2.1 Degrees-of-Freedom Interpretation

Consider the more general form of (4.3):

$$y_{ij} = \beta_0 + U_i + \beta_1 x_{ij} + \varepsilon_{ij}, \quad 1 \leq j \leq n_i, \ 1 \leq i \leq m. \tag{4.4}$$

Section 4.5 describes how to fit this model. As discussed in that section, we can write the vector of fitted values as

$$\hat{\mathbf{y}} = \mathbf{H}_0 \mathbf{y} + \mathbf{H}_U \mathbf{y} + \mathbf{H}_x \mathbf{y}$$

and then, analogously to (2.22), define the degrees of freedom for each component of (4.4):

$$df_0 = \text{tr}(\mathbf{H}_0), \quad df_U = \text{tr}(\mathbf{H}_U), \quad df_x = \text{tr}(\mathbf{H}_x).$$

Figure 4.2 Plot of
degrees of freedom
for random intercepts
(df_U) versus the
variance ratio $\hat{\sigma}_\varepsilon^2/\hat{\sigma}_U^2$
for the pig weight
example. The lines
correspond to the
estimated values via
REML, as discussed
in Section 4.5.4.

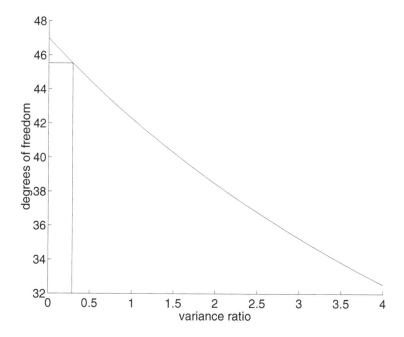

Not surprisingly,

$$df_0 = df_x = 1,$$

since each of these components involve a single parameter: β_0 and β_1, respectively. When $n_1 = \cdots = n_m$, we have

$$df_U = \frac{(m-1)n_1}{n_1 + \sigma_\varepsilon^2/\sigma_U^2}.$$

This shows that the effective number of parameters depends on the variance ratio

$$\sigma_\varepsilon^2/\sigma_U^2.$$

In the pig weight example,

$$df_U = 47\left(\frac{9}{9 + \sigma_\varepsilon^2/\sigma_U^2}\right).$$

Figure 4.2 shows this degrees of freedom plotted against $\sigma_\varepsilon^2/\sigma_U^2$. Using restricted maximum likelihood (described in Section 4.5.4), this ratio is estimated to be $\hat{\sigma}_\varepsilon^2/\hat{\sigma}_U^2 = 0.29$, corresponding to 46.5 degrees of freedom. We see from this that the random intercept model corresponds to a potential compromise between two possible models with fixed effects only.

(a) A single fixed intercept ($df_0 + df_U = 1$, $\sigma_U^2 = 0$); this is model (4.1).
(b) an intercept for each pig (48 parameters, so $df_0 + df_U = 48$, $\sigma_U^2 = \infty$); this is model (4.2).

In model (a), $\sigma_U = 0$ implies that $U_1, \ldots, U_{48} = 0$. Hence all the random effects drop out, and we are left with a fixed effects model.

When $\sigma_U^2 = \infty$ in model (b), the interpretation is that $U_1, \ldots, U_{48}$ are no longer random but rather are unknown fixed constants; the random effects have become fixed effects. Model (b) appears to have 49 parameters, $\beta_0, U_1, \ldots, U_{48}$, but in this model one constraint is needed to make the parameters well-defined. Often this constraint is chosen to be $U_1 + \cdots + U_{48} = 0$. Although the random intercept model is a potential compromise between (a) and (b), we see that in this example it is quite close to (b) because the random and fixed intercept models differ by only 1.5 degrees of freedom.

In this example the random intercept is closer to (b) than to (a) because the within-pig variability is considerably lower than the between-pig variability.

There is an analogy between the $\hat{U}_i$ and the $\hat{\beta}_{1k}$ of Section 3.5. In both cases, the estimates are shrunk in such a way that their contribution to the degrees of freedom of the fitted values is less than the number of coefficients. As will be clear by the end of this chapter, the fitting of both longitudinal data sets (such as the pig weight data) and nonlinear trends (such as for the LIDAR data) can be achieved through the same general approach.

4.3   Prediction

Mixed models contain fixed effects, random effects, and covariance matrix parameters. For model (4.3) the fixed effects are β_0 and β_1, the random effects are $U_1, \ldots, U_{48}$, and the covariance matrix parameters are σ_U^2 and σ_ε^2. The *parameters* in the model are $(\beta_0, \beta_1, \sigma_U^2, \sigma_\varepsilon^2)$, and their estimation can be achieved by using common statistical approaches such as maximum likelihood. This is treated for general linear mixed models in Sections 4.5.1 and 4.5.4. However, maximum likelihood does not apply to random effects. Instead we can form *predictions* of $U_1, \ldots, U_{48}$. The difference between prediction and estimation is that the target is random for the former but deterministic (nonrandom) for the latter. Some writers (Robinson 1991; Hayes and Haslett 1999) argue that this distinction is unnecessary and that the word "estimation" should be used for both types of targets. However, in accordance with the majority of relevant literature, we will here use the classical naming convention.

Prediction is a fundamental problem in statistics, although its treatment in textbooks is overshadowed by estimation. An excellent synopsis of prediction is provided by Chapter 9 of McCulloch and Searle (2001). We will summarize the main points here. Figure 4.3 shows the distributions of two random variables y and v, which are distributed according to

Prediction is at the heart of our semiparametric and nonparametric methods.

$$y = v + \varepsilon, \quad \text{where} \quad \begin{bmatrix} v \\ \varepsilon \end{bmatrix} \sim \mathrm{N}\left(\begin{bmatrix} 0 \\ 0 \end{bmatrix}, \begin{bmatrix} 1 & 0 \\ 0 & 4 \end{bmatrix} \right).$$

We observe only y. Based on this observation, what is a good prediction for the value of v? The *best predictor* (BP) of v is defined to be the $\tilde{v}$ for which

$$\mathrm{E}\{(\tilde{v} - v)^2\}$$

Figure 4.3 Simple
illustration of
prediction. The value
of y is observed but
v's value is not. The
best predictor of v is
$y/5$.

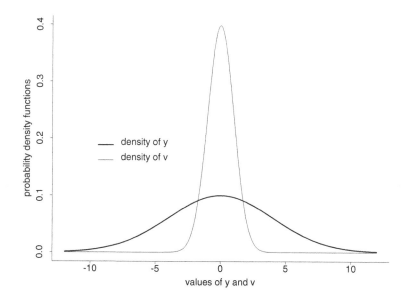

is minimized. For general (y, v) the solution is

$$\tilde{v} \equiv \mathrm{BP}(v) = \mathsf{E}(v|y);$$

in the current example, we obtain

$$\tilde{v} = y/5.$$

This is intuitively consistent with Figure 4.3, where v is seen to be a "shrunken" version of y.

In general: If $\mathbf{y}$ is the vector of observed data and $\mathbf{v}$ is a random vector, then best prediction corresponds to minimization of

$$\mathsf{E}\{\|\tilde{\mathbf{v}} - \mathbf{v}\|^2\},$$

and the solution is

$$\tilde{\mathbf{v}} \equiv \mathrm{BP}(\mathbf{v}) = \mathsf{E}(\mathbf{v}|\mathbf{y}).$$

4.3.1 *Best Linear Prediction (BLP)*

The best predictor is not necessarily a linear function of $\mathbf{y}$. A common simplification is to restrict the family of predictors to be linear. That is,

$$\tilde{\mathbf{v}} = \mathbf{A}\mathbf{y} + \mathbf{b}$$

for some matrix $\mathbf{A}$ and vector $\mathbf{b}$. The solution is called the *best linear predictor* (BLP) and can be shown to be

$$\tilde{\mathbf{v}} \equiv \mathrm{BLP}(\mathbf{v}) = \mathsf{E}(\mathbf{v}) + \mathbf{C}\mathbf{V}^{-1}\{\mathbf{y} - \mathsf{E}(\mathbf{y})\}, \qquad (4.5)$$

where

$$\mathbf{C} \equiv \mathsf{E}[\{\mathbf{v} - \mathsf{E}(\mathbf{v})\}\{\mathbf{y} - \mathsf{E}(\mathbf{y})\}^{\mathsf{T}}] \quad \text{and} \quad \mathbf{V} \equiv \mathrm{Cov}(\mathbf{y}).$$

If

$$\begin{bmatrix} \mathbf{v} \\ \mathbf{y} \end{bmatrix} \text{ is multivariate normal,}$$

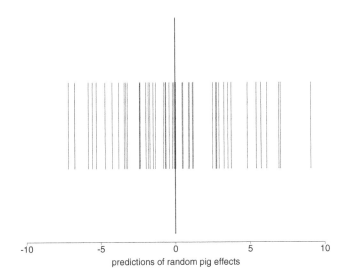

Figure 4.4
Predictions of
$U_1, \ldots, U_{48}$ for the
pig weight data. A
vertical line is plotted
for each $\tilde{U}_i$ value,
$1 \leq i \leq 48$.

then best prediction and best linear prediction coincide. That is,

$$BP(\mathbf{v}) = BLP(\mathbf{v}) = E(\mathbf{v}|\mathbf{y}) = E(\mathbf{v}) + \mathbf{CV}^{-1}\{\mathbf{y} - E(\mathbf{y})\}.$$

4.3.2 Application to Pig Weight Example

In model (4.3), let

$$\mathbf{U} = \begin{bmatrix} U_1 \\ \vdots \\ U_{48} \end{bmatrix} \quad \text{and} \quad \mathbf{y} = \begin{bmatrix} \text{weight}_{1,1} \\ \vdots \\ \text{weight}_{1,9} \\ \vdots \\ \text{weight}_{48,1} \\ \vdots \\ \text{weight}_{48,9} \end{bmatrix}.$$

Then

$$\begin{bmatrix} \mathbf{U} \\ \mathbf{y} \end{bmatrix} \text{ is multivariate normal}$$

and, for given β_0, β_1, σ_U^2, and σ_ε^2, the best predictor of U_i reduces to

$$\tilde{U}_i = \frac{n_i \sigma_U^2}{\sigma_\varepsilon^2 + n_i \sigma_U^2}(\bar{y}_{i.} - \beta_0 - \beta_1 \bar{x}_{i.})$$

$$= \frac{9\sigma_U^2}{\sigma_\varepsilon^2 + 9\sigma_U^2}(\overline{\text{weight}}_{i.} - \beta_0 - \beta_1 \overline{\text{week}}),$$

where $\bar{y}_{i.} = \overline{\text{weight}}_{i.}$ is the average weight of the ith pig and $\bar{x}_{i.} = 5$ is the average week value. See McCulloch and Searle (2001).

Figure 4.4 shows $\tilde{U}_1, \ldots, \tilde{U}_{48}$ after estimates of the variance components are plugged in. The variability in the intercepts among the 48 pigs is apparent, as is an estimated ranking of the pigs in this regard. Indeed, there is a branch of statistics devoted to *ranking* and *selection* of subjects that has roots in animal breeding and genetics.

4.4 The Linear Mixed Model (LMM)

Just as with the linear model, we can generalize mixed models to arbitrary design matrices. The covariance structure of the random effects vector can also be general. The resulting general *linear mixed model* is

$$\mathbf{y} = \mathbf{X}\boldsymbol{\beta} + \mathbf{Z}\mathbf{u} + \boldsymbol{\varepsilon}, \tag{4.6}$$

where

$$E\begin{bmatrix} \mathbf{u} \\ \boldsymbol{\varepsilon} \end{bmatrix} = \begin{bmatrix} \mathbf{0} \\ \mathbf{0} \end{bmatrix} \quad \text{and} \quad \text{Cov}\begin{bmatrix} \mathbf{u} \\ \boldsymbol{\varepsilon} \end{bmatrix} = \begin{bmatrix} \mathbf{G} & \mathbf{0} \\ \mathbf{0} & \mathbf{R} \end{bmatrix}.$$

Note that model (4.3) is a special case of (4.6) with

$$\mathbf{y} = \begin{bmatrix} \text{weight}_{1,1} \\ \vdots \\ \text{weight}_{1,9} \\ \vdots \\ \text{weight}_{48,1} \\ \vdots \\ \text{weight}_{48,9} \end{bmatrix}, \quad \mathbf{X} = \begin{bmatrix} 1 & \text{week}_1 \\ \vdots & \vdots \\ 1 & \text{week}_9 \\ \vdots & \vdots \\ 1 & \text{week}_1 \\ \vdots & \vdots \\ 1 & \text{week}_9 \end{bmatrix}, \quad \boldsymbol{\beta} = \begin{bmatrix} \beta_0 \\ \beta_1 \end{bmatrix},$$

$$\mathbf{Z} = \begin{bmatrix} \mathbf{1}_{9\times1} & \mathbf{0}_{9\times1} & \cdots & \mathbf{0}_{9\times1} \\ \mathbf{0}_{9\times1} & \mathbf{1}_{9\times1} & \cdots & \mathbf{0}_{9\times1} \\ \vdots & \vdots & \ddots & \vdots \\ \mathbf{0}_{9\times1} & \mathbf{0}_{9\times1} & \cdots & \mathbf{1}_{9\times1} \end{bmatrix}, \quad \mathbf{u} = \begin{bmatrix} U_1 \\ \vdots \\ U_{48} \end{bmatrix},$$

$$\mathbf{G} = \sigma_U^2 \mathbf{I}, \quad \text{and} \quad \mathbf{R} = \sigma_\varepsilon^2 \mathbf{I}.$$

As we will see in subsequent chapters, the general model (4.6) is extremely rich in that it includes a large number of special cases that are useful in practice. The next few sections discuss statistical inference within this general framework. We will then explain how such inference relates to semiparametric regression modeling.

4.5 Estimation and Prediction in LMM

We now treat estimation of $\boldsymbol{\beta}$, prediction of $\mathbf{u}$, and estimation of the parameters in $\mathbf{G}$ and $\mathbf{R}$.

4.5.1 *Estimation of Fixed Effects*

One way to derive an estimate of $\boldsymbol{\beta}$ is to rewrite (4.6) as

$$\mathbf{y} = \mathbf{X}\boldsymbol{\beta} + \boldsymbol{\varepsilon}^*, \quad \text{where} \quad \boldsymbol{\varepsilon}^* = \mathbf{Z}\mathbf{u} + \boldsymbol{\varepsilon}.$$

This is just a linear model with correlated errors, since

$$\text{Cov}(\boldsymbol{\varepsilon}^*) \equiv \mathbf{V} = \mathbf{ZGZ}^{\mathsf{T}} + \mathbf{R}.$$

For given $\mathbf{V}$, the classical textbook estimator of $\boldsymbol{\beta}$ (e.g., Rao 1973; Draper and Smith 1998) is

$$\tilde{\boldsymbol{\beta}} = (\mathbf{X}^{\mathsf{T}}\mathbf{V}^{-1}\mathbf{X})^{-1}\mathbf{X}^{\mathsf{T}}\mathbf{V}^{-1}\mathbf{y} \tag{4.7}$$

and is sometimes referred to as *generalized least squares* (GLS).

Expression (4.7) can be justified in a number of different ways. For $\mathbf{y}$ having a general distribution, (4.7) can be shown to be the *best linear unbiased estimator* (BLUE) for $\boldsymbol{\beta}$. Alternatively, if $\mathbf{y}$ is multivariate normal then the right-hand side of (4.7) is both the *maximum likelihood estimator* (MLE) and the *uniformly minimum variance unbiased estimator* (UMVUE). The latter is the estimator that has the best (smallest) possible variance of any unbiased estimator regardless of the parameter values.

4.5.2 *Prediction of Random Effects*

The random effects vector can be predicted via best linear prediction using (4.5). For given $\boldsymbol{\beta}$, we obtain

$$\tilde{\mathbf{u}} = \text{BLP}(\mathbf{u}) = \mathbf{GZ}^{\mathsf{T}}\mathbf{V}^{-1}(\mathbf{y} - \mathbf{X}\boldsymbol{\beta}). \tag{4.8}$$

In practice $\boldsymbol{\beta}$ would be replaced by an estimator such as $\tilde{\boldsymbol{\beta}}$ in (4.7), and the parameters in $\mathbf{G}$ and $\mathbf{V}$ would need to be estimated (see Section 4.6).

4.5.3 *Best Linear Unbiased Prediction (BLUP)*

A more unifying way to arrive at the results of the previous two subsections is through the notion of *best linear unbiased prediction* (BLUP). For arbitrary $n \times 1$ vectors $\mathbf{s}$ and $\mathbf{t}$, this involves the determination of linear $\tilde{\boldsymbol{\beta}}$ and $\tilde{\mathbf{u}}$ to minimize the prediction error

$$\mathrm{E}\{(\mathbf{s}^{\mathsf{T}}\mathbf{X}\tilde{\boldsymbol{\beta}} + \mathbf{t}^{\mathsf{T}}\mathbf{Z}\tilde{\mathbf{u}}) - (\mathbf{s}^{\mathsf{T}}\mathbf{X}\boldsymbol{\beta} + \mathbf{t}^{\mathsf{T}}\mathbf{Z}\mathbf{u})\}^2$$

subject to the unbiasedness condition

$$\mathrm{E}(\mathbf{s}^{\mathsf{T}}\mathbf{X}\tilde{\boldsymbol{\beta}} + \mathbf{t}^{\mathsf{T}}\mathbf{Z}\tilde{\mathbf{u}}) = \mathrm{E}(\mathbf{s}^{\mathsf{T}}\mathbf{X}\boldsymbol{\beta} + \mathbf{t}^{\mathsf{T}}\mathbf{Z}\mathbf{u}).$$

Then it can be shown (see e.g. Robinson 1991; Hayes and Haslett 1999) that the solutions for $\tilde{\boldsymbol{\beta}}$ and $\tilde{\mathbf{u}}$ are

$$\text{BLUP}(\boldsymbol{\beta}) \equiv \tilde{\boldsymbol{\beta}} = (\mathbf{X}^{\mathsf{T}}\mathbf{V}^{-1}\mathbf{X})^{-1}\mathbf{X}^{\mathsf{T}}\mathbf{V}^{-1}\mathbf{y},$$
$$\text{BLUP}(\mathbf{u}) \equiv \tilde{\mathbf{u}} = \mathbf{GZ}^{\mathsf{T}}\mathbf{V}^{-1}(\mathbf{y} - \mathbf{X}\tilde{\boldsymbol{\beta}}). \tag{4.9}$$

Note that the BLUP for $\boldsymbol{\beta}$ is identical to the generalized least-squares solution (4.7) and that the BLUP for $\mathbf{u}$ is the BLP with $\boldsymbol{\beta}$ replaced by $\text{BLUP}(\boldsymbol{\beta}) = \tilde{\boldsymbol{\beta}}$.

As described by Robinson (1991), there are several other ways to derive BLUP solutions. A simple (albeit somewhat ad hoc) way is *Henderson's justification*, which makes the distributional assumptions

$$\mathbf{y}|\mathbf{u} \sim N(\mathbf{X}\boldsymbol{\beta} + \mathbf{Z}\mathbf{u}, \mathbf{R}), \quad \mathbf{u} \sim N(\mathbf{0}, \mathbf{G}),$$

and maximizes the likelihood of the $(\mathbf{y}, \mathbf{u})$ over the unknowns $\boldsymbol{\beta}$ and $\mathbf{u}$. This leads to the criterion

$$(\mathbf{y} - \mathbf{X}\boldsymbol{\beta} - \mathbf{Z}\mathbf{u})^{\mathsf{T}}\mathbf{R}^{-1}(\mathbf{y} - \mathbf{X}\boldsymbol{\beta} - \mathbf{Z}\mathbf{u}) + \mathbf{u}^{\mathsf{T}}\mathbf{G}^{-1}\mathbf{u}. \tag{4.10}$$

This shows that BLUP estimation of $(\boldsymbol{\beta}, \mathbf{u})$ involves generalized least squares with a penalty term. It is easy to show from (4.10) that the BLUP of $(\boldsymbol{\beta}, \mathbf{u})$ can also be written as

$$\begin{bmatrix} \tilde{\boldsymbol{\beta}} \\ \tilde{\mathbf{u}} \end{bmatrix} = (\mathbf{C}^{\mathsf{T}}\mathbf{R}^{-1}\mathbf{C} + \mathbf{B})^{-1}\mathbf{C}^{\mathsf{T}}\mathbf{R}^{-1}\mathbf{y}, \tag{4.11}$$

where $\mathbf{C} \equiv [\mathbf{X} \ \mathbf{Z}]$ and

$$\mathbf{B} \equiv \begin{bmatrix} \mathbf{0} & \mathbf{0} \\ \mathbf{0} & \mathbf{G}^{-1} \end{bmatrix}.$$

The fitted values are then

$$\mathrm{BLUP}(\mathbf{y}) = \mathbf{X}\tilde{\boldsymbol{\beta}} + \mathbf{Z}\tilde{\mathbf{u}} = \mathbf{C}(\mathbf{C}^{\mathsf{T}}\mathbf{R}^{-1}\mathbf{C} + \mathbf{B})^{-1}\mathbf{C}^{\mathsf{T}}\mathbf{R}^{-1}\mathbf{y}.$$

This "ridge regression" formulation of BLUP shows the difference between $\tilde{\boldsymbol{\beta}}$ and $\tilde{\mathbf{u}}$ explicitly.

4.5.4 *Estimation of Covariance Matrices*

There is a large and varied literature on estimation of covariance matrices in mixed models. Dictated by computational issues, the earlier literature concentrated on strategies known as *minimum norm quadratic unbiased estimation* (MINQUE) and *minimum variance quadratic unbiased estimation* (MIVQUE) (e.g. Rao 1973). However, with the advent of better computing algorithms, *maximum likelihood* (ML) or *restricted maximum likelihood* (REML) have become the most common strategies for estimating the parameters in covariance matrices.

Restricted maximum likelihood also goes by the names *residual maximum likelihood, marginal maximum likelihood,* and *generalized maximum likelihood.*

First we describe ML. As in the previous section,

$$\mathbf{V} \equiv \mathrm{Cov}(\mathbf{y}) = \mathbf{Z}\mathbf{G}\mathbf{Z}^{\mathsf{T}} + \mathbf{R}.$$

Then the ML estimate of $\mathbf{V}$ is based on the model

$$\mathbf{y} \sim N(\mathbf{X}\boldsymbol{\beta}, \mathbf{V}).$$

The log-likelihood of $\mathbf{y}$ under this model is

$$\ell(\boldsymbol{\beta}, \mathbf{V}) = -\tfrac{1}{2}\{n \log(2\pi) + \log|\mathbf{V}| + (\mathbf{y} - \mathbf{X}\boldsymbol{\beta})^{\mathsf{T}}\mathbf{V}^{-1}(\mathbf{y} - \mathbf{X}\boldsymbol{\beta})\}, \tag{4.12}$$

so the ML estimate of $(\boldsymbol{\beta}, \mathbf{V})$ is the one that maximizes the right-hand side of this expression. If one first optimizes over $\boldsymbol{\beta}$ (which appears only in the last term) then we obtain that, for any fixed V, $\ell(\boldsymbol{\beta}, V)$ is maximized over $\boldsymbol{\beta}$ by

$$\tilde{\boldsymbol{\beta}} = (\mathbf{X}^{\mathsf{T}}\mathbf{V}^{-1}\mathbf{X})^{-1}\mathbf{X}^{\mathsf{T}}\mathbf{V}^{-1}\mathbf{y},$$

which corresponds to the best linear unbiased estimator given at (4.7). On substitution into (4.12) we obtain the *profile log-likelihood* for $\mathbf{V}$:

$$\ell_P(\mathbf{V}) = -\tfrac{1}{2}\{\log|\mathbf{V}| + (\mathbf{y} - \mathbf{X}\tilde{\boldsymbol{\beta}})^\mathsf{T}\mathbf{V}^{-1}(\mathbf{y} - \mathbf{X}\tilde{\boldsymbol{\beta}}) + n\log(2\pi)\}$$

$$= -\tfrac{1}{2}[\log|\mathbf{V}| + \mathbf{y}^\mathsf{T}\mathbf{V}^{-1}\{\mathbf{I} - \mathbf{X}(\mathbf{X}^\mathsf{T}\mathbf{V}^{-1}\mathbf{X})^{-1}\mathbf{X}^\mathsf{T}\mathbf{V}^{-1}\}\mathbf{y}] - \tfrac{n}{2}\log(2\pi).$$

$$\text{(4.13)}$$

The *profile log-likelihood* of a parameter is obtained from the log-likelihood by substitution of the ML estimators of the other parameters in the model.

ML estimates of the parameters in $\mathbf{V}$ can be found by maximizing (4.13) over
those parameters. In the pig weight example,

$$\mathbf{V} = \sigma_U^2\mathbf{Z}\mathbf{Z}^\mathsf{T} + \sigma_\varepsilon^2\mathbf{I},$$

so (4.13) is a function of the variance component pair $(\sigma_U^2, \sigma_\varepsilon^2)$. Their estimation
involves maximization of the bivariate function

$$\ell_P(\sigma_U^2\mathbf{Z}\mathbf{Z}^\mathsf{T} + \sigma_\varepsilon^2\mathbf{I})$$

over all $\sigma_U^2, \sigma_\varepsilon^2 \geq 0$. The answers are

$$\hat{\sigma}_{\varepsilon,\mathrm{ML}}^2 = 4.38 \quad \text{and} \quad \hat{\sigma}_{U,\mathrm{ML}}^2 = 14.8.$$

Derivation of the REML criterion is more complicated. It involves maximizing the likelihood of linear combinations of the elements of $\mathbf{y}$ that do not depend
on $\boldsymbol{\beta}$. Details can be found in, for example, Chapter 6 of Searle et al. (1992). The
resulting criterion function is the *restricted log-likelihood*,

$$\ell_R(\mathbf{V}) = \ell_P(\mathbf{V}) - \tfrac{1}{2}\log|\mathbf{X}^\mathsf{T}\mathbf{V}^{-1}\mathbf{X}|. \tag{4.14}$$

The main advantage of REML over ML is that REML takes into account the degrees of freedom for the fixed effects in the model. For example, in the special
case where a random sample $X_1, \ldots, X_n$ is collected from the $N(\mu, \sigma^2)$ distribution, with $\bar{X} = n^{-1}\sum_{i=1}^{n} X_i$ we have

$$\hat{\sigma}_{\mathrm{ML}}^2 = \frac{1}{n}\sum_{i=1}^{n}(X_i - \bar{X})^2, \qquad \hat{\sigma}_{\mathrm{REML}}^2 = \frac{1}{n-1}\sum_{i=1}^{n}(X_i - \bar{X})^2.$$

The $n-1$ in the denominator of $\hat{\sigma}_{\mathrm{REML}}^2$ accounts for the estimation of μ via $\bar{X}$.
For small sample sizes REML is expected to be more accurate than ML, but for
large samples there is little difference between the two approaches.

There has been a considerable amount of work devoted to computation of covariance matrix estimates (e.g., Lindstrom and Bates 1988; Wolfinger, Tobias,
and Sall 1994) and ensuing software development. The procedure PROC MIXED
in the SAS computing system and the function lme() in the S-PLUS package
both compute REML and ML covariance matrix estimates.

4.6   Estimated BLUP (EBLUP)

The BLUPs of $\boldsymbol{\beta}$ and $\mathbf{u}$ given at (4.9) depend on

$$\mathbf{G} = \mathrm{Cov}(\mathbf{u}) \quad \text{and} \quad \mathbf{R} = \mathrm{Cov}(\boldsymbol{\varepsilon}),$$

especially through

$$\mathbf{V} = \text{Cov}(\mathbf{y}) = \mathbf{ZGZ}^\mathsf{T} + \mathbf{R}.$$

As described in the previous section, the parameters in these covariance matrices are typically estimated via ML or REML; in practice, the BLUPs are usually replaced by

$$\hat{\boldsymbol{\beta}} = (\mathbf{X}^\mathsf{T}\hat{\mathbf{V}}^{-1}\mathbf{X})^{-1}\mathbf{X}^\mathsf{T}\hat{\mathbf{V}}^{-1}\mathbf{y},$$

$$\hat{\mathbf{u}} = \hat{\mathbf{G}}\mathbf{Z}^\mathsf{T}\hat{\mathbf{V}}^{-1}(\mathbf{y} - \mathbf{X}\hat{\boldsymbol{\beta}})$$

where $\hat{\mathbf{G}}$ and $\hat{\mathbf{V}}$ are obtained by plugging in the ML or REML estimates of their parameters.

We will refer to $\hat{\boldsymbol{\beta}}$ and $\hat{\mathbf{u}}$ as *estimated* BLUPs, or EBLUPs, of $\boldsymbol{\beta}$ and $\mathbf{u}$. Similarly, the EBLUP of

$$\text{BLUP}\{\text{E}(\mathbf{y}|\mathbf{u})\} = \mathbf{X}\tilde{\boldsymbol{\beta}} + \mathbf{Z}\tilde{\mathbf{u}}$$

is

$$\text{EBLUP}\{\text{E}(\mathbf{y}|\mathbf{u})\} \equiv \hat{\mathbf{y}} \equiv \mathbf{X}\hat{\boldsymbol{\beta}} + \mathbf{Z}\hat{\mathbf{u}}.$$

Estimated BLUPs therefore have two sources of variability: that from estimation of $(\boldsymbol{\beta}, \mathbf{u})$, and that from estimation of $\mathbf{G}$ and $\mathbf{V}$. Ideally, both should be taken into account when making inference about the quantity of interest. As described in the next section, this is a somewhat delicate matter.

4.7   Standard Error Estimation

From the BLUP expressions in Section 4.5, it follows that

$$\text{Cov}(\tilde{\boldsymbol{\beta}}) = (\mathbf{X}^\mathsf{T}\mathbf{V}^{-1}\mathbf{X})^{-1}$$

and so the natural estimate of the standard deviation of the ith entry of the EBLUP $\hat{\beta}_i$ is

$$\widehat{\text{st.dev.}}(\hat{\beta}_i) = \sqrt{i\text{th diagonal entry of } (\mathbf{X}^\mathsf{T}\hat{\mathbf{V}}^{-1}\mathbf{X})^{-1}}. \tag{4.15}$$

Note that such an estimate ignores the variability due to estimation of $\mathbf{V}$. For larger samples this extra variability will be negligible. However, it can make a difference for smaller samples. As pointed out by McCulloch and Searle (2001, p. 258), the variance of $\hat{\beta}_i$ is largely intractable although there have been some attempts (Kackar and Harville 1984; Prasad and Rao 1990) at using approximations to quantify the extra variability in EBLUPs. The mixed model packages use (4.15), and we will use this estimator throughout much of this book in the hope that the samples are sufficiently large.

However, it is possible to handle these "intractable" calculations through a Bayesian approach and Markov chain Monte Carlo methods. The details are postponed to Chapter 16. Treating the variance components as known – even though they really have been estimated – is what Bayesians call an empirical Bayes method. By taking instead a fully Bayesian approach, the extra variability in the EBLUPs due to estimation of variance components can be taken into account.

Table 4.1 Summary of REML/EBLUP fit of (4.3) for the pig weight data.

Component	Coeff.	St. dev.	Z-Ratio
intercept	19.4	0.603	32.1
week	6.21	0.0391	159
	$\hat{\sigma}_U^2 = 15.1$	$\hat{\sigma}_\varepsilon^2 = 4.39$	

To estimate the precision of BLUPs involving **u**, we also need

$$\mathrm{Cov}\left(\begin{bmatrix} \tilde{\beta} - \beta \\ \tilde{u} - u \end{bmatrix}\right) = \mathrm{Cov}\left(\begin{bmatrix} \tilde{\beta} \\ \tilde{u} - u \end{bmatrix}\right).$$

Using (4.11), it can be shown that

$$\mathrm{Cov}\left(\begin{bmatrix} \tilde{\beta} \\ \tilde{u} - u \end{bmatrix}\right) = (\mathbf{C}^\mathsf{T}\mathbf{R}^{-1}\mathbf{C} + \mathbf{B})^{-1}, \tag{4.16}$$

where (as before)

$$\mathbf{C} \equiv [\mathbf{X} \ \mathbf{Z}] \quad \text{and} \quad \mathbf{B} \equiv \begin{bmatrix} \mathbf{0} & \mathbf{0} \\ \mathbf{0} & \mathbf{G}^{-1} \end{bmatrix}.$$

In some contexts it is useful to estimate the *conditional* covariance matrix

$$\mathrm{Cov}\left(\begin{bmatrix} \tilde{\beta} - \beta \\ \tilde{u} \end{bmatrix}\middle| u\right) = \mathrm{Cov}\left(\begin{bmatrix} \tilde{\beta} \\ \tilde{u} \end{bmatrix}\middle| u\right).$$

From (4.11),

$$\mathrm{Cov}\left(\begin{bmatrix} \tilde{\beta} \\ \tilde{u} \end{bmatrix}\middle| u\right) = (\mathbf{C}^\mathsf{T}\mathbf{R}^{-1}\mathbf{C} + \mathbf{B})^{-1}\mathbf{C}^\mathsf{T}\mathbf{R}^{-1}\mathbf{C}(\mathbf{C}^\mathsf{T}\mathbf{R}^{-1}\mathbf{C} + \mathbf{B})^{-1}. \tag{4.17}$$

These results suggest the approximations

$$\mathrm{Cov}\left(\begin{bmatrix} \hat{\beta} \\ \hat{u} - u \end{bmatrix}\right) \simeq (\mathbf{C}^\mathsf{T}\hat{\mathbf{R}}^{-1}\mathbf{C} + \hat{\mathbf{B}})^{-1}$$

and

$$\mathrm{Cov}\left(\begin{bmatrix} \hat{\beta} \\ \hat{u} \end{bmatrix}\middle| u\right) \simeq (\mathbf{C}^\mathsf{T}\hat{\mathbf{R}}^{-1}\mathbf{C} + \hat{\mathbf{B}})^{-1}\mathbf{C}^\mathsf{T}\hat{\mathbf{R}}^{-1}\mathbf{C}(\mathbf{C}^\mathsf{T}\hat{\mathbf{R}}^{-1}\mathbf{C} + \hat{\mathbf{B}})^{-1}.$$

Standard errors may also be estimated for the covariance matrix parameters, although the details are omitted. Searle et al. (1992) contains some details. Confidence intervals for both fixed effects and covariance matrix parameters (i.e., variance components) are described by Pinheiro and Bates (2000).

4.7.1   *Summary of Fit to Pig Weights*

Table 4.1 summarizes the fit of (4.3) based on REML estimation of the variance components and EBLUP.

A graphical summary of the fit is shown in Figure 4.5. There is a strongly significant positive growth but with considerable between-pig variability in the intercepts.

Figure 4.5 Graphical summary of REML/BLUP fit of (4.3) for the pig weight data. The line is estimated mean weight. The shaded region corresponds to plus and minus two standard deviations. The curve at the left is a density estimate of the estimated random intercepts.

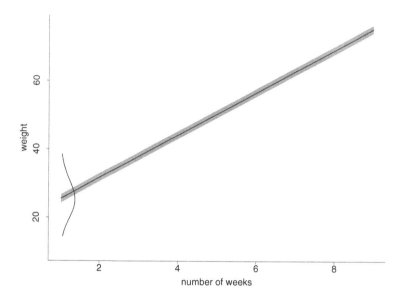

4.8 Hypothesis Testing

There is a rich literature on hypothesis testing in the linear mixed model frame-work. Verbeke and Molenberghs (1997) and Khuri, Mathew, and Sinha (1998) provide surveys. We will restrict discussion to some of the simpler methods here.

4.8.1 *Normal Theory Tests*

First consider the problem of hypothesis testing for β_i, the ith entry of $\boldsymbol{\beta}$. Ideally, this can be done through a result of the form

$$z_i \equiv \frac{\hat{\beta}_i - \beta_i}{\widehat{\text{st.dev.}(\hat{\beta}_i)}} \overset{\text{approx.}}{\sim} \text{N}(0, 1). \tag{4.18}$$

Specifically, for the hypothesis set-up

$$H_0 : \beta_i = 0 \quad \text{versus} \quad H_1 : \beta_i \neq 0, \tag{4.19}$$

the approximate p-value is given by the tail area:

$$p\text{-value} \simeq 2\{1 - \Phi(|z_{0,i}|)\},$$

where Φ is the cumulative distribution function of the standard normal distribution, and $z_{0,i}$ is given by (4.18) with $\beta_i = 0$. However, the theoretical justification of (4.18) for general mixed models is somewhat elusive owing to the dependence in **y** imposed by the random effects. Theoretical back-up for (4.18) exists in certain special cases, such as those arising in analysis of variance and longitudinal data analysis (e.g. Miller 1977). For some mixed models, including many used in the subsequent chapters of this book, justification of (4.18) remains an open problem.

4.8.2 *Likelihood Ratio Tests*

We saw in Sections 4.5.1 and 4.5.4 that parameters in mixed models can be estimated by maximum likelihood; hence, the *likelihood ratio* procedure can be used to test hypotheses. We will first give a brief description of the likelihood ratio test procedure for general parametric models.

Let $\mathcal{L}(\boldsymbol{\theta}; \mathbf{y})$ be the likelihood of the parameter vector $\boldsymbol{\theta}$ based on the data in $\mathbf{y}$. The likelihood ratio statistic for testing a null restricted model against an alternative unrestricted model is

$$\text{LR}(\mathbf{y}) = \mathcal{L}(\hat{\boldsymbol{\theta}}_0; \mathbf{y})/\mathcal{L}(\hat{\boldsymbol{\theta}}; \mathbf{y}),$$

where $\hat{\boldsymbol{\theta}}_0$ and $\hat{\boldsymbol{\theta}}$ are the maximum likelihood estimates of $\boldsymbol{\theta}$ under the null model and unrestricted model, respectively. It is more common to work with

$$-2 \log\{\text{LR}(\mathbf{y})\} = -2\{\ell(\hat{\boldsymbol{\theta}}_0; \mathbf{y}) - \ell(\hat{\boldsymbol{\theta}}; \mathbf{y})\}, \tag{4.20}$$

where $\ell(\boldsymbol{\theta}; \mathbf{y}) = \log \mathcal{L}(\boldsymbol{\theta}; \mathbf{y})$ is the log-likelihood.

The classical result for determining the significance of the observed value of $\ell(\boldsymbol{\theta}; \mathbf{y})$ is one that states, under H_0,

$$-2 \log\{\text{LR}(\mathbf{y})\} \overset{\text{approx.}}{\sim} \chi_\nu^2; \tag{4.21}$$

here the right-hand side is the chi-squared distribution with ν degrees of freedom, where

$$\nu = \text{number of independent parameters in unrestricted model}$$
$$- \text{number of independent parameters in null model.}$$

For hypothesis test (4.19) $\nu = 1$, so (4.21) provides an alternative way to test this hypothesis. Since a χ_1^2 random variable is the square of a standard normal random variable, we have

$$p\text{-value} \simeq 1 - \Phi\left(\sqrt{-2 \log \text{LR}(\mathbf{y})}\right). \tag{4.22}$$

Once again the dependence in $\mathbf{y}$ means that justification of (4.21), and hence (4.22), is dependent on the type of correlation structure induced by the $\mathbf{G}$ and $\mathbf{R}$ matrices.

Hypothesis tests for covariance matrix parameters may also be of interest. Consider, for example, the random intercept straight line model for repeated measures regression data:

$$y_{ij} = \beta_0 + U_i + \beta_1 x_{ij} + \varepsilon_{ij}, \quad 1 \le j \le n_i, \ 1 \le i \le m, \ U_i \sim N(0, \sigma_U^2). \tag{4.23}$$

One may wish to determine whether the intercepts of the individuals are significantly different from one another – that is, whether the submodel

$$y_{ij} = \beta_0 + \beta_1 x_{ij} + \varepsilon_{ij}, \quad 1 \le j \le n_i, \ 1 \le i \le m, \tag{4.24}$$

is adequate compared with (4.23). This boils down to testing the hypotheses

$$H_0 : \sigma_U^2 = 0 \quad \text{versus} \quad H_1 : \sigma_U^2 > 0. \tag{4.25}$$

For (4.25) this would suggest that $-2\log\{LR(\mathbf{y})\}$ be compared with percentiles from the χ_1^2 distribution. However, the theory behind (4.21) assumes that the parameter of interest is not on the boundary of its parameter space. Since the parameter space for σ_U^2 is $[0, \infty)$, this assumption is violated. In fact, under certain independence assumptions to be discussed shortly, the asymptotic distribution when H_0 is true is such that there is a 50 : 50 chance that

$$\hat{\sigma}_{u,\mathrm{ML}}^2 = 0.$$

This type of behavior leads to

The notation
$X \overset{\text{approx.}}{\sim} \frac{1}{2}\chi_s^2 + \frac{1}{2}\chi_t^2$
means that the random variable X has a approximate density function equal to a 50 : 50 mixture of the χ_s^2 and χ_t^2 densities. This is different from the density of the average of independent random variables from each of these densities.

$$-2\log\{LR(\mathbf{y})\} \overset{\text{approx.}}{\sim} \tfrac{1}{2}\chi_0^2 + \tfrac{1}{2}\chi_1^2 \qquad (4.26)$$

for (4.25), where χ_0^2 means a point mass at zero and the right-hand side of (4.26) is shorthand for a 50 : 50 mixture between a χ_0^2 and a χ_1^2 distribution (Self and Liang 1987). Use of (4.26) rather than (4.21) leads to p-values being halved.

If, instead, one were interested in testing the adequacy of

$$y_{ij} = \beta_0 + \varepsilon_{ij}, \quad 1 \le j \le n_i, \ 1 \le i \le m,$$

compared with (4.23), then the hypotheses would be

$$H_0 : \sigma_U^2 = \beta_1 = 0 \quad \text{versus} \quad H_1 : \sigma_U^2 > 0 \text{ or } \beta_1 \ne 0;$$

under H_0, we would have

$$-2\log\{LR(\mathbf{y})\} \overset{\text{approx.}}{\sim} \tfrac{1}{2}\chi_1^2 + \tfrac{1}{2}\chi_2^2.$$

More generally, if H_0 constrains one variance component and s regression coefficients to be zero, then

$$-2\log\{LR(\mathbf{y})\} \overset{\text{approx.}}{\sim} \tfrac{1}{2}\chi_s^2 + \tfrac{1}{2}\chi_{s+1}^2. \qquad (4.27)$$

If there are two variance components of interest then the distribution theory for $-2\log\{LR(\mathbf{y})\}$ becomes much more complicated (see e.g. Self and Liang 1987; Stram and Lee 1994; Silvapulle 1996; Verbeke and Molenberghs 1997).

The classical large-sample theory for likelihood ratio tests assumes independence of the $\mathbf{y}$ vector under all values of the parameters, or more precisely, that the $\mathbf{y}$ vector can be partitioned into subvectors that are independent. This assumption does not hold in general for mixed models, at least not under the alternative. However, for the simple longitudinal model (4.23), independence between subjects allows for extension of the classical theory and validation of (4.26) and (4.27) for a large number of subjects. The idea is that these approximations assume that the number of independent observations approaches infinity, which is true if we take the subjects to be the observations and the number of subjects increases to infinity (Stram and Lee 1994). As for the normal theory tests described in the previous section, the asymptotic distribution theory for general mixed models is more difficult and, in some instances, yet to be worked out.

One case that has been studied carefully in Crainiceanu, Ruppert, and Vogelsang (2002) and Crainiceanu and Ruppert (2002) is the balanced one-way analysis of variance with random treatment (or subject) effects. This model is given by equation (4.23) with β_1 known to equal 0 and $n_i = n$ for some fixed n and all i. If n is fixed and m tends to infinity, then the Self and Liang (1987) assumptions are met and (4.26) does hold. In contrast, if n goes to infinity with m fixed, then we have a fixed number of subjects (or treatments) and the number of observations per subject goes to infinity. In this case, the Self and Liang (1987) assumptions do not hold and (4.26) fails to hold. The asymptotic probability of zero – that is, the probability attached to the χ_0^2 component – is not $\frac{1}{2}$ but rather is greater than $\frac{1}{2}$ and tends very slowly to $\frac{1}{2}$ as m goes to infinity (Crainiceanu et al. 2002). In fact, this probability is about 0.65 when m is 10 and 0.55 when m is 100. Moreover, the component that is nonzero is not χ_1^2 but something that tends to be smaller than χ_1^2 (Crainiceanu and Ruppert 2002). Thus, the likelihood ratio test statistic tends to be *smaller* than under the Self and Liang asymptotics, so that using those asymptotics to obtain *p*-values gives tests that are conservative (i.e., have nominal *p*-values larger than the true *p*-value). Conservative tests have smaller type 1 error probabilities than stated, which is not a problem. However, they have the disadvantage of having less power at the alternative than a test with correct type 1 error probability.

As we will see in the next section, penalized splines can be viewed as BLUPs in a certain mixed model. Approximation (4.26) is very poor when applied to penalized splines (Crainiceanu and Ruppert 2002). In general, the asymptotics of Self and Liang (1987) do not apply to the semiparametric models we discuss in this book. This means that asymptotics cannot be used to find *p*-values, at least not until alternative asymptotics are derived.

As we discuss in later sections, critical values of likelihood ratio tests can be determined satisfactorily by Monte Carlo simulations. The idea is to set the values of all fixed effect and variance component parameters equal to their estimates under the null distribution and then to simulate the distribution of the likelihood ratio test under the null model at the parameters and with the covariates equal to their observed values. More precisely, we simulate a large number of independent data sets (say, 10,000 to 100,000) with fixed effects parameters at their estimated values and with the ε-values and random effects generated according to their estimated variances, both estimations under the null hypothesis. The likelihood ratio test statistic (4.20) is calculated for each simulated data set. The *p*-value of the test is the proportion of simulated values of the test statistic that exceed the value at the real data.

4.8.3 *Restricted Likelihood Ratio Tests*

Instead of using the likelihood function to form test statistics, one could do so using the maximum restricted log-likelihood $\ell_R(\mathbf{V})$ defined at (4.14). Since REML estimates of $\mathbf{V}$ are less biased, the accuracy of the test might be improved. There is some discussion on this approach in Verbeke and Molenberghs (1997). They mention that restricted likelihood can only be used to compare models with the same mean structure – that is, the same fixed effects model. The reason for this is that restricted likelihood is the likelihood of the residuals after fitting the fixed effects and so is not appropriate when there is more than one fixed effects model under consideration.

As discussed in Crainiceanu et al. (2002) and Crainiceanu and Ruppert (2002), restricted likelihood ratio tests have the same complex asymptotic theory as likelihood ratio tests. For this reason, we advocate computing p-values by simulation as just discussed at the end of Section 4.8.2.

4.9   Penalized Splines as BLUPs

In Chapter 3 we considered the ordinary nonparametric regression model

$$y_i = f(x_i) + \varepsilon_i, \quad 1 \leq i \leq n, \tag{4.28}$$

and showed how f could be estimated by penalized splines. In this section we show that this estimate can be written as the BLUP of a mixed model.

For clarity we will treat the linear case and suppose that the errors satisfy $\text{Cov}(\boldsymbol{\varepsilon}) = \sigma_\varepsilon^2 \mathbf{I}$. The linear spline model for f is

$$f(x_i) = \beta_0 + \beta_1 x_i + \sum_{k=1}^{K} u_k (x_i - \kappa_k)_+. \tag{4.29}$$

Let

$$\boldsymbol{\beta} = \begin{bmatrix} \beta_0 \\ \beta_1 \end{bmatrix} \quad \text{and} \quad \mathbf{u} = \begin{bmatrix} u_1 \\ \vdots \\ u_K \end{bmatrix}$$

be the coefficients of the polynomial functions and truncated line functions, respectively. Corresponding to these vectors, define

$$\mathbf{X} = \begin{bmatrix} 1 & x_1 \\ \vdots & \vdots \\ 1 & x_n \end{bmatrix} \quad \text{and} \quad \mathbf{Z} = \begin{bmatrix} (x_1 - \kappa_1)_+ & \cdots & (x_1 - \kappa_K)_+ \\ \vdots & \ddots & \vdots \\ (x_n - \kappa_1)_+ & \cdots & (x_n - \kappa_K)_+ \end{bmatrix}.$$

The penalized spline fitting criterion (3.7), when divided by σ_ε^2, can then be written as

$$\frac{1}{\sigma_\varepsilon^2} \| \mathbf{y} - \mathbf{X}\boldsymbol{\beta} - \mathbf{Z}\mathbf{u} \|^2 + \frac{\lambda^2}{\sigma_\varepsilon^2} \| \mathbf{u} \|^2.$$

Notice that this can be made to equal the BLUP criterion given at (4.10) by treating the $\mathbf{u}$ as a set of random coefficients with

$$\text{Cov}(\mathbf{u}) = \sigma_u^2 \mathbf{I}, \quad \text{where} \quad \sigma_u^2 = \sigma_\varepsilon^2 / \lambda^2.$$

Putting all of this together yields the mixed model representation of the regression spline

$$\mathbf{y} = \mathbf{X}\boldsymbol{\beta} + \mathbf{Z}\mathbf{u} + \boldsymbol{\varepsilon}, \quad \text{Cov} \begin{bmatrix} \mathbf{u} \\ \boldsymbol{\varepsilon} \end{bmatrix} = \begin{bmatrix} \sigma_u^2 \mathbf{I} & \mathbf{0} \\ \mathbf{0} & \sigma_\varepsilon^2 \mathbf{I} \end{bmatrix}. \tag{4.30}$$

Note that the fitted values $\tilde{\mathbf{f}}$ can be rewritten as

$$\tilde{\mathbf{f}} = \mathbf{C}(\mathbf{C}^{\mathsf{T}}\mathbf{C} + \lambda^2 \mathbf{D})^{-1}\mathbf{C}^{\mathsf{T}}\mathbf{y}, \tag{4.31}$$

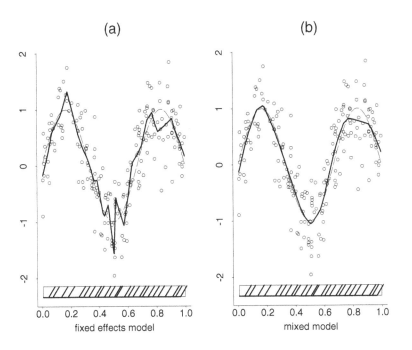

(a)   (b)

fixed effects model   mixed model

Figure 4.6
Comparison
between treating the
coefficients of the
knots as fixed effects
versus random effects.
The solid curve is the
estimated curve, while
the dashed curve is the
true function.

where
$$\mathbf{C} = [\mathbf{X} \ \mathbf{Z}], \quad \mathbf{D} = \mathrm{diag}(0, 0, 1, 1, \ldots, 1), \quad \lambda^2 = \sigma_\varepsilon^2 / \sigma_u^2,$$

matching (3.12) for $p = 1$.

Figure 4.6 shows how the mixed model approach to fitting the regression splines leads to a smooth result. In this example, data are simulated according to the situation where $f(x) = \sin(3\pi x)$, $0 \le x \le 1$, and $\sigma_\varepsilon = 0.4$. In Figure 4.6(a) we see the result when

$$y_i = \beta_0 + \beta_1 x_i + \sum_{k=1}^{K} u_k (x_i - \kappa_k)_+ + \varepsilon_i$$

is fit using ordinary least squares. Notice that it overfits the data rather than smoothing it. The fit in Figure 4.6(b) corresponds to treating (4.29) as a mixed model with

$$u_k \text{ i.i.d. } \mathrm{N}(0, \sigma_u^2).$$

Ordinary least squares corresponds to $\sigma_u = \infty$, where the u_k are unrestricted. Taking σ_u to be finite – in this case, $\sigma_u = 3\sigma_\varepsilon = 1.2$ – leads to smaller estimates of the u_k and the effect of the $(x_i - \kappa_k)_+$ being diminished. A smooth fit results.

This representation of the penalized spline as a BLUP in a mixed model is useful because it allows smoothing to be done using mixed model methodology and software. This will be exemplified in the following chapters. It also lends itself (via e.g. Robinson 1991) to a host of other derivations, including one as a *Bayesian estimator*. As explained there, if $\boldsymbol{\beta}$ is regarded as a parameter with a uniform and improper prior, then $\tilde{\mathbf{f}}$ corresponds to the posterior mode. Robinson (1991) also demonstrates how the *Kalman filter* can be used to compute BLUPs. See also the discussion by Spall (1991) and Robinson's rejoinder.

Bayesian estimation is a large branch of statistical methodology based on the notion of incorporating prior beliefs about parameters of interest. Chapter 16 deals with Bayesian estimation.

A prior distribution is *improper* if its total probability equals infinity rather than one.

The *Kalman filter* is a fast algorithm for fitting optimal solutions to a certain class of linear statistical models. An introduction to the Kalman filter is provided by Maybeck (1979).

When σ_u^2 and σ_ε^2 are replaced by estimators, such as those obtained from ML and REML, the final vector of fitted values is

$$\hat{\mathbf{f}} = \mathbf{X}\hat{\boldsymbol{\beta}} + \mathbf{Z}\hat{\mathbf{u}}.$$

For arbitrary $x \in \mathbb{R}$, the estimate of $f(x)$ is

$$\hat{f}(x) = \hat{\beta}_0 + \hat{\beta}_1 x + \sum_{k=1}^{K} \hat{u}_k (x - \kappa_k)_+,$$

where $\hat{\beta}_0$, $\hat{\beta}_1$, and $\hat{u}_k$ $(1 \leq k \leq K)$ are EBLUPs.

4.10   Bibliographical Notes

Mixed modeling is a massive and growing branch of statistics. Here we have simply summarized the aspects that are most relevant to the subsequent chapters in this book.

An excellent introduction to general design linear mixed models is Robinson (1991). REML estimation of covariance matrices is due to Patterson and Thompson (1971). An important reference for computational issues in mixed models is Harville (1977).

In recent years, several books on mixed models – some with an emphasis on longitudinal data analysis – have been published. These include Searle et al. (1992), Vonesh and Chinchilli (1997), Pinheiro and Bates (2000), Verbeke and Molenberghs (2000), and McCulloch and Searle (2001). Brumback, Ruppert, and Wand (1999) discuss the mixed model representation of penalized splines, though this representation had been developed earlier for the special case of smoothing splines; see their paper for references to the latter.

4.11   Summary of Formulas

Linear mixed model

$$\mathbf{y} = \mathbf{X}\boldsymbol{\beta} + \mathbf{Z}\mathbf{u} + \boldsymbol{\varepsilon}$$

where

$$E\begin{bmatrix} \mathbf{u} \\ \boldsymbol{\varepsilon} \end{bmatrix} = \begin{bmatrix} \mathbf{0} \\ \mathbf{0} \end{bmatrix} \quad \text{and} \quad \mathrm{Cov}\begin{bmatrix} \mathbf{u} \\ \boldsymbol{\varepsilon} \end{bmatrix} = \begin{bmatrix} \mathbf{G} & \mathbf{0} \\ \mathbf{0} & \mathbf{R} \end{bmatrix}$$

$$\mathbf{V} \equiv \mathrm{Cov}(\mathbf{y}) = \mathbf{Z}\mathbf{G}\mathbf{Z}^{\mathsf{T}} + \mathbf{R}$$

Best linear unbiased prediction (BLUP)

$$\mathrm{BLUP}(\boldsymbol{\beta}) \equiv \tilde{\boldsymbol{\beta}} = (\mathbf{X}^{\mathsf{T}}\mathbf{V}^{-1}\mathbf{X})^{-1}\mathbf{X}^{\mathsf{T}}\mathbf{V}^{-1}\mathbf{y}$$

$$\mathrm{BLUP}(\mathbf{u}) \equiv \tilde{\mathbf{u}} = \mathbf{G}\mathbf{Z}^{\mathsf{T}}\mathbf{V}^{-1}(\mathbf{y} - \mathbf{X}\tilde{\boldsymbol{\beta}})$$

Alternatively,

$$\begin{bmatrix} \tilde{\beta} \\ \tilde{u} \end{bmatrix} = (\mathbf{C}^{\mathsf{T}}\mathbf{R}^{-1}\mathbf{C} + \mathbf{B})^{-1}\mathbf{C}^{\mathsf{T}}\mathbf{R}^{-1}\mathbf{y}$$

where

$$\mathbf{C} = [\mathbf{X} \ \mathbf{Z}] \quad \text{and} \quad \mathbf{B} = \begin{bmatrix} \mathbf{0} & \mathbf{0} \\ \mathbf{0} & \mathbf{G}^{-1} \end{bmatrix}$$

Likelihood-based estimation of V

$$\mathbf{y} \sim \mathrm{N}(\mathbf{X}\beta, \mathbf{V})$$

Profile log-likelihood for ML estimation of $\mathbf{V}$ is

$$\ell_P(\mathbf{V}) = -\tfrac{1}{2}[\log|\mathbf{V}| + \mathbf{y}^{\mathsf{T}}\mathbf{V}^{-1}\{\mathbf{I} - \mathbf{X}(\mathbf{X}^{\mathsf{T}}\mathbf{V}^{-1}\mathbf{X})^{-1}\mathbf{X}^{\mathsf{T}}\mathbf{V}^{-1}\}\mathbf{y}] - \tfrac{n}{2}\log(2\pi)$$

Restricted profile log-likelihood for REML estimation of $\mathbf{V}$ is

$$\ell_R(\mathbf{V}) = \ell_P(\mathbf{V}) - \tfrac{1}{2}\log|\mathbf{X}^{\mathsf{T}}\mathbf{V}^{-1}\mathbf{X}|$$

Standard error estimation

$$\mathrm{Cov}(\tilde{\beta}) = (\mathbf{X}^{\mathsf{T}}\mathbf{V}^{-1}\mathbf{X})^{-1}$$

$$\mathrm{Cov}\left(\begin{bmatrix} \tilde{\beta} \\ \tilde{u} \end{bmatrix} \middle| \mathbf{u}\right) = (\mathbf{C}^{\mathsf{T}}\mathbf{R}^{-1}\mathbf{C} + \mathbf{B})^{-1}\mathbf{C}^{\mathsf{T}}\mathbf{R}^{-1}\mathbf{C}(\mathbf{C}^{\mathsf{T}}\mathbf{R}^{-1}\mathbf{C} + \mathbf{B})^{-1}$$

BLUP representation of linear penalized spline
Scatterplot data are (x_i, y_i), $1 \le i \le n$. Knots are $\kappa_1, \ldots, \kappa_K$.

$$\mathbf{y} = \begin{bmatrix} y_1 \\ \vdots \\ y_n \end{bmatrix}, \quad \mathbf{X} = \begin{bmatrix} 1 & x_1 \\ \vdots & \vdots \\ 1 & x_n \end{bmatrix}, \quad \mathbf{Z} = \begin{bmatrix} (x_1 - \kappa_1)_+ & \cdots & (x_1 - \kappa_K)_+ \\ \vdots & \ddots & \vdots \\ (x_n - \kappa_1)_+ & \cdots & (x_n - \kappa_K)_+ \end{bmatrix}$$

$$\mathbf{G} = \sigma_u^2\mathbf{I}, \quad \mathbf{R} = \sigma_\varepsilon^2\mathbf{I}, \quad \lambda^2 = \sigma_\varepsilon^2/\sigma_u^2$$

5

Automatic Scatterplot Smoothing

5.1 Introduction

In Chapter 3 we showed how one can smooth a scatterplot using spline functions. However, the methods described there were not *automatic* in that the user must specify a number of quantities: the degree of the piecewise polynomials, the knot locations, and the smoothing parameter λ. To varying extents, each of these has an effect on the quality of the smooth. It is natural, therefore, to ask if the data can guide the choice of these quantities and so lead to reasonably automatic scatterplot smooths.

Inspection of Figure 5.1 (which is identical to Figure 3.11) shows that the choice of λ has a profound influence on the fit. In fact, λ can be chosen to give any one of a spectrum of fits between the unconstrained regression spline fit and the least-squares line. In this particular example, $\lambda = 30$ could be chosen by

Figure 5.1 Linear penalized spline regression fits to LIDAR data for λ-values of 0, 10, 30, and 1000 (24 knots are used).

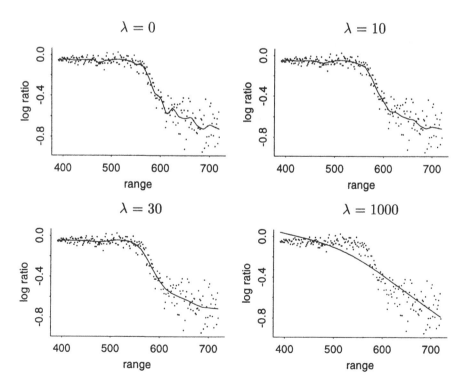

eye as reasonable. However, some applications require many scatterplot smooths
and it is not feasible to have the amount of smoothing chosen subjectively by a
human. Rather, it is advantageous that the data themselves "choose" an appropriate amount of smoothing.

Apart from the smoothing parameter, the degree of the polynomial basis and
the number and positioning of the knots need to be specified. As we will demonstrate in Section 5.5, these choices are much less crucial than the smoothing
parameter, and reasonable default values are relatively easy to devise. Hence, the
first part of this chapter in concerned solely with smoothing parameter selection;
it is assumed that the degree and knot locations are fixed.

By the end of the chapter we will have given full prescriptions for choosing the
smoothing parameter, degree, and knot locations from the data. Such *automatic*
scatterplot smoothing is the focus of this chapter.

5.2   The Likelihood Approach

An attractive consequence of the BLUP representation of the penalized spline
smoother is that it can be fit using mixed model software, with ML or REML
used to select the amount of smoothing. REML is the more commonly used of
the two, so we will concentrate on that here.

Recall from Section 4.5.4 that, for a linear mixed model

$$\mathbf{y} = \mathbf{X}\boldsymbol{\beta} + \mathbf{Z}\mathbf{u} + \boldsymbol{\varepsilon},$$

the REML criterion for estimating covariance matrix parameters is

$$\ell_R(\mathbf{V}) = -\tfrac{1}{2}[n \log(2\pi) + \log|\mathbf{V}| + \log|\mathbf{X}^{\mathsf{T}}\mathbf{V}^{-1}\mathbf{X}|$$
$$+ \mathbf{y}^{\mathsf{T}}\mathbf{V}^{-1}\{\mathbf{I} - \mathbf{X}(\mathbf{X}^{\mathsf{T}}\mathbf{V}^{-1}\mathbf{X})^{-1}\mathbf{X}^{\mathsf{T}}\mathbf{V}^{-1}\}\mathbf{y}],$$

where $\mathbf{V} = \text{Cov}(\mathbf{y})$. For the pth-degree penalized spline model,

$$\mathbf{V} = \sigma_u^2 \mathbf{Z}\mathbf{Z}^{\mathsf{T}} + \sigma_\varepsilon^2 \mathbf{I}$$

and the smoothing parameter λ is

$$\lambda = (\sigma_\varepsilon^2/\sigma_u^2)^{1/2p}.$$

Therefore, if the REML criterion is minimized over $\sigma_u^2, \sigma_\varepsilon^2 \geq 0$ to give estimates
$(\hat{\sigma}_{u,\text{REML}}^2, \hat{\sigma}_{\varepsilon,\text{REML}}^2)$, then the selected bandwidth is

$$\hat{\lambda}_{\text{REML}} = (\hat{\sigma}_{\varepsilon,\text{REML}}^2/\hat{\sigma}_{u,\text{REML}}^2)^{1/2p}.$$

The same idea, of course, can be used to select smoothing parameters via ML.

Figure 5.2 shows fits to the LIDAR data obtained using REML and ML. The
two fits are here almost indistinguishable because

$$\hat{\lambda}_{\text{REML}} = 38.7 \quad \text{and} \quad \hat{\lambda}_{\text{ML}} = 40.8,$$

which corresponds to 8.36 and 8.09 degrees of freedom, respectively. In our ex-
perience, $\hat{\lambda}_{\text{REML}}$ and $\hat{\lambda}_{\text{ML}}$ are usually quite close. See, for example, the results
from a simulation study presented in Figure 5.6.

Figure 5.2 Automatic penalized linear spline fits to LIDAR data based on REML and ML selection of the smoothing parameter and using 24 knots. The two fits are visually indistinguishable for these data.

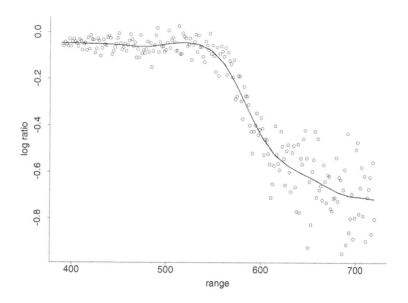

5.3   The Model Selection Approach

The maximum likelihood–based smoothing parameter selection methods depend on the mixed model representation of penalized splines. Many other smoothing methods, such as those described in Section 3.15, do not have such a representation. Hence more generic methods, based on classical model selection ideas, are typically used to select the amount of smoothing. We will describe the most common of these approaches here.

5.3.1   *Cross-Validation (CV)*

As we discussed in Section 2.6, one of the most common measures for the "goodness of fit" of a regression curve to a scatterplot is the *residual sum of squares*:

$$\text{RSS} = \sum_{i=1}^{n}(y_i - \hat{y}_i)^2 = \|\mathbf{y} - \hat{\mathbf{y}}\|^2. \tag{5.1}$$

However, since RSS is minimized at the interpolant ($\hat{y}_i = y_i$, $1 \le i \le n$), minimization of this criterion will lead to the smooth that is closest to interpolation. For penalized spline or local polynomial regression this corresponds to a zero smoothing parameter. As illustrated by the fit in the upper left panel of Figure 5.1, this is usually unacceptable. This difficulty is similar in nature to the problem that we described in parametric model selection in Section 2.6 – the observation y_i being used as part of its own predictor. Therefore, a small amount of smoothing, which gives high weight to y_i, *appears* optimal for prediction according to RSS.

We saw in Section 2.6 that *cross-validation* (CV) gets around this problem. Let $\hat{f}(x; \lambda)$ denote the nonparametric regression estimate at a point x with smoothing parameter λ. Then we can rewrite the RSS formula (5.1) as

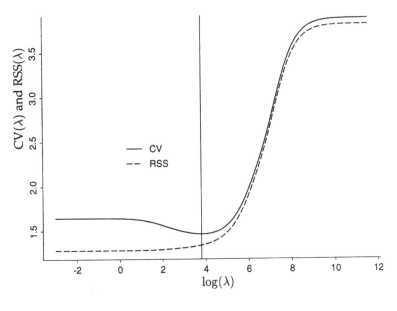

Figure 5.3 Cross-validation and RSS curves for the LIDAR data using 24-knot linear regression splines.

$$\text{RSS}(\lambda) = \sum_{i=1}^{n} \{y_i - \hat{f}(x_i; \lambda)\}^2.$$

The *cross-validation* criterion is

$$\text{CV}(\lambda) = \sum_{i=1}^{n} \{y_i - \hat{f}_{-i}(x_i; \lambda)\}^2,$$

where $\hat{f}_{-i}$ denotes the nonparametric regression estimator applied to the data but with (x_i, y_i) deleted. This "leaving one out" strategy is a way of guarding against the wiggly answer that RSS(λ) gives. The CV choice of λ is the one that minimizes CV(λ) over $\lambda \geq 0$. We denote this minimizer by $\hat{\lambda}_{CV}$.

For the LIDAR data, Figure 5.3 shows the CV and RSS functions for 24-knot penalized linear spline regression plotted against log(λ). CV(λ) is minimized on this grid at $\lambda = 44.7$, which is close to the values determined by ML and REML and is indicated by a vertical line in the plot. Also, $\lambda = 30$ (log(λ) = 3.4), which we saw gives a visually appealing curve in Figure 5.1 and comes close to minimizing CV; whereas $\lambda = 10$ (log(λ) = 2.3), which is a slight undersmooth in Figure 5.1, yields a somewhat larger CV value. Furthermore, $\lambda = 1000$ (log(λ) = 6.9) is obviously a severe oversmooth in Figure 5.1 and gives a rather large CV value in Figure 5.3. In this case, cross-validation appears to give a reasonable result.

In this example 24 knots are used, so OLS gives a somewhat undersmoothed but not disastrously poor estimate. As $\lambda \to 0$, CV(λ) converges to 1.63, which is only somewhat above the minimum-CV value of 1.46. As $\lambda \to \infty$, the estimate converges to the ordinary least-squares fit to a straight line model with severe bias. Thus, as $\lambda \to \infty$, CV(λ) converges to the rather large value of 3.88.

In Figure 5.3 one sees that RSS is monotonically increasing in λ, as theory predicts, so that minimizing RSS does indeed lead to near-interpolation.

5.3.1.1   *Computation of CV*

When calculating cross-validation, some care is needed to avoid high computational cost. If one directly programs

$$\mathrm{CV}(\lambda) = \sum_{i=1}^{n} \{y_i - \hat{f}_{-i}(x_i; \lambda)\}^2,$$

then n versions of $\hat{f}_{-i}(x; \lambda)$ are required, which leads to an order-n^2 algorithm. Fortunately, there now exist fast order-n algorithms for computation of $\mathrm{CV}(\lambda)$ for most common smoothing techniques (Hutchinson and de Hoog 1985).

Let $\mathbf{S}_\lambda$ be the smoother matrix associated with $\hat{f}$. The vector of fitted values is

$$\begin{bmatrix} \hat{f}(x_1; \lambda) \\ \vdots \\ \hat{f}(x_n; \lambda) \end{bmatrix} = \mathbf{S}_\lambda \mathbf{y}$$

so that

$$\hat{f}(x_i; \lambda) = \sum_{j=1}^{n} S_{\lambda, ij} y_j,$$

where $S_{\lambda, ij}$ is the (i, j) entry of $\mathbf{S}_\lambda$. For many smoothers,

$$\hat{f}_{-i}(x_i; \lambda) = \frac{\sum_{j \neq i} S_{\lambda, ij} y_j}{\sum_{j \neq i} S_{\lambda, ij}}. \tag{5.2}$$

Even if (5.2) does not hold exactly, it usually holds approximately. Moreover, we could take (5.2) as a *definition* of $\hat{f}_{-i}(x_i; \lambda)$ for use in cross-validation. Also, all smoothers used routinely have the sensible property that if $y_i \equiv 1$ then $\hat{y}_i \equiv 1$, which implies that

$$\sum_{j=1}^{n} S_{\lambda, ij} = 1 \quad \text{for all } i, \tag{5.3}$$

so that the denominator in (5.2) is equal to $1 - S_{\lambda, ii}$. Using this fact, (5.2), and some algebra, one can show that

$$\begin{aligned} \mathrm{CV}(\lambda) &= \sum_{i=1}^{n} \left(\frac{y_i - \hat{f}(x_i; \lambda)}{1 - S_{\lambda, ii}} \right)^2 \\ &= \sum_{i=1}^{n} \left(\frac{\{(\mathbf{I} - \mathbf{S}_\lambda)\mathbf{y}\}_i}{1 - S_{\lambda, ii}} \right)^2 \\ &= \sum_{i=1}^{n} \left(\frac{y_i - \hat{y}_i}{1 - S_{\lambda, ii}} \right)^2. \end{aligned} \tag{5.4}$$

The beauty of (5.4) is that CV can be computed using only ordinary residuals and the diagonal elements of the smoother matrix.

5.3.2   *Generalized Cross-Validation (GCV)*

Efficient algorithms for computation of $\mathrm{CV}(\lambda)$ were developed in the mid-1980s (Hutchinson and de Hoog 1985). Before that time, the difficulties surrounding

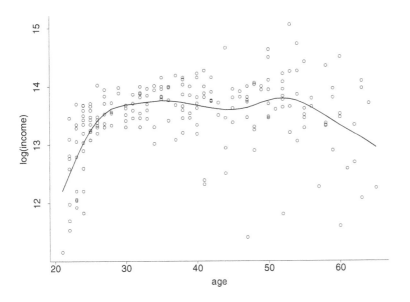

Figure 5.4
Scatterplot of the
age and log(income)
data with CV-based
penalized spline
smooth.

computation of the cross-validation criterion led to the proposal of a simplified version where the

$$\mathbf{S}_{\lambda, ii}$$

are replaced by their average

$$\frac{1}{n} \sum_{i=1}^{n} S_{\lambda, ii} = \frac{1}{n} \operatorname{tr}(\mathbf{S}_{\lambda}).$$

This criterion is known as *generalized cross-validation,* or GCV for short:

$$\begin{aligned}
\operatorname{GCV}(\lambda) &= \sum_{i=1}^{n} \left(\frac{\{(\mathbf{I} - \mathbf{S}_{\lambda})\mathbf{y}\}_i}{1 - n^{-1} \operatorname{tr}(\mathbf{S}_{\lambda})} \right)^2 \\
&= \frac{\operatorname{RSS}(\lambda)}{\{1 - n^{-1} \operatorname{tr}(\mathbf{S}_{\lambda})\}^2}.
\end{aligned} \tag{5.5}$$

Of course, GCV is not a generalization of CV. However, we will observe convention and continue to use "generalized" where we would prefer to use "approximate". We let $\hat{\lambda}_{\mathrm{GCV}}$ denote the smoothing parameter that minimizes GCV(λ). In the smoothing spline context, GCV was proposed by Craven and Wahba (1979).

The GCV curve for the LIDAR data is so close to the CV curve that the two are difficult to distinguish on a plot. Moreover, on the 40-point grid of λ-values used in Figure 5.3, GCV is minimized at the same point as CV.

5.3.2.1 *Age and Income Data*

Figure 5.4 is a scatterplot of the log(income) versus age for a sample of 205 Canadian workers, all of whom were educated to grade 13.

Consider the problem of modeling log(income) as a function of age. The scatterplot smooth in the figure is an estimate of the conditional expectation of log(income) given age using 24-knot linear regression splines with the smoother

The Canadian workers data were used by Ullah (1985), who identifies their source as a 1971 Canadian Census Public Use Tape.

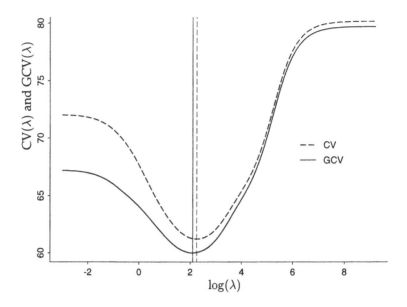

Figure 5.5 Cross-validation and generalized cross-validation curves for age–income data using 24-knot linear regression splines. The vertical lines correspond to the minima for each function.

parameter chosen via CV. The dip around age = 45 years in the spline smooth in the figure might be of interest to economists. If real, this dip represents a mid-career decline in mean salary.

Figure 5.5 shows GCV and CV curves for the age–income data. We see that, as with the LIDAR data, both criteria choose the same amount of smoothing. However, the two curves of GCV and CV versus λ differ somewhat more here than for the LIDAR data.

As with the LIDAR data, there is a suggestion of heteroscedasticity because log(income) appears more variable for older individuals compared to those under 40. For illustration, we will use a simple analysis that ignores this heteroscedasticity. See Chapter 14 for methods of analyzing heteroscedastic data.

5.3.3 *Mallows's C_p Criterion*

In Section 2.6 we briefly mentioned Mallows's C_p criterion (Mallows 1973),

$$C_p = \text{RSS}(p) + 2\hat{\sigma}_\varepsilon^2 p,$$

as a means of choosing among multiple regression models. The motivation for this criterion can be derived through nonparametric regression ideas. Suppose that the true model is

$$\mathbf{y} = \mathbf{f} + \boldsymbol{\varepsilon}, \quad \text{where } \text{Cov}(\boldsymbol{\varepsilon}) = \sigma_\varepsilon^2 \mathbf{I},$$

and that the estimator of $\mathbf{f}$ is a linear smoother

$$\hat{\mathbf{f}} = \mathbf{L}\mathbf{y}.$$

Then, as shown in Section 3.11, the mean summed squared error of $\hat{\mathbf{f}}$ can be written as

$$\text{MSSE}(\hat{\mathbf{f}}) = \|(\mathbf{L} - \mathbf{I})\mathbf{f}\|^2 + \sigma_\varepsilon^2 \,\text{tr}(\mathbf{L}\mathbf{L}^\mathsf{T}).$$

Using first (3.31) and then (3.33), we have

$$E(RSS) = MSSE(\hat{f}) + \sigma_\varepsilon^2(n - 2df_{fit}),$$

where $df_{fit} = tr(\mathbf{L})$ as defined in Section 3.13. It follows that if $\hat{\sigma}_\varepsilon^2$ is an unbiased estimate of σ_ε^2, then

$$RSS + 2\hat{\sigma}_\varepsilon^2 df_{fit}$$

is an unbiased estimator of

$$MSSE(\hat{f}) + n\sigma_\varepsilon^2.$$

Since $n\sigma_\varepsilon^2$ does not depend on $\mathbf{L}$, minimization of C_p is approximately the same as minimization of $MSSE(\hat{f})$.

For scatterplot smoothers such as penalized splines, putting $\mathbf{L} = \mathbf{S}_\lambda$ leads to the criterion

$$C_p(\lambda) \equiv RSS(\lambda) + 2\hat{\sigma}_\varepsilon^2 df_{fit}(\lambda)$$

for choosing λ. We let $\hat{\lambda}_{C_p}$ denote the smoothing parameter that minimizes $C_p(\lambda)$. The estimate $\hat{\sigma}_\varepsilon^2$ requires a choice of df_{fit}.

In Section 5.3.3.1, we will discuss two ways to choose df_{fit} when calculating $\hat{\mathbf{f}}$ to obtain residuals for estimating σ_ε^2: (a) use a large value of df_{fit} so there is little smoothing and little bias; (b) use GCV (or CV). Because C_p is used as an alternative to GCV and CV, (a) will be used here.

5.3.3.1 *Estimation of σ_ε^2*

Since the residuals estimate the true errors, a crude estimate of σ_ε^2 is RSS/n. This statistic underestimates σ_ε^2 because the residuals are less variable than the true errors. The same problem was encountered in parametric regression in Section 2.4. The solution there was to divide RSS by $n - p$ rather than n. The same solution is appropriate here, with $n - p$ replaced by $df_{res}(\lambda)$ as defined in Section 3.13. Therefore, we take

$$\hat{\sigma}_\varepsilon^2 = \frac{RSS(\lambda)}{df_{res}(\lambda)}. \tag{5.6}$$

How should λ be chosen for estimation of σ_ε^2? Using λ that minimizes CV or GCV is appropriate, but one might instead choose λ somewhat smaller than the minimizers of these criteria. The reason is that CV and GCV balance the bias and the variance $\hat{f}$. However, for the purpose of estimating σ_ε^2, bias in $\hat{f}$ is to be avoided even at the expense of extra variability of $\hat{f}$. This is because the bias in $\hat{f}$ inflates $\hat{\sigma}_\varepsilon^2$ in a manner that cannot be easily corrected. Variance in $\hat{f}$ deflates $\hat{\sigma}_\varepsilon^2$, but this is precisely the effect that can be corrected by dividing RSS by df_{res} instead of n.

From this argument we see that, when estimating σ_ε^2, a very small amount of smoothing is acceptable. When using penalized regression splines, even $\lambda = 0$ could be used to estimate σ_ε^2 – provided there are not so many knots that the OLS fit can no longer be computed in a numerically stable manner.

5.3.3.2 *Relationship between GCV and C_p*

Recall that

$$GCV(\lambda) = \frac{RSS(\lambda)}{\{1 - n^{-1}df_{fit}(\lambda)\}^2}.$$

If f is a smooth
function, then the
one-term Taylor
series approximation
to $f(x + h)$ is
$f(x) + hf'(x)$.

By a one-term Taylor series approximation,

$$\{n - df_{\text{fit}}(\lambda)\}^{-2} \simeq n^{-2}\{1 + 2df_{\text{fit}}(\lambda)/n\}.$$

Therefore, by letting

$$\hat{\sigma}_\varepsilon^2(\lambda) \equiv \frac{\text{RSS}(\lambda)}{n - df_{\text{fit}}(\lambda)} \simeq \frac{\text{RSS}(\lambda)}{n}$$

we obtain

$$\text{GCV}(\lambda) \simeq \text{RSS}(\lambda) + 2\hat{\sigma}_\varepsilon^2(\lambda)df_{\text{fit}}(\lambda).$$

Thus, GCV(λ) is approximately equal to $C_p(\lambda)$. The major difference between the two criteria is that GCV estimates σ_ε^2 using RSS(λ) whereas $C_p(\lambda)$ requires a prior estimate of σ_ε^2. This makes GCV somewhat more attractive.

5.3.4 *Other Model Selection Criteria*

Apart from GCV and C_p, there are several other selection criteria that trade off RSS against $df_{\text{fit}}(\lambda)$ in various ways. One popular one is *Akaike's information criterion* (AIC) (Akaike 1973):

$$\text{AIC}(\lambda) \equiv \log\{\text{RSS}(\lambda)\} + 2df_{\text{fit}}(\lambda)/n. \tag{5.7}$$

The search for better-performing model selection criteria for selecting smoothing parameters is ongoing. For selection of a bandwidth for kernel regression, Hurvich, Simonoff, and Tsai (1998) proposed a modification of AIC called *corrected* AIC:

$$\text{AIC}_C(\lambda) \equiv \log\{\text{RSS}(\lambda)\} + \frac{2\{df_{\text{fit}}(\lambda) + 1\}}{n - df_{\text{fit}}(\lambda) - 2}.$$

They give a mathematical justification and conduct a large simulation study that shows good comparative performance of this criterion for selecting the bandwidth in kernel regression.

5.4   Caveats of Automatic Parameter Selection

The preceding sections may give the impression that the choice of a good smoothing parameter can be achieved by simply implementing one of the algorithms described there. Unfortunately, this is not the case. Like all estimators, automatic smoothing parameter selectors are subject to variability and will not necessarily estimate the "best" smoothing parameter very well.

Since the early 1980s, a great deal of theoretical and simulation-based investigation into the performance of automatic smoothing has taken place. Most of the theory has been done by asymptotics in the local polynomial context (Section 3.15.1), where a theoretically optimal smoothing parameter is taken to be the target. However, it has been shown (Härdle, Hall, and Marron 1988) that, as the sample size increases to infinity, smoothing parameter selectors such as GCV and C_p converge very slowly to the optimum. This translates to the variability of the

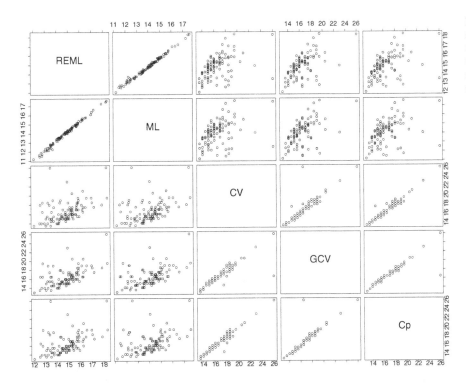

Figure 5.6 Pairwise scatterplots of df_{fit} for REML, ML, CV, GCV, and C_p from small simulation study.

best smoothing parameter being quite high. Therefore, one should not implicitly trust the answer given by one of these methods.

Simulation studies are useful for gaining an appreciation of the practical performance of smoothing parameter selectors and for making comparisons. Some large studies were published in the 1990s – such as Park and Turlach (1992), Cao, Cuevas, and González-Manteiga (1994), Chiu (1996), and Jones, Marron, and Sheather (1996a) – although each of these concerned the related problem of probability density estimation.

To illustrate the practical performance of automatic smoothing parameter selectors in the context of penalized spline regression, we ran a small simulation study in which (x_i, y_i), $1 \leq i \leq 200$, were generated from

$$y_i = f(x_i) + \varepsilon_i, \quad \varepsilon_i \overset{\text{ind.}}{\sim} \mathrm{N}(0, 0.01),$$

with the x_i uniformly distributed on the interval $(0, 1)$. The true regression function is

$$f(x) = 1.5\phi\left(\frac{x - 0.35}{0.15}\right) - \phi\left(\frac{x - 0.8}{0.04}\right),$$

where $\phi(x) = (2\pi)^{-1/2}e^{-x^2/2}$ is the standard normal density function. There were 100 simulated data sets.

Figure 5.6 shows pairwise scatterplots of df_{fit} for REML, ML, CV, GCV, and C_p. Immediately it is seen that the selectors separate into two groups that give roughly the same results for each sample: the likelihood-based selectors REML

Figure 5.7
(a) $df_{\text{fit}}(\hat{\lambda}_{\text{REML}})$ versus $df_{\text{fit}}(\hat{\lambda}_{\text{GCV}})$ and (b) $\log_{10}(\text{RMSE})$ for REML versus $\log_{10}(\text{RMSE})$ for GCV for the small simulation study described in the text. The 45° line allows comparison of the values for each sample.

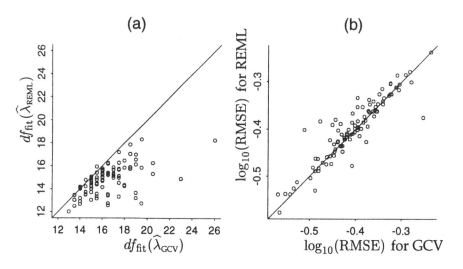

and ML; and the model selection–based selectors CV, GCV, and C_p. Thus, we will restrict further comparison to REML and GCV.

Figure 5.7(a) is a reproduction of the $df_{\text{fit}}(\hat{\lambda}_{\text{REML}})$ versus $df_{\text{fit}}(\hat{\lambda}_{\text{GCV}})$ scatterplot with a 45° line added. Notice that GCV tends to choose more degrees of freedom than REML for this example. In Figure 5.7(b) we compare the two selectors via the root mean squared error (RMSE) goodness-of-fit measure,

$$\text{RMSE} = \sqrt{\sum_{i=1}^{200}\{\hat{f}(x_i) - f(x_i)\}^2},$$

on the $\log_{10}$ scale. Notice that the RMSE values for GCV and REML are quite similar, with neither being systematically larger than the other. In summary, REML will produce smoother fits than GCV because it chooses lower values of df_{fit} than GCV. This means that in the fitted curves REML will have more bias and less variance than GCV. However, the two methods of selecting the amount of smoothing will be about equally accurate in terms of MSE; they trade off bias and variance differently but achieve nearly the same values of MSE = bias^2 + variance.

Figure 5.8 shows the regression function estimates corresponding to the 50th and 90th percentiles of the RMSE values for REML and GCV. This reveals that REML does not handle the dip around $x = 0.8$ as well as GCV does. However, the GCV-based fits tend to be more wiggly. Out of the 100 simulations, GCV has a lower RMSE on 57 occasions. Moreover, a Wilcoxon test shows the mean RMSE differences to be significantly less than zero and so, based on this measure, GCV comes out on top for this example.

Simulation studies with many more settings are required to better understand the relative practical performance of smoothing parameter selectors described in this chapter. For scatterplot smoothing such studies are somewhat piecemeal

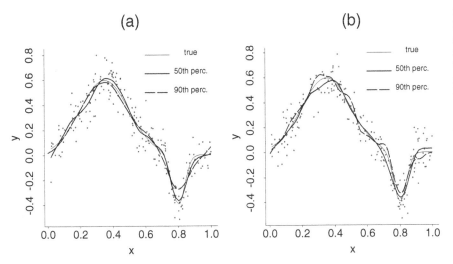

Figure 5.8 Estimates based on (a) REML and (b) GCV corresponding to the 50th and 90th percentiles of the RMSE distributions for the simulation study described in the text.

throughout the literature, and general recommendations are difficult to extract at this stage.

5.5   Choosing the Knots and Basis Functions

Two regression spline fits that differ by their numbers of knots or by their basis functions can be compared if each has a λ value that was automatically chosen "optimally" (say, by CV). We can also compare the amount of smoothing that they impose by comparing their df_{fit} values. Thus, by using CV and df_{fit}, we can study the effects of varying the number of knots or the degree of a regression spline. This same principle was applied earlier to compare regression spline fitting to polynomial regression for the LIDAR data; see Section 3.13.

5.5.1   *Varying the Number of Knots*

First we study the effect of varying the number of knots by using CV and df_{fit}. In Figure 5.9, CV(λ) is plotted against λ for 24-knot linear splines and for 48-knot linear splines. Only the region of df_{fit} values that come close to minimizing CV is shown. One sees that in this region there is little difference between using 24 knots and 48 knots, at least in terms of CV. For these data there really is no essential difference between 24 and 48 knots. In fact, plots of the 24-knot and 48-knot minimum-CV smooths are virtually indistinguishable by eye.

Of course, if there are too few knots then a good fit may not be achievable. Figure 5.10 shows minimum-CV linear spline fits with 6, 12, and 24 knots. The 12-knot fit is only slightly different from the 24-knot fit and still fits the data well. The 6-knot fit is less flexible and fits the data less well. The 6-knot spline also has a noticeable piecewise linear or "kinky" appearance. The 12- and 24-knot

Figure 5.9 Plot of
CV(λ) versus $df_{\text{fit}}(\lambda)$
for the LIDAR data
for 24-knot and
48-knot linear splines.
The vertical lines pass
through the minima
of the CV(λ) curves
which are virtually
indistinguishable in
this case.

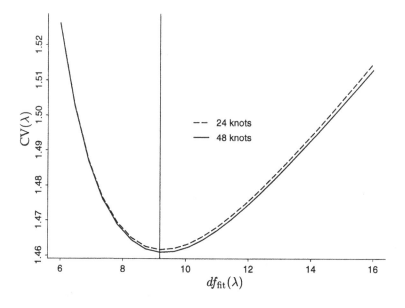

Figure 5.10 Plot of
minimum-CV linear
splines with 6, 12,
and 24 knots for the
LIDAR data.

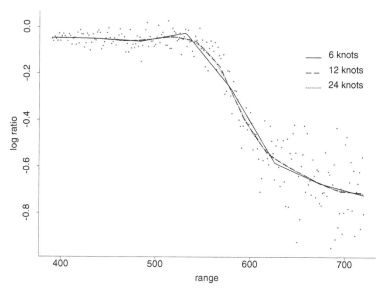

splines are, of course, piecewise linear, but they appear smooth because the jump
at each of their knots is small.

5.5.2 *Varying the Degree of the Regression Spline*

What happens to a regression spline smooth if the degree of the spline is changed?
The answer depends on how many knots are being used and how smooth the true
function f is.

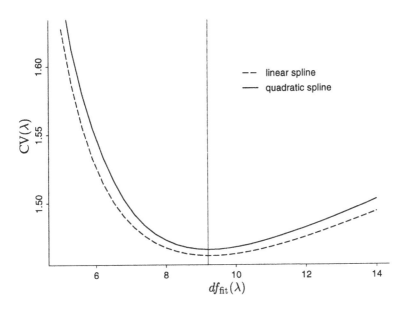

Figure 5.11 Plot of CV for 24-knot linear and quadratic splines for the LIDAR data. The vertical lines pass through the minima of the CV(λ) curves, which are virtually indistinguishable in this case.

Suppose that f is smooth (e.g., f'' is continuous). If one is using a linear spline with enough knots so that increasing the number of knots has no appreciable effect on the penalized fit, then increasing the degree of the spline is also unlikely to have a noticeable effect. As an example, consider 24-knot linear and quadratic spline fits to the LIDAR data. The CV functions for both fits are shown in Figure 5.11. One can see that, for a fixed value of df_{fit}, the two fits have nearly equal CV values. The range of possible values of df_{fit} is shifted 1 unit to the right when the degree is increased by 1, but CV-optimal values of df_{fit} are well within both ranges. Figure 5.12 shows minimum-CV linear and quadratic fits that are nearly indistinguishable.

However, if one uses fewer knots then quadratic regression splines can fit the LIDAR data better than linear splines. Figure 5.13 shows the minimum-CV linear and quadratic 6-knot fits. One sees that the kinky appearance of the linear spline disappears when one uses a higher-degree fit.

5.5.3   *Default Choices for Knot Locations*

In Section 5.5.1 we saw that, provided the set of knots was relatively "dense" with respect to the x_i, the result hardly changed. The idea is to choose enough knots to resolve the essential structure in the underlying regression function. But for more elaborate penalized spline models (to be studied in later chapters) there are computational advantages to keeping the number of knots relatively low. A reasonable default is to choose the knots to ensure that there are a fixed number of unique observations, say 4–5, between each knot. For large data sets this can lead to an excessive number of knots, so a maximum number of allowable knots (say, 20–40 total) is recommended.

Figure 5.12 Plot of
the minimum-CV
24-knot linear and
quadratic penalized
spline fits for the
LIDAR data.

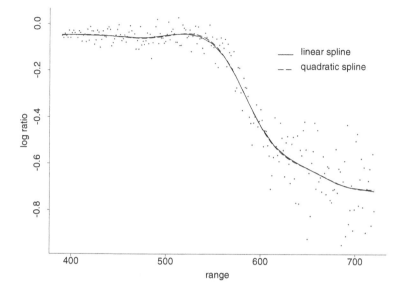

Figure 5.13 Plot of
minimum-CV 6-knot
linear and quadratic
splines.

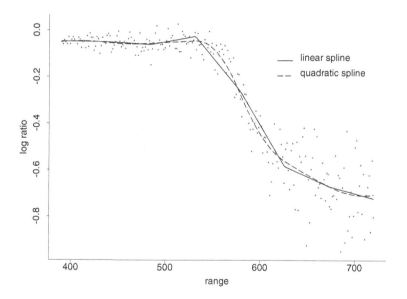

Our current default for knot locations is:

$$\kappa_k = \left(\tfrac{k+1}{K+2}\right)\text{th sample quantile of the unique } x_i \tag{5.8}$$

for $k = 1, \ldots, K$.

A simple default choice of K that usually works well is

$$K = \min\left(\tfrac{1}{4} \times \text{number of unique } x_i, \, 35\right). \tag{5.9}$$

Although (5.8) and (5.9) work well in most of the examples we come across, they
do not use any information in the data except the sample size. There are times
where one needs a more sophisticated algorithm that uses the data to choose K.

Alternatively, the user may choose K based on visual inspection of the scatterplot to determine the complexity of f relative to the noise in the data.

For example, if the regression function seems to have a lot of fine detail then K should be increased. For this reason, we have developed two automatic algorithms for determining K. These are studied in the next section.

5.6 Automatic Selection of the Number of Knots

This section discusses two algorithms for data-based selection of the number of knots, K. The *myopic* algorithm searches a sequence of trial values of K and stops when there is no improvement in GCV. The *full-search* algorithm searches the entire sequence of trial values and uses the one that minimizes GCV. The myopic algorithm works very well for most problems, but it can be fooled into stopping prematurely.

5.6.1 *Myopic Algorithm*

The myopic algorithm uses a sequence of trial values of K (5, 10, 20, 40, 80, and 120), except that only values of K in this sequence that are less than $n_{uniq} - p - 1$ are used, where n_{uniq} is the number of unique x_i.

The algorithm for selecting the number of knots is as follows. First, the penalized spline fit is computed for K equal to 5 and 10. In each case λ is chosen to minimize GCV(λ) for that number of knots. If GCV at $K = 10$ is greater than 0.98 times GCV at $K = 5$, then one concludes that further increases in K are unlikely to decrease GCV and hence uses $K = 5$ or 10, whichever has the smallest GCV. Otherwise, one computes the penalized spline fit with $K = 20$ and compares GCV for $K = 10$ with GCV for $K = 20$ – in the same way one compared GCV for $K = 5$ and 10. One stops and uses $K = 10$ or 20 (whichever gives the smaller GCV) if GCV at $K = 20$ exceeds 0.98 times GCV at $K = 10$. Otherwise, one computes the penalized spline at $K = 40$, and so forth. The algorithm is called "myopic" because it never looks beyond the value of K where it stops.

5.6.2 *Full-Search Algorithm*

The full-search algorithm computes GCV, minimized over λ, at all values of K in our trial sequence. The value of K (in that sequence) that minimizes GCV is selected.

The myopic algorithm has the advantage that it usually takes far less computation than the full-search algorithm. However, penalized splines can be computed so rapidly that this advantage is not compelling.

The one drawback to the myopic algorithm is that it can stop "before it really gets started". More precisely, for regression functions with enough complexity, neither $K = 5$ nor 10 will fit the data satisfactorily; it may happen that 5 knots is just as good as 10. In this case, the myopic policy will stop at 10 knots whereas the full-search policy will select a much greater number of knots and achieve a much better fit. An example where this phenomenon occurs is the cyclic function example in Section 5.6.3. This problem with the myopic policy may be occurring in the fossil data example given in Section 5.6.4, though with real data it is impossible to know the "right answer".

5.6.3 A Simulation Study

A small simulation study of the two algorithms used $n = 250$ observations per data set. The x_i were equally spaced on $[0, 1]$. Two regression functions were studied: the first, called the "bump function", was

$$f(x) = \frac{1}{0.1 + x} + 8 \exp\{-400(x - 0.5)^2\};$$

the second, called the "cyclic function", was

$$f(x) = 10 \sin(20\pi x^{1.3}).$$

The standard deviation σ was equal to 2 for the bump function and to 4 for the cyclic function. There were 300 simulated data sets for each regression function. On each data set, the myopic and full-search algorithms were applied. Penalized splines using each of the fixed values of K searched by the automatic algorithms (i.e., 5, 10, 20, 40, 80, 120) were also calculated.

Figure 5.14 shows the results for the bump function. Panel (a) shows the function, a typical data set, and the full search estimator for that data set. In panel (b) there is a plot of relative MASE (mean average squared error) versus K for the penalized splines with K fixed. Here MASE is the average over the 300 simulated data sets of

$$\text{ASE} = n^{-1} \sum_{i=1}^{n} \{\hat{f}(x_i; \hat{\lambda}) - f(x_i)\}^2.$$

Relative MASE is MASE divided by MASE minimized over fixed K. For the bump function, the minimum occurs at $K = 10$. There are also horizontal lines to indicate MASE for the two automatic algorithms. Panel (c) shows histograms of the values of K chosen by the algorithms. Panel (d) plots ASE for $K = 40$ versus for $K = 10$.

MASE is very high for $K = 5$, is minimized by $K = 10$, and is near a minimum for $K = 20, 40, 80,$ and 120. Neither automatic algorithm chooses $K = 5$ for any of the 300 data sets. Both tend to choose the best K, $K = 10$. Not surprisingly, the myopic algorithm tends to choose smaller values of K than the full-search algorithm, and for this reason the myopic algorithm has a smaller MASE in this example than the full-search algorithm.

The results for the cyclic function are shown in Figure 5.15. Note that in panel (b), the vertical axis is on the log scale because the MASE for $K = 5, 10,$ and 20 is an order of magnitude greater than for $K = 80$ or 120. The full-search

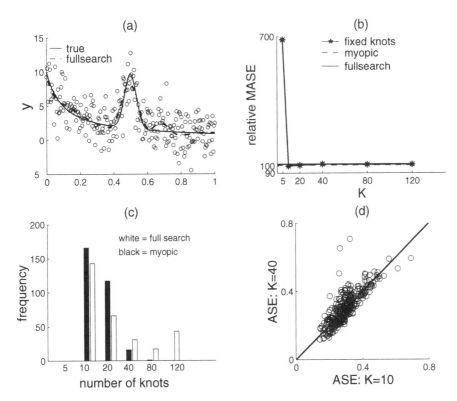

Figure 5.14 Bump function. (a) Typical data set, true regression function, and estimate from the full-search algorithm. (b) Semilog plot of relative MASE as a function of K with horizontal lines through MASE for the myopic and full-search algorithms. (c) Histograms of K as chosen by the myopic (black on left) and full-search (white on right) algorithms. (d) Plot of average squared errors for the 300 samples for a small and a large value of K.

algorithm is successful in that it chooses 80 or 120 knots in each of the 300 simulated data sets. The myopic algorithm chooses 5 or 10 knots in about 80% of the data sets. The problem here is that 10 knots is no better than 5, and the myopic algorithm stops prematurely. Interestingly, the myopic algorithm never chooses 20 knots: once the algorithm decides not to stop at 10 knots, it finds a moderate improvement going from 10 to 20 knots and then a huge improvement going from 20 to 40 knots. In panel (d) one sees that the ASE for 40 knots is rather consistently slightly larger than for 80 knots, an effect of larger bias for 40 knots compared to 80.

5.6.4 *Fossil Data*

When applied to the fossil data, the myopic algorithm chooses 5 knots and the full-search algorithm chooses 80 knots. See Figure 5.16. The fits with 10 and 20 knots are also shown. The 40-knot fit is similar to the full-search fit with 80 knots. Clearly the number of knots does affect the fit in the region between 95 and 105 million years. A dip in this region is seen in all fits except the 5-knot fit. As with any example of real data, the "true" regression function is unknown and we cannot be certain which estimate is best. Since the myopic algorithm chooses 5 knots, one might conjecture that it has stopped prematurely. However, Chaudhuri and Marron (1999) use their feature significance methodology (see Section 6.9)

Figure 5.15 Cyclic
function. (a) Typical
data set, true
regression function,
and estimate from the
full-search algorithm.
(b) Semilog plot
of relative MASE
as a function of
K with horizontal
lines through MASE
for the myopic and
full-search algorithms.
(c) Histograms of
K as chosen by the
myopic (black on left)
and full-search (white
on right) algorithms.
(d) Plot of average
squared errors for
the 300 samples for
a small and a large
value of K.

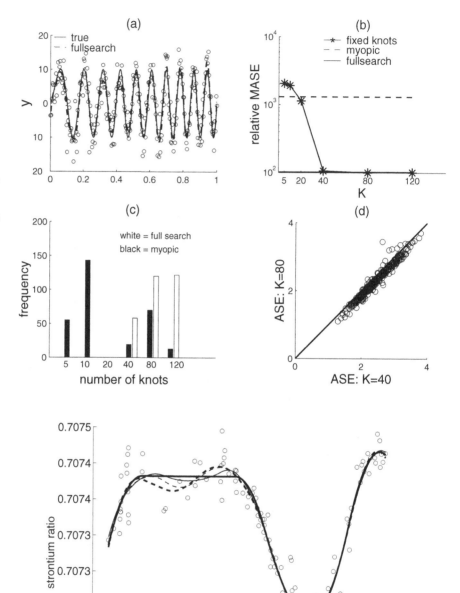

Figure 5.16 Myopic,
full-search, and two
fixed-knots fits to the
fossil data.

to conclude that the data do *not* support the hypothesis of a dip at about 97 million years. Thus, there is little or no evidence that the 5-knot fit without the dip selected by the myopic algorithm is inferior to the other fits; in fact, if the dip is spurious then the 5-knot fit could be considered better than the other estimators.

5.7 Bibliographical Notes

Automatic smoothing parameter selection was the focus of a great deal of research in the 1980s and early 1990s. Much of it was concerned with the kernel density estimation setting; Jones, Marron, and Sheather (1996a,b) provide a good survey. See also Chiu (1996). Ansley and Kohn (1985) and Wahba (1985) investigated the use of REML for automatic smoothing spline fitting. The myopic algorithm for selecting K was proposed by Ruppert and Carroll (2000). The myopic and full-search algorithms were studied in detail by Ruppert (2002). His simulation study is much like that in Section 5.6.3 but more extensive.

In contrast to the situation with penalized estimation, with OLS knot selection is crucially important. Therefore, it is not surprising that there is a large literature on knot selection when fitting is by OLS. Smith (1982) appears to be the first paper in this area, and Friedman (1991) and Smith and Kohn (1996) are important contributions. See Stone et al. (1997) and Hansen et al. (2003) and references therein for a good introduction to this work.

5.8 Summary of Formulas

REML smoothing parameter selection

For pth-degree truncated splines,

$$\hat{\lambda}_{\text{REML}} = (\hat{\sigma}_{\varepsilon,\text{REML}}^2 / \hat{\sigma}_{u,\text{REML}}^2)^{1/2p}$$

where $(\hat{\sigma}_{u,\text{REML}}^2, \hat{\sigma}_{\varepsilon,\text{REML}}^2)$ minimizes

$$\ell_R(\mathbf{V}) = -\tfrac{1}{2}[n\log(2\pi) + \log|\mathbf{V}| + \log|\mathbf{X}^\mathsf{T}\mathbf{V}^{-1}\mathbf{X}|$$
$$+ \mathbf{y}^\mathsf{T}\mathbf{V}^{-1}\{\mathbf{I} - \mathbf{X}(\mathbf{X}^\mathsf{T}\mathbf{V}^{-1}\mathbf{X})^{-1}\mathbf{X}^\mathsf{T}\mathbf{V}^{-1}\}\mathbf{y}]$$

with $\mathbf{V} \equiv \sigma_u^2\mathbf{Z}\mathbf{Z}^\mathsf{T} + \sigma_\varepsilon^2$.

ML smoothing parameter selection

$$\hat{\lambda}_{\text{ML}} = (\hat{\sigma}_{\varepsilon,\text{ML}}^2 / \hat{\sigma}_{u,\text{ML}}^2)^{1/2p}$$

where $(\hat{\sigma}_{u,\text{ML}}^2, \hat{\sigma}_{\varepsilon,\text{ML}}^2)$ minimizes

$$\ell(\mathbf{V}) \equiv -\tfrac{1}{2}[n\log(2\pi) + \log|\mathbf{V}| + \mathbf{y}^\mathsf{T}\mathbf{V}^{-1}\{\mathbf{I} - \mathbf{X}(\mathbf{X}^\mathsf{T}\mathbf{V}^{-1}\mathbf{X})^{-1}\mathbf{X}^\mathsf{T}\mathbf{V}^{-1}\}\mathbf{y}]$$

Cross-validation (CV)

$\hat{\lambda}_{\text{CV}}$ minimizes

$$\text{CV}(\lambda) \equiv \sum_{i=1}^{n}\{y_i - \hat{f}_{-i}(x_i;\lambda)\}^2$$

where $\hat{f}_{-i}(x_i;\lambda)$ is the scatterplot smooth based on the data with (x_i, y_i) omitted.

Generalized cross-validation (GCV)

$\hat{\lambda}_{\text{GCV}}$ minimizes

$$\text{GCV}(\lambda) \equiv \frac{\text{RSS}(\lambda)}{\{1 - n^{-1}df_{\text{fit}}(\lambda)\}^2}$$

Mallows's C_p

$\hat{\lambda}_{C_p}$ minimizes

$$C_p(\lambda) \equiv \mathrm{RSS}(\lambda) + 2\hat{\sigma}_\varepsilon^2 df_{\mathrm{fit}}(\lambda)$$

for some estimate $\hat{\sigma}_\varepsilon^2$ of $\sigma_\varepsilon^2 \equiv \mathrm{var}(\varepsilon_i)$.

Akaike's information criterion (AIC)

$\hat{\lambda}_{\mathrm{AIC}}$ minimizes

$$\mathrm{AIC}(\lambda) \equiv \log\{\mathrm{RSS}(\lambda)\} + 2df_{\mathrm{fit}}(\lambda)/n$$

Corrected AIC (AIC_C)

$\hat{\lambda}_{\mathrm{AIC}_C}$ minimizes

$$\mathrm{AIC}_C(\lambda) \equiv \log\{\mathrm{RSS}(\lambda)\} + \frac{2\{df_{\mathrm{fit}}(\lambda) + 1\}}{n - df_{\mathrm{fit}}(\lambda) - 2}.$$

6

Inference

6.1 Introduction

The methodology in Chapters 3 and 5 solves the problem of fitting a smooth curve to a scatterplot. More formally, for the nonparametric regression model

$$y_i = f(x_i) + \varepsilon_i, \quad \mathsf{E}(\varepsilon_i) = 0,$$

it estimates the function $f(x) = \mathsf{E}(y|x)$ without the stringency of a parametric model.

For a particular value x of the predictor, the value of a scatterplot smooth at x, $\hat{f}(x)$, is a *point estimate* of $f(x)$. Natural follow-up questions are:

- What is the estimated *standard deviation* of $\hat{f}(x)$?
- What is a *95% confidence interval* for $f(x)$?

These problems fall within the realm of statistical inference for the unknown quantity $f(x)$ and are simply in keeping with those used routinely in parametric modeling. However, in this *function estimation* context there are a number of new *global* inferential questions that arise, such as:

- Is f linear or nonlinear?
- Is the dip apparent in $\hat{f}$ "really there"?
- Is f monotonically increasing?

In this chapter we will describe techniques for addressing both point and global inferential questions.

6.2 Variability Bands

Figure 6.1 is a penalized spline smooth of the fossil scatterplot described on page 68. For each value of age, the height of the curve is a point estimate of

$$f(\text{age}) = \mathsf{E}(\text{strontium ratio}|\text{age}).$$

For example,

$$\hat{f}(100) = 0.7074118.$$

However, a common convention in statistical analysis is to report a standard deviation estimate:

Figure 6.1 Smooth
of the fossil data.
For each value of
age, the vertical
height of the curve
is a point estimate
of E(strontium
ratio|age). The
shaded bar at
age = 100 is an
approximate 95%
confidence interval
for E(strontium
ratio|100).

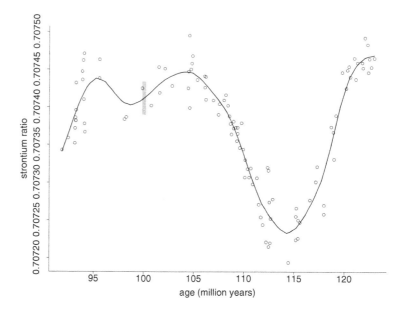

Figure 6.2 Variability
band for a smooth
of the fossil data.
The shaded region
represents plus and
minus twice the
estimated standard
deviation at each value
of age.

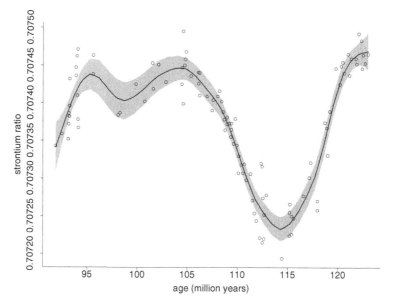

$$\hat{f}(100) = 0.7074118 \ (0.000011).$$

Here 0.000011 is the estimated standard deviation of $\hat{f}(100)$, which we will denote by $\widehat{\text{st.dev.}}\{\hat{f}(100)\}$. If $\hat{f}(100)$ is approximately normally distributed (see next section), then an approximate 95% confidence interval for $f(100)$ is

$$\hat{f}(100) \pm 2 \times \widehat{\text{st.dev.}}\{\hat{f}(100)\} = (0.7073898, 0.7074339);$$

this interval is shown by the shaded vertical strip around age = 100 in Figure 6.1.

Obviously it would be useful to do this for all values of age in the range of the data set. The result is shown in Figure 6.2. The "cloud" around the scatterplot smooth corresponds to plus and minus twice the estimated standard deviation at each age value; we will call this a *variability band*.

Variability bands are straightforward to calculate if the smooth is linear in the response vector **y**, as is the case for a fixed smoothing parameter in the penalized spline or the local polynomial estimates described in Chapter 3. For the special case of penalized splines, the mathematics for calculating variability bands differs depending on whether a mixed model representation is being used. In this section and the next we will not use a mixed model representation, leaving that to Section 6.4.

Let x be a general value of the predictor variable. Then the estimate at $f(x)$ is

$$\hat{f}(x) = \boldsymbol{\ell}_x^{\mathsf{T}} \mathbf{y} \tag{6.1}$$

for some $n \times 1$ vector $\boldsymbol{\ell}_x$. Ignoring, for now, the dependence of $\boldsymbol{\ell}_x$ on estimated smoothing parameters, we have

$$\mathrm{var}\{\hat{f}(x)\} = \boldsymbol{\ell}_x^{\mathsf{T}} \mathrm{Cov}(\mathbf{y})\boldsymbol{\ell}_x.$$

If it is reasonable to assume homoscedasticity via

$$\mathrm{Cov}(\mathbf{y}) = \sigma_\varepsilon^2 \mathbf{I},$$

then

$$\widehat{\mathrm{st.dev.}}\{\hat{f}(x)\} = \hat{\sigma}_\varepsilon \|\boldsymbol{\ell}_x\|$$

for some suitable estimate $\hat{\sigma}_\varepsilon^2$ of σ_ε^2. Such a strategy was used to produce Figure 6.2. The variability bands correspond to

$$\hat{f}(x) \pm 2 \times \widehat{\mathrm{st.dev.}}\{\hat{f}(x)\} \quad \text{for } 91.8 \le x \le 123.$$

Strictly speaking, an adjustment should be made to $\widehat{\mathrm{st.dev.}}\{\hat{f}(x)\}$ to account for

(1) variability in $\hat{\sigma}_\varepsilon$ as an estimate of σ_ε,
(2) bias due to curvature, and
(3) variability in $\boldsymbol{\ell}_x$ due to smoothing parameter estimation.

In the next section we argue that replacement of the "2" by $t(0.975; df_{\mathrm{res}})$ is a reasonable strategy for addressing (1), although for typical scatterplot smoothing this correction will not make much difference. For example, for the fossil data $n = 106$ and $df_{\mathrm{res}} = 90.2$, so $t(0.975; df_{\mathrm{res}}) = 1.987 \approx 2$.

Sources (2) and (3) are somewhat more delicate and are best appreciated through the mixed model representation of penalized splines and Bayesian modeling. Thus, we postpone further discussion of these issues until Section 6.4 and Chapter 16.

6.3 Confidence and Prediction Intervals

Consider, first, the homoscedastic *normal errors* model

$$y_i = f(x_i) + \varepsilon_i, \quad \varepsilon_i \overset{\text{ind.}}{\sim} N(0, \sigma_\varepsilon^2),$$

and estimation of $f(x)$ through a general linear smoother

$$\hat{f}(x) = \boldsymbol{\ell}_x^{\mathsf{T}} \mathbf{y}. \tag{6.2}$$

For fixed values of the smoothing parameter $\boldsymbol{\ell}_x$,

$$\hat{f}(x) \sim \mathrm{N}(\mathrm{E}\{\hat{f}(x)\}, \sigma_\varepsilon^2 \|\boldsymbol{\ell}_x\|^2)$$

and so

$$\frac{\hat{f}(x) - \mathrm{E}\{\hat{f}(x)\}}{\sigma_\varepsilon \|\boldsymbol{\ell}_x\|} \sim \mathrm{N}(0, 1). \tag{6.3}$$

If σ_ε is replaced by an estimate $\hat{\sigma}_\varepsilon$ then, for small n, the normal approximation is poor. In parametric regression models such studentized statistics have a t-distribution with degrees of freedom equal to $n - p$, where p is the number of parameters in the model. In scatterplot smoothing we have the approximation:

$$\frac{\hat{f}(x) - \mathrm{E}\{\hat{f}(x)\}}{\hat{\sigma}_\varepsilon \|\boldsymbol{\ell}_x\|} \overset{\text{approx.}}{\sim} t_{[df_{\text{res}}]} \tag{6.4}$$

(Hastie and Tibshirani 1990), where $[x]$ is the closest integer to x.

Central limit theorems for smoothers require certain assumptions on $\boldsymbol{\ell}_x$. Several papers, starting with Schuster (1972), have worked out the details.

If the errors are not necessarily Gaussian then, by a central limit theorem for the smoother $\hat{f}(x)$, for large samples we have

$$\frac{\hat{f}(x) - \mathrm{E}\{\hat{f}(x)\}}{\hat{\sigma}_\varepsilon \|\boldsymbol{\ell}_x\|} \overset{\text{approx.}}{\sim} \mathrm{N}(0, 1).$$

Confidence intervals like the t-intervals of Section 2.4 can be based on (6.4) and (6.3). The resulting intervals are

$$\hat{f}(x) \pm \begin{cases} t\left(1 - \frac{\alpha}{2}; df_{\text{res}}\right)\hat{\sigma}_\varepsilon \|\boldsymbol{\ell}_x\| & \text{for small } n, \\ z\left(1 - \frac{\alpha}{2}\right)\hat{\sigma}_\varepsilon \|\boldsymbol{\ell}_x\| & \text{for large } n. \end{cases} \tag{6.5}$$

Note that these intervals cover $\mathrm{E}\{\hat{f}(x)\}$ with $100(1 - \alpha)\%$ confidence rather than $f(x)$. Interpretation of them as approximate confidence intervals for $f(x)$ requires approximate unbiasedness of $\hat{f}(x)$. Often the plausibility of unbiasedness can be assessed from inspection of the fitted curve to the scatterplot, but in high-noise situations and more complex settings this may be difficult. Theory for local polynomial regression (e.g. Tsybakov 1986) shows that bias is inherent in nonparametric regression when the amount of smoothing is optimal, and that it tends to be higher at peaks and valleys in the regression curve. Because of this, it is difficult to give a theoretical "stamp of approval" for (6.5) being a $100(1 - \alpha)\%$ confidence interval for $f(x)$. Nevertheless, it is often the case that approximate unbiasedness is reasonably assumed and hence variability bands can be interpreted as approximate confidence intervals. A form of bias correction is discussed in Section 6.4.

A prediction interval for a new observation at x, analogous to parametric interval (2.16), is

$$\hat{f}(x) \pm \begin{cases} t\left(1 - \frac{\alpha}{2}; df_{\text{res}}\right)\hat{\sigma}_\varepsilon \sqrt{1 + \|\boldsymbol{\ell}_x\|^2} & \text{for small } n, \\ z\left(1 - \frac{\alpha}{2}\right)\hat{\sigma}_\varepsilon \sqrt{1 + \|\boldsymbol{\ell}_x\|^2} & \text{for large } n. \end{cases} \tag{6.6}$$

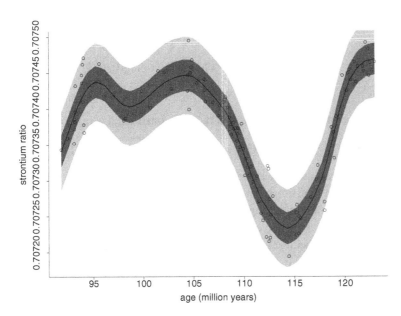

Figure 6.3
Confidence intervals
for $f(x_i)$ (dark
band) and prediction
intervals for new y
(light band).

Figure 6.3 shows the 95% confidence intervals (6.5) and prediction intervals (6.6). Bias is the same in prediction and confidence intervals, but variability is much greater in the former. For this reason, generally bias can safely be ignored in (6.6). The prediction intervals should contain about 95% of the observations. Because of the large sample size, the confidence intervals are relatively narrow. The prediction intervals are wider because most of the uncertainty in prediction is due to variation of a new observation about its mean, not to uncertainty regarding that mean.

Although the bands depicted in Figure 6.3 are useful for conveying the amount of sample variability inherent in a scatterplot smooth, it is important to realize that they can only be interpreted in a *pointwise* fashion. They cannot be used to infer the existence of any features, such as bumps or dips, that depend on the entire curve. Section 6.9 describes inference for feature existence.

6.4 Inference for Penalized Splines

Inference for penalized splines is a delicate matter because the variability estimates depend on whether or not the mixed model formulation is being used and (if not) how the random coefficients are handled.

Consider first the linear penalized spline *without* mixed model representation:

$$y_i = \beta_0 + \beta_1 x_i + \sum_{k=1}^{K} \beta_{1k}(x_i - \kappa_k)_+ + \varepsilon_i, \qquad (6.7)$$

where none of the coefficients $\beta_0, \beta_1, \beta_{11}, \ldots, \beta_{1K}$ are considered random and $\varepsilon_i \overset{\text{ind.}}{\sim} N(0, \sigma^2)$. This is the representation used in Chapter 3. In this case

$$\hat{f}(x) = \boldsymbol{\ell}_x^{\mathsf{T}} \mathbf{y},$$

where

$$\ell_x = \mathbf{C}(\mathbf{C}^\mathsf{T}\mathbf{C} + \lambda^2\mathbf{D})^{-1}\mathbf{C}_x^\mathsf{T}$$

with $\mathbf{C}_x \equiv [1 \ x \ (x - \kappa_1)_+ \ \ldots \ (x - \kappa_K)_+]$, $\mathbf{D} \equiv \mathrm{diag}(0, 0, 1, \ldots, 1)$, and

$$\mathbf{C} \equiv [\mathbf{C}_{x_i}]_{1 \le i \le n}.$$

Approximate confidence intervals can then be constructed using (6.5) with

$$\|\ell_x\| = \sqrt{\mathbf{C}_x(\mathbf{C}^\mathsf{T}\mathbf{C} + \lambda^2\mathbf{D})^{-1}\mathbf{C}^\mathsf{T}\mathbf{C}(\mathbf{C}^\mathsf{T}\mathbf{C} + \lambda^2\mathbf{D})^{-1}\mathbf{C}_x^\mathsf{T}}. \tag{6.8}$$

In Chapter 5 we used the mixed model formulation of penalized splines as a convenient fiction to estimate smoothing parameters. The mixed model is a reasonable (though not compelling) Bayesian prior for a smooth curve, and ML or REML estimates of variance components give estimates of the smoothing parameter that generally behave well. Can we push this idea further and use the mixed model for inference? We will see that the answer is "yes". Mixed model theory gives confidence intervals that are similar to (6.5) but a little wider. The extra width is a good thing, since it comes about because the mixed model theory takes bias into account whereas (6.5) does not.

Consider the mixed model representation of (6.7):

$$y_i = \beta_0 + \beta_1 x_i + \sum_{k=1}^{K} u_k(x_i - \kappa_k)_+ + \varepsilon_i,$$

which can be written as

$$\mathbf{y} = \mathbf{X}\boldsymbol{\beta} + \mathbf{Z}\mathbf{u} + \boldsymbol{\varepsilon}, \quad \mathrm{Cov}\begin{bmatrix} \mathbf{u} \\ \boldsymbol{\varepsilon} \end{bmatrix} = \begin{bmatrix} \sigma_u^2\mathbf{I} & 0 \\ 0 & \sigma_\varepsilon^2\mathbf{I} \end{bmatrix}, \tag{6.9}$$

where

$$\mathbf{X} = [1 \ x_i]_{1 \le i \le n}, \quad \mathbf{Z} = \left[(x_i - \kappa_k)_+\right]_{\substack{1 \le k \le K \\ 1 \le i \le n}}.$$

Let

$$\mathbf{X}_x = [1 \ x], \quad \mathbf{Z}_x = \left[(x - \kappa_k)_+\right]_{1 \le k \le K},$$

and

$$\tilde{f}(x) \equiv \mathbf{X}_x\tilde{\boldsymbol{\beta}} + \mathbf{Z}_x\tilde{\mathbf{u}},$$

where $\tilde{\boldsymbol{\beta}}$ and $\tilde{\mathbf{u}}$ are the BLUPs of $\boldsymbol{\beta}$ and $\mathbf{u}$. Then $\tilde{f}(x)$ is the BLUP of

$$f(x) \equiv \mathbf{X}_x\boldsymbol{\beta} + \mathbf{Z}_x\mathbf{u}.$$

We let

$$\hat{f}(x) \equiv \mathbf{X}_x\hat{\boldsymbol{\beta}} + \mathbf{Z}_x\hat{\mathbf{u}}$$

denote the corresponding EBLUP of $f(x)$.

Note that, within this framework, the target function $f(x)$ is random owing to randomness in $\mathbf{u}$. Variability estimates differ depending on whether randomness in $\mathbf{u}$ is taken into account. One argument, which leads to the same variability estimates as the ridge regression formulation, is that randomness of $\mathbf{u}$ is a device used to model curvature, while $\boldsymbol{\varepsilon}$ accounts for variability about the curve.

According to this argument, variance calculations should be done with respect to the conditional distribution $\mathbf{y}|\mathbf{u}$ rather than the unconditional distribution of $\mathbf{y}$. Variability bands can then be obtained from

$$\text{var}\{\tilde{f}(x)|\mathbf{u}\} = [\mathbf{X}_x \ \mathbf{Z}_x]\,\text{Cov}\left(\begin{bmatrix}\tilde{\boldsymbol{\beta}}\\ \tilde{\mathbf{u}}\end{bmatrix}\Big|\mathbf{u}\right)[\mathbf{X}_x \ \mathbf{Z}_x]^{\mathsf{T}}$$

$$= \mathbf{C}_x\,\text{Cov}\left(\begin{bmatrix}\tilde{\boldsymbol{\beta}}\\ \tilde{\mathbf{u}}\end{bmatrix}\Big|\mathbf{u}\right)\mathbf{C}_x^{\mathsf{T}}.$$

Then, from (4.17) of Section 4.7,

$$\text{Cov}\left(\begin{bmatrix}\tilde{\boldsymbol{\beta}}\\ \tilde{\mathbf{u}}\end{bmatrix}\Big|\mathbf{u}\right) = \sigma_\varepsilon^2\left(\mathbf{C}^{\mathsf{T}}\mathbf{C} + \frac{\sigma_\varepsilon^2}{\sigma_u^2}\mathbf{D}\right)^{-1}\mathbf{C}^{\mathsf{T}}\mathbf{C}\left(\mathbf{C}^{\mathsf{T}}\mathbf{C} + \frac{\sigma_\varepsilon^2}{\sigma_u^2}\mathbf{D}\right)^{-1}. \tag{6.10}$$

Hence

$$\widehat{\text{st.dev.}}\{\tilde{f}(x)|\mathbf{u}\} = \hat{\sigma}_\varepsilon\sqrt{\mathbf{C}_x\left(\mathbf{C}^{\mathsf{T}}\mathbf{C} + \frac{\hat{\sigma}_\varepsilon^2}{\hat{\sigma}_u^2}\mathbf{D}\right)^{-1}\mathbf{C}^{\mathsf{T}}\mathbf{C}\left(\mathbf{C}^{\mathsf{T}}\mathbf{C} + \frac{\hat{\sigma}_\varepsilon^2}{\hat{\sigma}_u^2}\mathbf{D}\right)^{-1}\mathbf{C}_x^{\mathsf{T}}}, \tag{6.11}$$

which matches (6.8). Also, if $\boldsymbol{\varepsilon} \sim \text{N}(\mathbf{0}, \sigma_\varepsilon^2\mathbf{I})$ then

$$\tilde{f}(x)|\mathbf{u} \sim \text{N}[\text{E}\{\tilde{f}(x)|\mathbf{u}\}, \text{var}\{\tilde{f}(x)|\mathbf{u}\}].$$

It follows that

$$\frac{\tilde{f}(x) - \text{E}\{\tilde{f}(x)|\mathbf{u}\}}{\widehat{\text{st.dev.}}\{\tilde{f}(x)|\mathbf{u}\}}\Big|\mathbf{u} \sim \text{N}(0, 1),$$

and an approximate $100(1 - \alpha)\%$ confidence interval for $\text{E}\{\tilde{f}(x)|\mathbf{u}\}$ is

$$\hat{f}(x) \pm z\left(1 - \tfrac{\alpha}{2}\right)\widehat{\text{st.dev.}}\{\hat{f}(x)|\mathbf{u}\}. \tag{6.12}$$

If there is no appreciable bias then $\text{E}\{\tilde{f}(x)|\mathbf{u}\} \approx f(x)$, and this interval can be interpreted as a confidence interval for $f(x)$. However, in this mixed model framework we can get a handle on bias. Note that this conditional bias is

$$\text{E}\{\tilde{f}(x) - f(x)|\mathbf{u}\} = \mathbf{X}_x\{\text{E}(\tilde{\boldsymbol{\beta}}|\mathbf{u}) - \boldsymbol{\beta}\} + \mathbf{Z}_x\{\text{E}(\tilde{\mathbf{u}}|\mathbf{u}) - \mathbf{u}\}$$

$$= -\frac{\sigma_\varepsilon^2}{\sigma_u^2}\mathbf{C}_x\left(\mathbf{C}^{\mathsf{T}}\mathbf{C} + \frac{\sigma_\varepsilon^2}{\sigma_u^2}\mathbf{D}\right)^{-1}\begin{bmatrix}\mathbf{0}\\ \mathbf{u}\end{bmatrix},$$

which is nonzero. But, since $\text{E}(\mathbf{u}) = \mathbf{0}$, the unconditional bias is

$$\text{E}\{\tilde{f}(x) - f(x)\} = 0.$$

Thus, on average over the distribution of $\mathbf{u}$, $\tilde{f}(x)$ is unbiased for $f(x)$. To account for bias in the confidence intervals, the ridge regression variance $\text{var}\{\tilde{f}(x)|\mathbf{u}\}$ should be replaced by the conditional mean squared error

$$\text{E}[\{\tilde{f}(x) - f(x)\}^2|\mathbf{u}] = \text{var}\{\tilde{f}(x)|\mathbf{u}\} + [\text{E}\{\tilde{f}(x) - f(x)|\mathbf{u}\}]^2$$

and then averaged over the $\mathbf{u}$ distribution. Noting that $\text{var}\{\tilde{f}(x)|\mathbf{u}\}$ is constant (not dependent on $\mathbf{u}$), we obtain

$$E\big(E[\{\tilde{f}(x)-f(x)\}^2|\mathbf{u}]\big)=\mathrm{var}\{\tilde{f}(x)|\mathbf{u}\}+E\big([E\{\tilde{f}(x)-f(x)|\mathbf{u}\}]^2\big).$$

But the left-hand side is just

$$E[\{\tilde{f}(x)-f(x)\}^2]=\mathrm{var}\{\tilde{f}(x)-f(x)\}$$

$$=\mathrm{var}\left\{\mathbf{C}_x\begin{bmatrix}\tilde{\boldsymbol{\beta}}-\boldsymbol{\beta}\\\tilde{\mathbf{u}}-\mathbf{u}\end{bmatrix}\right\}$$

$$=\mathbf{C}_x\,\mathrm{Cov}\begin{bmatrix}\tilde{\boldsymbol{\beta}}\\\tilde{\mathbf{u}}-\mathbf{u}\end{bmatrix}\mathbf{C}_x^{\mathsf{T}}.$$

Therefore, the ridge regression and bias adjusted variability estimates differ: the former uses

$$\mathrm{Cov}\left(\begin{bmatrix}\tilde{\boldsymbol{\beta}}\\\tilde{\mathbf{u}}\end{bmatrix}\Big|\mathbf{u}\right)=\mathrm{Cov}\left(\begin{bmatrix}\tilde{\boldsymbol{\beta}}\\\tilde{\mathbf{u}}-\mathbf{u}\end{bmatrix}\Big|\mathbf{u}\right)$$

whereas the latter uses

$$\mathrm{Cov}\left(\begin{bmatrix}\tilde{\boldsymbol{\beta}}\\\tilde{\mathbf{u}}-\mathbf{u}\end{bmatrix}\right)=\sigma_\varepsilon^2\left(\mathbf{C}^{\mathsf{T}}\mathbf{C}+\frac{\sigma_\varepsilon^2}{\sigma_u^2}\mathbf{D}\right)^{-1}.\tag{6.13}$$

Note that (6.13) is a special case of (4.16). This suggests

$$\widehat{\mathrm{st.dev.}}\{\hat{f}(x)-f(x)\}=\hat{\sigma}_\varepsilon\sqrt{\mathbf{C}_x\left(\mathbf{C}^{\mathsf{T}}\mathbf{C}+\frac{\sigma_\varepsilon^2}{\sigma_u^2}\mathbf{D}\right)^{-1}\mathbf{C}_x^{\mathsf{T}}}.$$

Also, under certain assumptions,

$$\frac{\hat{f}(x)-f(x)}{\widehat{\mathrm{st.dev.}}\{\hat{f}(x)-f(x)\}}\sim N(0,1),$$

and an approximate $100(1-\alpha)\%$ confidence interval for $f(x)$ is

$$\hat{f}(x)\pm\begin{cases}t\left(1-\frac{\alpha}{2};df_{\mathrm{res}}\right)\hat{\sigma}_\varepsilon\,\widehat{\mathrm{st.dev.}}\{\hat{f}(x)-f(x)\}&\text{for small }n,\\z\left(1-\frac{\alpha}{2}\right)\hat{\sigma}_\varepsilon\,\widehat{\mathrm{st.dev.}}\{\hat{f}(x)-f(x)\}&\text{for large }n.\end{cases}\tag{6.14}$$

Interval (6.14) will be somewhat longer than interval (6.12) because (6.14) accounts for both components of error (variance and squared bias) whereas (6.12) accounts only for variance and covers $E\{\tilde{f}(x)|\mathbf{u}\}$, not $f(x)$.

A revealing comparison between the two intervals can be made using the smoother matrix (Hastie and Tibshirani 1990, p. 60). If $\tilde{\mathbf{f}}$ and $\mathbf{f}$ are the vectors of $\tilde{f}(x_i)$ and $f(x_i)$, respectively, then

$$\mathrm{Cov}(\tilde{\mathbf{f}}-\mathbf{f}|\mathbf{u})=\sigma_\varepsilon^2\mathbf{S}\mathbf{S}^{\mathsf{T}},\qquad\mathrm{Cov}(\tilde{\mathbf{f}}-\mathbf{f})=\sigma_\varepsilon^2\mathbf{S},$$

where

$$\mathbf{S}\equiv\mathbf{C}\left(\mathbf{C}^{\mathsf{T}}\mathbf{C}+\frac{\sigma_\varepsilon^2}{\sigma_u^2}\mathbf{D}\right)\mathbf{C}^{\mathsf{T}}$$

is the smoother matrix associated with $\tilde{\mathbf{f}}$ ($\tilde{\mathbf{f}}=\mathbf{S}\mathbf{y}$). Therefore, at the x_i, the ridge regression variability bands use diagonal entries of $\mathbf{S}\mathbf{S}^{\mathsf{T}}$ while the bias adjusted bands use diagonal entries of $\mathbf{S}$.

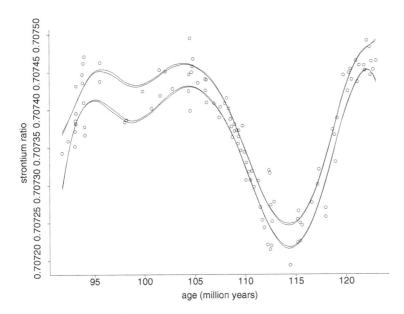

Figure 6.4 Fossil data with confidence bands that account for squared bias (outer) and do not account for squared bias (inner).

Figure 6.4 shows two sets of 95% confidence bands for the fossil data, the inner one given by (6.12) and the outer one given by (6.14). Note that the differences between the two are greatest in regions of higher curvature around 100 and 115 million years of age. This is where bias in curve estimates is greatest.

Often there is only a small difference between (6.14) and (6.12), but the difference can be nontrivial. We prefer (6.14) because it is an interval for $f(x)$, the true object of interest. We discussed (6.12) carefully, since that interval is in common use and there is a need to understand why (6.14) is superior.

The bias adjustment described here is equivalent to that developed for smoothing splines by Wahba (1983) and Nychka (1988). Their derivation was based on a Bayesian perspective and so they used the name *Bayesian* confidence intervals. Because the mixed model can be viewed as a partially Bayesian model – with the distribution of the random effects as their prior but with the fixed effects and variance components not having priors – these intervals can be also viewed as non-Bayesian intervals that account for bias. Nychka (1988) argues that the Bayesian intervals have a frequentist interpretation in having coverage probabilities that, on average over x, are $1 - \alpha$. However, as shown in Ruppert and Carroll (2000), if there are regions of sharp curvature in an otherwise flat regression function, then the coverage probability can be far below $1 - \alpha$ in the regions of high curvature and greater than $1 - \alpha$ elsewhere. As demonstrated by Ruppert and Carroll (2000) and Cummins, Filloon, and Nychka (2001), spatially adaptive splines can be used to correct this problem; see Chapter 17.

Finally, we note that none of the interval estimates given thus far in this chapter account for variability in smoothing parameter estimation. In penalized splines with mixed model representation, this corresponds to variability in $\hat{\sigma}_u$ and $\hat{\sigma}_\varepsilon$. In Section 4.7 we discussed this issue and its difficulties for general mixed models, so the same comments apply to penalized spline scatterplot smoothing. It is common practice to ignore this source of variability in the hope that the sample sizes are sufficiently large that the extra variability will be negligible, but in Chapter 16 we show how this extra variability can be taken into account by fully Bayesian inference.

6.5 Simultaneous Confidence Bands

Each of the confidence intervals presented so far in this chapter are *pointwise*. For example, we can use Figure 6.4 to separately make the statements

$(0.70739, 0.70743)$ is an approximate 95% confidence interval for $f(100)$

and

$(0.70732, 0.70735)$ is an approximate 95% confidence interval for $f(110)$.

However, as is well known in multiple comparison circles, it is a fallacy to say that

$f(100)$ is contained in $(0.70739, 0.70743)$ and *simultaneously* $f(110)$ is contained in $(0.70732, 0.70735)$ with 95% confidence.

How might we modify these intervals so that such statements are valid? Let $\mathcal{X}$ denote the set of x values of interest. Often $\mathcal{X}$ will be the smallest interval containing each of the x_i (e.g., for the fossil data $\mathcal{X} = (91.3, 123)$). Mathematically, the pointwise $100(1 - \alpha)\%$ confidence bands $\{(L(x), U(x)) : x \in \mathcal{X}\}$ approximately satisfy

$$P\{L(x) \leq f(x) \leq U(x)\} \geq 1 - \alpha \quad \text{for all } x \in \mathcal{X}. \quad (6.15)$$

In Section 6.4 we argued that, for large n,

$$L(x) = \hat{f}(x) - z\left(1 - \tfrac{\alpha}{2}\right) \widehat{\text{st.dev.}}\{\hat{f}(x) - f(x)\},$$

$$U(x) = \hat{f}(x) + z\left(1 - \tfrac{\alpha}{2}\right) \widehat{\text{st.dev.}}\{\hat{f}(x) - f(x)\}$$

approximately satisfies (6.15). In contrast to (6.15), a $100(1 - \alpha)\%$ *simultaneous* confidence band must satisfy

$$P\{L(x) \leq f(x) \leq U(x) \text{ for all } x \in \mathcal{X}\} \geq 1 - \alpha.$$

Penalized splines lend themselves to fairly straightforward simulation-based simultaneous confidence bands as we now describe. Suppose that we want a simultaneous confidence band for f over a grid of M x-values

$$g = (g_1, \ldots, g_M).$$

Define

$$\mathbf{f}_g \equiv \begin{bmatrix} f(g_1) \\ \vdots \\ f(g_M) \end{bmatrix}$$

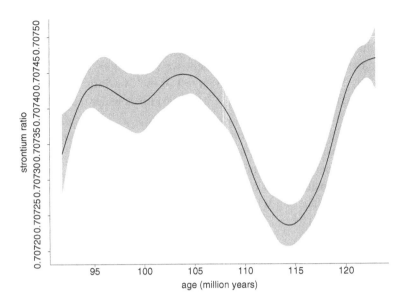

Figure 6.5
95% simultaneous
confidence band for
the fossil data.

to be the true function over g and let $\hat{\mathbf{f}}_g$ be the corresponding EBLUP based on
linear penalized splines in the mixed model framework. Note that

$$\hat{\mathbf{f}}_g - \mathbf{f}_g = \mathbf{C}_g \begin{bmatrix} \hat{\beta} - \beta \\ \hat{\mathbf{u}} - \mathbf{u} \end{bmatrix},$$

where, for linear splines,

$$\mathbf{C}_g = [1 \quad \mathbf{g} \quad (\mathbf{g} - \kappa_1 \mathbf{1})_+ \quad \cdots \quad (\mathbf{g} - \kappa_K \mathbf{1})_+]$$

and

$$\begin{bmatrix} \hat{\beta} - \beta \\ \hat{\mathbf{u}} - \mathbf{u} \end{bmatrix} \overset{\text{approx.}}{\sim} \mathbf{N}\left\{ \mathbf{0}, \hat{\sigma}_\varepsilon^2 \left(\mathbf{C}^\mathsf{T}\mathbf{C} + \frac{\hat{\sigma}_\varepsilon^2}{\hat{\sigma}_u^2}\mathbf{D} \right)^{-1} \right\}. \tag{6.16}$$

A $100(1 - \alpha)\%$ simultaneous confidence band for $\mathbf{f}_g$ is

$$\hat{\mathbf{f}}_g \pm m_{1-\alpha} \begin{bmatrix} \widehat{\text{st.dev.}}\{\hat{f}(g_1) - f(g_1)\} \\ \vdots \\ \widehat{\text{st.dev.}}\{\hat{f}(g_M) - f(g_M)\} \end{bmatrix},$$

where $m_{1-\alpha}$ is the $(1 - \alpha)$ quantile of the random variable

$$\sup_{x \in \mathcal{X}} \left| \frac{\hat{f}(x) - f(x)}{\widehat{\text{st.dev.}}\{\hat{f}(x) - f(x)\}} \right| \approx \max_{1 \le \ell \le M} \left| \frac{\left(\mathbf{C}_g \begin{bmatrix} \hat{\beta} - \beta \\ \hat{\mathbf{u}} - \mathbf{u} \end{bmatrix} \right)_\ell}{\widehat{\text{st.dev.}}\{\hat{f}(g_\ell) - f(g_\ell)\}} \right|. \tag{6.17}$$

The quantile $m_{1-\alpha}$ can be approximated using simulation. One simulates a realization of (6.16) and then computes the corresponding value of (6.17). This
process is repeated a large number of times, say $N = 10{,}000$. The N simulated values of (6.17) are sorted from smallest to largest, and the one with rank
$\lceil (1 - \alpha)N \rceil$ is used as $m_{1-\alpha}$.

Figure 6.5 shows simultaneous confidence bands based on this simulation-
based method. The approximation to the 95% quantile of (6.17) based on a simulation of size $N = 10{,}000$ was

$\sup_{x \in \mathcal{X}} g(x)$ is the
supremum or *least
upper bound* on the
set $\{g(x) : x \in \mathcal{X}\}$
and often corresponds
to the maximum value
obtained by $g(x)$
over $\mathcal{X}$.

$$m_{0.95}^{\text{sim}} = 3.17.$$

With N this large, $m_{0.95}$ can be approximated very accurately. On five independent simulations of 10,000 draws each, we obtained $m_{0.95} \simeq 3.172, 3.198, 3.172, 3.201, 3.199$. The smallest and largest of the five differ by less than 1%. The global intervals are about $1.62 (= 3.17/1.96)$ times wider than the pointwise intervals.

Simulation-based approximation of $m_{1-\alpha}$, though accurate, is computationally somewhat expensive, but this is not a significant problem because the computation involves a matter of seconds. However, analytic approximations to $m_{1-\alpha}$ are also worth considering. Loader (1999, sec. 9.2) surveys asymptotics for simultaneous confidence bands based on *upcrossing theory* (Rice 1939). For linear smoothers as defined by (6.2),

$$P\left\{\left|\frac{\hat{f}(x) - f(x)}{\widehat{\text{st.dev.}\{\hat{f}(x) - f(x)\}}}\right| > c \text{ for all } x \in \mathcal{X}\right\} \simeq \frac{\kappa_0}{\pi} e^{-c^2/2} + 2\{1 - \Phi(c)\},$$

(6.18)

where

$$\kappa_0 \equiv \int_{\mathcal{X}} \frac{\sqrt{\|\boldsymbol{\ell}_x\|^2 \|\boldsymbol{\ell}_x'\|^2 - (\boldsymbol{\ell}_x^{\mathsf{T}} \boldsymbol{\ell}_x')^2}}{\|\boldsymbol{\ell}_x\|^2} \, dx.$$

Here $\boldsymbol{\ell}_x' \equiv (d/dx)\boldsymbol{\ell}_x$, with the differentiation applied elementwise. Such an approximation was used by Knafl, Sacks, and Ylvisaker (1985) to derive simultaneous confidence bands for a general class of regression functions.

The approximation to $m_{1-\alpha}$ implied by (6.18) is

$$m_{1-\alpha}^{\text{UCI}} \equiv \{c > 0 : (\kappa_0/\pi)e^{-c^2/2} + 2\{1 - \Phi(c)\} - \alpha = 0\}.$$

However, $2\{1 - \Phi(c)\}$ is quite small for $c > 2$ and often has little effect on the result. If this term is ignored then the following closed-form upcrossing probability-based approximation ensues:

$$m_{1-\alpha}^{\text{UCII}} = \sqrt{2 \log\{\kappa_0/(\alpha\pi)\}}.$$

For differentiable smoothers, the vectors $\boldsymbol{\ell}_x$ and $\boldsymbol{\ell}_x'$ are straightforward and inexpensive to compute over a fine grid, and quadrature can be used to obtain an accurate approximation to κ_0. For the smooth of the fossil data using quintic radial splines (see Section 3.7.3) but with the same df_{fit} value as that used to produce Figure 6.5, one obtains

$$m_{0.95}^{\text{UCI}} = 3.11 \quad \text{and} \quad m_{0.95}^{\text{UCII}} = 3.10,$$

which are about 97–98% the size of those obtained via simulation.

Despite the simultaneous coverage enjoyed by the confidence bands described in this section, they are prone to misinterpretation. For example, it is not valid to infer that the true relationship is linear just because a line can be drawn within the band. Hastie and Tibshirani (1990, sec. 3.8.2) provide detailed discussion on this matter. They point out that the confidence sets for $\mathbf{f}_g$ exist in M-dimensional space, and the simultaneous confidence bands are just a projection of this space.

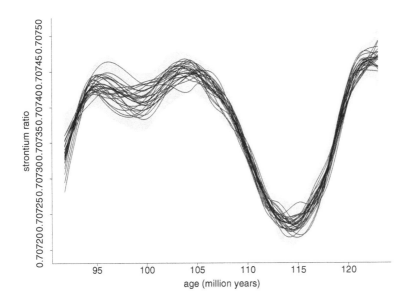

Figure 6.6
Embellishment of
Figure 6.5. Smooths
corresponding to 25
draws from (6.16)
have been added.

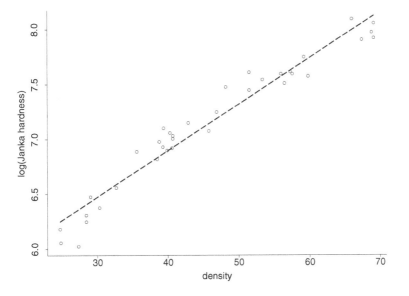

Figure 6.7 Logarithm
of Janka hardness
plotted against density
for 36 Australian
eucalypt hardwood
timbers.

Figure 6.6 provides some insight. It shows 25 fits corresponding to the simulation.
The bands simply give approximate bounds on the maxima and minima of the
curves, but they do not say anything about other structure apparent in the 25 curves.

6.6   Testing the Adequacy of Parametric Models

Figure 6.7 shows the logarithm of *Janka hardness* of a sample of Australian timbers against the density of the timber. Janka hardness is a structural property of
the timber, but it is difficult to measure and so a regression model linking it to

The Janka hardness
data are from
Regression Analysis
(Williams 1959).

Figure 6.8 Figure 6.7 with penalized spline fit added. The linear fit is rejected by a significance test, but it is close to the nonparametric fit and might be adequate for some purposes – one should not confuse statistical significance with practical significance.

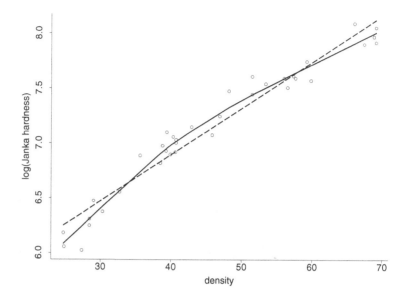

density is desirable. The least-squares line is shown on the plot. But is the relationship really linear? Closer inspection of Figure 6.7 shows that there is a hint of curvature about the line that suggests inadequacy of the linear model.

In Figure 6.8, a penalized spline regression fit is added to the plot. It indicates some degree of nonlinearity. But how much evidence is there for this more complex model compared with the linear model? This can be assessed by testing the hypotheses:

$$H_0 : \mathsf{E}\{\log(\texttt{hardness})|\texttt{density}\} = \beta_0 + \beta_1\texttt{density},$$
$$H_1 : \mathsf{E}\{\log(\texttt{hardness})|\texttt{density}\} = f(\texttt{density}) \tag{6.19}$$

for some "smooth" function f.

In this section we will describe some procedures for testing the adequacy of a particular parametric model against the nonparametric alternative. These are sometimes referred to as *lack-of-fit* tests.

6.6.1 Restricted Likelihood Ratio Tests

Consider the mixed model representation of the linear spline model for $f(x)$:

$$f(x) = \beta_0 + \beta_1 x + \sum_{k=1}^{K} u_k(x - \kappa_k)_+, \quad u_k \text{ i.i.d. } N(0, \sigma_u^2). \tag{6.20}$$

In this case, hypotheses (6.19) reduce to

$$H_0 : \sigma_u^2 = 0 \quad \text{versus} \quad H_1 : \sigma_u^2 > 0, \tag{6.21}$$

so we can appeal to the likelihood ratio paradigm summarized in Section 4.8. Since the fixed effect model is the same under the null and alternative hypotheses, restricted likelihood ratio tests are appropriate for (6.21) – and our simulations have shown them to be better behaved than ordinary likelihood ratio tests.

In the case of (6.20) with normally distributed homoscedastic errors having variance σ_ε^2, the response vector is such that

$$\mathbf{y} \sim N(\mathbf{X}\boldsymbol{\beta}, \mathbf{V}), \quad \text{where } \mathbf{V} = \sigma_u^2 \mathbf{Z}\mathbf{Z}^\mathsf{T} + \sigma_\varepsilon^2 \mathbf{I}.$$

The restricted log-likelihood is

$$
\begin{aligned}
& -2\ell_R(\sigma_u^2, \sigma_\varepsilon^2; \mathbf{y}) \\
& \quad = n\log(2\pi) + \log|\mathbf{V}| + (\mathbf{y} - \mathbf{X}\boldsymbol{\beta})^\mathsf{T}\mathbf{V}^{-1}(\mathbf{y} - \mathbf{X}\boldsymbol{\beta}) + \log|\mathbf{X}^\mathsf{T}\mathbf{V}^{-1}\mathbf{X}|.
\end{aligned}
$$

The restricted likelihood ratio statistic is then

$$-2\log\mathrm{LR}_R(\mathbf{y}) = -2\{\ell_R(0, \hat{\sigma}_{\varepsilon,0}^2; \mathbf{y}) - \ell_R(\hat{\sigma}_u^2, \hat{\sigma}_\varepsilon^2; \mathbf{y})\},$$

where $\hat{\sigma}_{\varepsilon,0}^2$ minimizes $-2\ell_R(0, \sigma_\varepsilon^2; \mathbf{y})$ and $(\hat{\sigma}_u^2, \hat{\sigma}_\varepsilon^2)$ minimizes $-2\ell_R(\sigma_u^2, \sigma_\varepsilon^2; \mathbf{y})$.

We conclude that the null hypothesis of linearity should be rejected if the observed $-2\log\mathrm{LR}_R(\mathbf{y})$ is in the upper tail of its null distribution. We now discuss approximations to the null distribution of $-2\log\mathrm{LR}_R(\mathbf{y})$.

6.6.1.1   *Null Distribution of the Likelihood Ratio*

Since we are treating the penalized spline model as a mixed model, the discussion in Section 4.8.2 is relevant. As mentioned in that section, classical asymptotics may fail when $\mathbf{y}$ cannot be partitioned into a large number of independent subvectors. Crainiceanu and Ruppert (2002) show that this problem is especially severe when a polynomial null hypothesis is tested against a spline alternative. For this reason, we strongly recommend against using the standard asymptotics that assume independence, such as in Self and Liang (1987). Asymptotics that take dependence into account may be a useful area for further study, but for now we have only preliminary results that do not seem ready to present.

As discussed in Section 4.8.2, one can use simulation to determine the null distribution of the likelihood ratio test statistic. This is the approach that we recommend.

For the Janka hardness example, the restricted likelihood ratio test of linearity versus a linear spline has a p-value of 0.000010. This value is based on 1,000,000 simulations. Computational time was 98.7 seconds for 10,000 simulations.

6.6.2   *F-Test Approach*

The parametric F-tests of Section 2.4.7 can be generalized to semiparametric regression, though they are only approximate when the models are not parametric. As we described in Section 2.4.7.1, the F-statistic for parametric regression models can be defined in terms of R^2 values as

$$F = \frac{R_{\text{larger}}^2 - R_{\text{smaller}}^2}{(1 - R_{\text{larger}}^2)(p_{\text{larger}} - p_{\text{smaller}})/(n - p_{\text{larger}})},$$

where, for each model,

$$R^2 = \text{square of correlation coefficient between } \mathbf{y} \text{ and } \hat{\mathbf{y}}$$

and p_{smaller} and p_{larger} are the numbers of parameters in each model.

Figure 6.9
Scatterplots of
y versus ŷ, with
45° line, and
corresponding
squared correlation
coefficients (R^2
values) for the linear
and smooth function
models fitted to the
Janka hardness data.
The smooth function
is estimated via a
penalized spline with
REML smoothing
parameter choice.

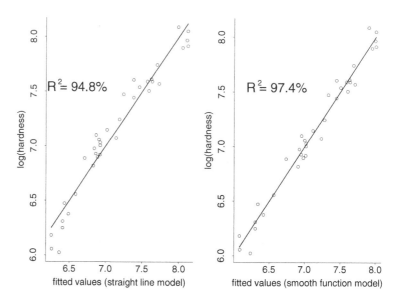

In the semiparametric context it can be argued (Hastie and Tibshirani 1990, sec. 3.9) that an appropriate F-statistic is

$$F = \frac{R^2_{\text{larger}} - R^2_{\text{smaller}}}{(1 - R^2_{\text{larger}})(df_{\text{res, smaller}} - df_{\text{res, larger}})/df_{\text{res, larger}}},$$

with $df_{\text{res, smaller}}$ and $df_{\text{res, larger}}$ the values of df_{res} (as defined in Section 3.14) for the respective models' fits.

Under the null hypothesis, F will have an approximate F-distribution with

$$df_{\text{res, smaller}} - df_{\text{res, larger}} \quad \text{and} \quad df_{\text{res, larger}}$$

degrees of freedom. In general, neither of these degrees of freedom will be integers. Some software packages provide values of the cumulative distribution function of the F-distribution function for noninteger degrees-of-freedom values, so use of these (rather than tables) is recommended. If only tables are available then linear interpolation should be used.

Obviously this test procedure depends on how much smoothing is done to obtain the two fits – in other words, the values of $df_{\text{res, larger}}$ and $df_{\text{res, smaller}}$. Ideally, these values are chosen via a reasonable automatic smoothing criterion such as REML or GCV.

Figure 6.9 shows the result of applying the F-test procedure to the Janka hardness example. The left panel corresponds to the linear model, while the right panel corresponds to a REML-based penalized spline fit. In this case we have

$$R^2_{\text{smaller}} = 0.948, \quad df_{\text{res, smaller}} = 34,$$
$$R^2_{\text{larger}} = 0.974, \quad df_{\text{res, larger}} = 31.3194.$$

Under H_0 (linearity), the F-statistic is approximately distributed as an $F(2.6805, 31.3194)$ random variable. The observed F-statistic is

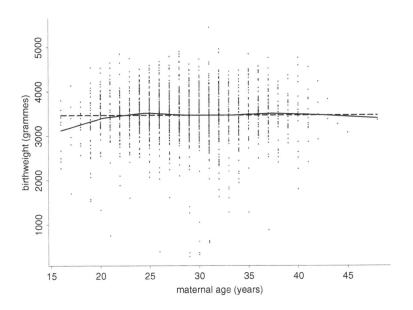

Figure 6.10 Plot of birthweight versus maternal age for data described in Section 8.4.1. The solid line is an ML-based penalized spline fit. The dashed line is a horizontal line through the mean of the birthweights and represents maternal age having no effect.

$$F = \frac{0.974 - 0.948}{(1 - 0.974)(34 - 31.3194)/31.3194} = 11.919,$$

which leads to an approximate p-value of 0.0000387. Once again, there is strong evidence against linearity.

Unpublished work by C. Crainiceanu shows that F-tests using the F-distribution for critical values are often liberal, meaning the stated p-value is smaller than the true p-value and the type 1 error probability is larger than stated. In fact, in the example of Section 6.7.1, use of the F-distribution appears to give a very liberal test compared to using simulation when computing the p-value, a topic we discuss next.

6.6.3 *Simulation for p-Values*

The F-test statistic does not have an exact F-distribution, even when the values of the smoothing parameters are fixed in advance. It seems reasonable that the F-approximation is worse when the smoothing parameters depend on the data, as will typically be the case in practice. An alternative to using the F-distribution is to use simulation. Computation for p-values is discussed in Section 4.8.2 for likelihood ratio tests, and the same technique can be applied to other test statistics such as the F-test.

6.7   Testing for No Effect

The likelihood ratio idea can also be used to test whether a particular predictor variable has no effect on the response. This is illustrated in Figure 6.10 for some data from the example described in Section 8.4.1. The response variable is birthweight and the predictor is maternal age. The sample size is $n = 1630$. As seen from inspection of Figure 6.10, the relationship between these two variables seems tenuous.

The solid curve is a penalized spline fit, while the dashed line is just the overall mean of `birthweight`. In this case the hypotheses are

$$H_0 : \mathrm{E}(\texttt{birthweight}|\texttt{maternal.age}) = \text{constant},$$
$$H_1 : \mathrm{E}(\texttt{birthweight}|\texttt{maternal.age}) = f(\texttt{maternal.age}).$$

(6.22)

If f is modeled as a linear penalized spline,

$$f(x) = \beta_0 + \beta_1 x + \sum_{k=1}^{K} u_k (x - \kappa_k)_+, \quad u_k \overset{\text{ind.}}{\sim} N(0, \sigma_u^2),$$

then the hypotheses become

$$H_0 : \beta_1 = \sigma_u^2 = 0 \quad \text{versus} \quad H_1 : \beta_1 \neq 0 \text{ or } \sigma_u^2 > 0.$$

Using simulation to compute the p-value as discussed in Section 4.8.2, we find the p-value to be 0.12 using 10 knots.

6.7.1 *F-Test for No Effect*

The F-test approach can also be extended to testing for no effect. For the maternal age–birthweight example, we obtain (based on the ML fits)

$$R^2_{\text{larger}} = 0.011356 \quad \text{and} \quad df_{\text{res, larger}} = 1623.99.$$

For the constant model,

$$R^2_{\text{larger}} = 0 \quad \text{and} \quad df_{\text{res, smaller}} = n - 1 = 1629.$$

The observed F-statistic is

$$F = \frac{0.011356 - 0}{(1 - 0.011356)(1629 - 1623.99)/1623.99} = 3.724,$$

and its approximate null distribution is F with 5.01 and 1623.99 degrees of freedom. From this we obtain

$$p\text{-value} = 0.0023,$$

which is rather different from the conclusions of the likelihood ratio test. However, although the F-approximation is widely used, we have found it to be inaccurate. Moreover, the p-value of 0.0023 is rather different from the conclusions of the likelihood ratio test *and* from an exact p-value based on simulation. In other words, the F-approximation is not to be trusted.

In fact, using maximum likelihood to estimate the smoothing parameters, we get a simulation-based p-value of 0.13, whereas using GCV yields a p-value of 0.050.

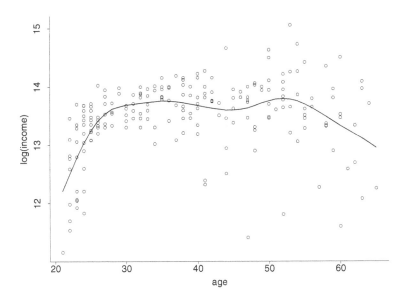

Figure 6.11 Scatter plot of age and income data with smooth.

The difference between the simulation-based *p*-values using ML and GCV is a bit disturbing, since 0.050 is borderline significant by the conventional 0.05 standard while 0.13 is not significant by even the rather liberal 0.10 standard. The difference is due to a substantial difference between the amount of smoothing chosen by GCV and ML. The ML fit is shrunk to nearly a straight line, whereas the GCV is smoothed less – see Figure 6.10.

In summary, our experience is that simulations should be used to obtain accurate *p*-values. Simulation is not too expensive computationally. In fact, the simulations to compute a *p*-value used 10,000 iterations and took about 1.6 minutes in MATLAB on a rather old 600-MHz personal computer.

6.8   Inference Using First Derivatives

The derivatives of the regression function f can be of interest as well as f itself. Consider the simple linear model

$$\mathsf{E}(y|x) = \beta_0 + \beta_1 x.$$

Generally, there is less interest in β_0 than in β_1. Why? Because β_1 is the effect of x on y as measured by a rate of change – that is, it is the derivative of $\mathsf{E}(y)$ with respect to x. In contrast, the additive constant β_0 tells us nothing about the effect of x on y (though it is, of course, a crucial part of a prediction model).

When we plot $\hat{f}(x)$, our eye often looks at its slope and how it changes with x. In our mind, we note where the slope is positive, where it is negative, and where it is essentially zero. Let us look again at the age and income data, shown now in Figure 6.11.

The interesting features of this plot can be expressed in terms of the first derivative $f'(x)$. It appears that f' is large and positive at young ages, meaning

Figure 6.12 Age and income data: estimate of the first derivative with confidence intervals.

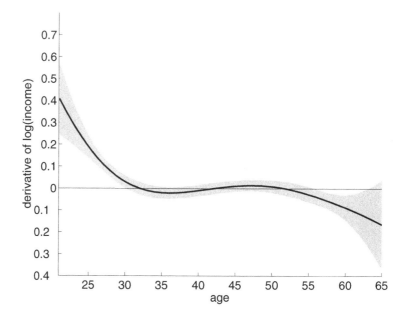

that young workers see their incomes rise rapidly. But eventually f reaches a plateau, suggesting that middle-age workers do not experience much (if any) rise in income. There is some suggestion in the data that f' is negative at older ages, meaning that older workers actually may see a drop in income. Of course, these data are cross-sectional and not longitudinal; one must be very cautious about interpreting cross-sectional data in such a longitudinal fashion (see Diggle et al. 2002).

Figure 6.12 shows the first derivative of the scatterplot smooth of Figure 6.11, along with a pointwise 95% confidence band. It helps us answer some basic questions:

- How fast is income rising at young ages?
- When does income start to plateau?
- Does income really decline at older ages (but before 65) or is the apparent decline not statistically significant?

Notice that the first derivative is significantly above zero until age 30 and then stays quite near zero until age 55. Between ages 55 and 60 it is significantly negative. After age 55, the estimated derivative becomes increasingly negative, but the large boundary variance overwhelms the estimate and it is not significantly different from zero at age 65, the right boundary. Note that the confidence intervals are pointwise, not simultaneous, so one should be cautious about concluding that f' is ever negative. Also, the confidence intervals are based on a homoscedasticity assumption although the data suggest higher variance of log.income at higher ages. To construct better confidence intervals, one should avoid the homoscedasticity assumption and construct intervals based upon an estimate of the conditional variance of log.income given age. Chapter 14 describes this extension.

Besides telling us about the shape of f, derivatives of f are important in themselves, as can be seen in examples from several fields. In the field of pollution monitoring, for the LIDAR example $-f'(\texttt{range})$ is proportional to the concentration of mercury at a given value of $\texttt{range}$. Nonparametric regression is used by engineers to smooth data from the Monte Carlo study of turbulence; in such studies, partial derivatives with respect to spatial variables are of fundamental interest.

6.8.1 *Derivative Estimation via Penalized Splines*

For derivative estimation via penalized splines, it is recommended that higher-degree polynomial basis functions be used to ensure that the resulting derivative estimates are smooth. We will start by describing first derivative estimation, for which quadratic splines are the simplest basis leading to continuous fits.

Let $\hat{f}$ be a quadratic penalized spline fit:

$$\hat{f}(x) = \hat{\beta}_0 + \hat{\beta}_1 x + \hat{\beta}_2 x^2 + \sum_{k=1}^{K} \hat{u}_k (x - \kappa_k)_+^2.$$

This is a piecewise quadratic function that can be differentiated over each piece to obtain the piecewise linear estimate of f':

$$\hat{f}'(x) = \hat{\beta}_1 + 2\hat{\beta}_2 x + \sum_{k=1}^{K} 2\hat{u}_k (x - \kappa_k)_+.$$

Operationally, a derivative estimate at location x can be obtained from the quadratic fit coefficients $\hat{\boldsymbol{\beta}} = [\hat{\beta}_0 \ \hat{\beta}_1 \ \hat{\beta}_2]^{\mathsf{T}}$ and $\hat{\mathbf{u}} = [\hat{u}_1, \ldots, \hat{u}_K]^{\mathsf{T}}$ by setting

$$\mathbf{X}'_x = [0 \ 1 \ 2x] \quad \text{and} \quad \mathbf{Z}'_x = \left[2(x - \kappa_k)_+\right]_{1 \le k \le K}.$$

Then

$$\hat{f}'(x) = \mathbf{X}'_x \hat{\boldsymbol{\beta}} + \mathbf{Z}'_x \hat{\mathbf{u}}.$$

Also,

$$\mathrm{var}\{\hat{f}'(x) - f'(x)\} \simeq \mathbf{C}'_x \, \mathrm{Cov}\left(\begin{bmatrix} \tilde{\boldsymbol{\beta}} \\ \tilde{\mathbf{u}} - \mathbf{u} \end{bmatrix}\right) \mathbf{C}'^{\mathsf{T}}_x$$

$$= \sigma_\varepsilon^2 \mathbf{C}'_x \left(\mathbf{C}^{\mathsf{T}}\mathbf{C} + \frac{\sigma_\varepsilon^2}{\sigma_u^2} \mathbf{D}\right)^{-1} \mathbf{C}'^{\mathsf{T}}_x, \qquad (6.23)$$

where $\mathbf{C}'_x = [\mathbf{X}'_x \ \mathbf{Z}'_x]$ and $\mathbf{D} = \mathrm{diag}(0, 0, 0, 1, \ldots, 1)$.

If $\hat{f}$ is based on cubic radial basis functions,

$$\hat{f}(x) = \hat{\beta}_0 + \hat{\beta}_1 x + \sum_{k=1}^{K} \hat{u}_k |x - \kappa_k|^3,$$

then the first derivative can be estimated by

$$\hat{f}'(x) = \hat{\beta}_1 + \sum_{k=1}^{K} 3\hat{u}_k (x - \kappa_k)|x - \kappa_k|.$$

and inference can be handled using matrix algebra similar to that used for quadratic fits – for example, (6.23).

The extension to other basis functions and higher derivatives is straightforward. Note that the degree of the spline used to estimate f should exceed the order derivative by at least 1 in order to avoid piecewise constant estimates.

6.8.2 *Choosing the Smoothing Parameter*

The smoothing parameter that minimizes the mean squared error of $\hat{f}'$ as an estimate of f' will not be the same as that minimizing the mean squared error of $\hat{f}$ as an estimate of f. Nonetheless, we find in practice that using REML or GCV to select the smoothing parameter is generally an effective strategy.

Derivative estimates are typically more noisy than estimates of f, and this fact suggests that derivative estimates should be smoothed more than estimators of f. In the context of local polynomial estimation, it is known that the best bandwidth for estimating f' converges to zero at a slower rate than the optimal bandwidth for f. Again, this is a reason why $\hat{f}'$ might be smoothed more than $\hat{f}$. Nonetheless, smoothing parameter selection can be a problem because asymptotics often do not take effect until sample sizes are enormous – much larger than the sample sizes in our examples. Our experience with finite-sample problems is that the optimal smoothing parameter for estimation of f' or even f'' is generally close to that which is optimal for f. Furthermore, if one adopts the mixed model formulation of penalized splines, then the smoothing parameter depends only on the estimated variance components, not on which order of derivative is being estimated. Yet in one case study Jarrow, Ruppert, and Yu (2003) did find problems when using GCV to choose the smoothing parameter when estimating a derivative. They developed an alternative to GCV based on Ruppert's EBBS (empirical bias bandwidth selection) methodology.

Thus, our recommendation is to use REML or GCV to select a smoothing parameter appropriate for estimating f and then to use the same smoothing parameter value when estimating the derivative. This is our strategy in both examples of the next section.

6.8.3 *LIDAR Data*

Figure 6.13 shows $-\hat{f}'$ (i.e., the negative of the estimated first derivative) for the LIDAR data. As mentioned previously, $-f'(\text{range})$ is proportional to the concentration of mercury at a given value of range. The estimate is the derivative of a 15-knot penalized cubic spline. The top plot shows the estimate with the penalty parameter chosen by GCV. The "bump" where the estimate is significantly positive reveals a plume of mercury. This is an example where a single penalty parameter λ does not achieve the best possible estimate. The value of λ chosen by GCV is rather small, because GCV minimizes the bias around the bump. Minimizing this bias is a good thing, but it has the undesirable side effect of undersmoothing to both sides of the bump where the function is relatively

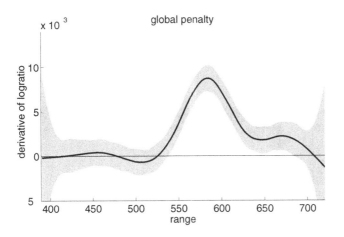

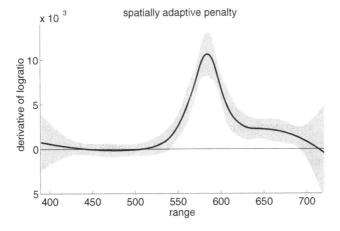

Figure 6.13 LIDAR
data: estimate of
first derivative with
confidence intervals.
The top plot uses a
global penalty and
the bottom plot uses
a spatially adaptive
penalty.

flat and where a large value of λ is appropriate. The problem is *not* with GCV
but rather with the constraint that λ be constant over all values of the predictor
variable range.

In Chapter 17, a method is presented that allows the penalty parameter to vary
as a smooth function of the predictor variable. This "local" or "spatially adaptive" penalty parameter is chosen by GCV (though in principle REML could be
used), and for this data set GCV is more successful in choosing the right amount
of smoothing when it is not constrained to a "global" penalty parameter. The
bottom plot in Figure 6.13 shows the estimate with a spatially adaptive penalty
parameter. We see that, compared to the estimate in the top plot, the spatially
adaptive estimate has a sharper estimate of the peak and smoother estimates of
the regions where f' is close to zero. The large boundary variance of the global
penalty estimator is reduced by using a local penalty, so the confidence intervals
near the boundaries are much narrower in the bottom plot.

A simultaneous confidence band for the derivative f' can be constructed in exactly the same way as described in Section 6.5 by replacing the basis functions
with their derivatives. Figure 6.14 shows the 95% pointwise and simultaneous

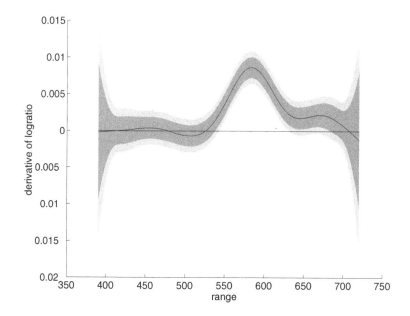

Figure 6.14 LIDAR data: estimate of first derivative with pointwise (dark) and global (light) confidence limits. Global penalty.

confidence bands for $-f'$ using the LIDAR data; $M = 100$ grid points were used, and $m_{0.95}$ was computed using $N = 10{,}000$ simulations.

Since $-f'$ is proportional to the concentration of mercury, one can conclude with 95% confidence that there is mercury at all values of range where the global interval is entirely above zero. This example is used only for illustration. Because of the heteroscedasticity, the intervals should be constructed assuming a nonconstant variance function; see Chapter 14.

6.9 Testing for Existence of a Feature

Consider, again, the fit of the fossil data shown (with variability band) in Figure 6.15. A question of interest is: Do the bumps at age = 95 and age = 105 and the dip at age = 115 represent structure that is "really there"? By this we mean: If several other laboratories collected a different realization of the same data, would they tend to have the same features in the same position?

Features such as bumps and dips in regression curves are often of practical interest. For example, the dip in Figure 6.11 for workers in their mid-40s corresponds to a mid-career decline in income. In this section we describe some recent methodology known as *significance zero crossings of derivatives* (SiZer) due to Chaudhuri and Marron (1999) that assesses feature significance systematically.

Figure 6.16 shows a derivative estimate with corresponding 95% simultaneous confidence band. The regions over which the variability band is positive correspond to those where the regression function is significantly increasing. Regions for which the variability band is below the zero line correspond to those where the regression function is significantly decreasing. If the variability band covers a portion of the zero line then nothing can be concluded about the slope of the

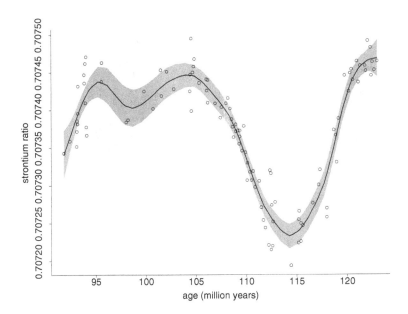

Figure 6.15
Penalized spline smooth of the fossil data with variability band.

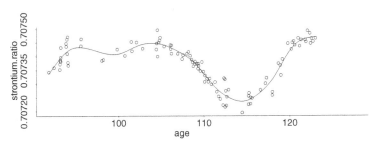

Figure 6.16 *Top:* Penalized spline smooth of the fossil data. *Bottom:* Penalized spline estimate of first derivative with bar at the base showing significant zero crossings.

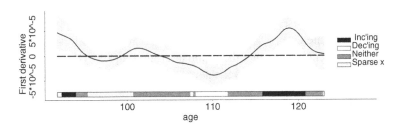

mean function. The bar at the base of the plot summarizes this through a simple graphic that uses black for increasing, white for decreasing, and dark grey for neither increasing nor decreasing. Light grey is used in regions where the data are too sparse to draw any conclusions. The fact that, around age = 115, the band goes from white to dark grey to black means that the dip there is statistically significant. Similarly, the large hump from around age = 95 to age = 110 is significant, since the band goes from black to grey to white. However, there is no such behavior immediately about age = 95 and age = 105, so the secondary bumps at age = 95 and 105 with the associated dip around age = 100

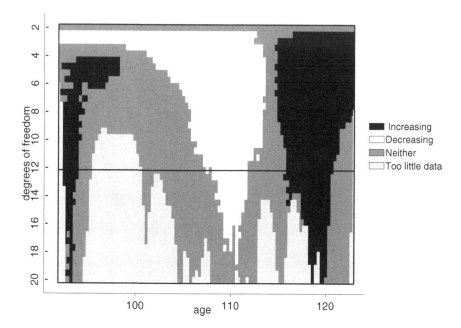

Figure 6.17 Map of significance of zero crossings against several degrees of freedom values. The horizontal line corresponds to degrees of freedom chosen by REML.

are, perhaps, an aberration. A drawback of Figure 6.16 is that it depends on the amount of smoothing. For example, higher amounts of smoothing can wash away features that are apparent in less smoothed plots. Chaudhuri and Marron (1999) propose a "SiZer map" of significant zero crossing bars across a range of smoothing parameters. Figure 6.17 shows such a map for the fossil data. Chaudhuri and Marron (1999) also discuss adjustments for simultaneous confidence bands across degrees of freedom, but they advise that using simultaneous confidence bands for individual degrees of freedom in SiZer maps often yields a reasonable approximation. Figure 6.17 uses this approximation, although further research in this direction is warranted.

From Figure 6.17 we see that the bump at around age $= 115$ is significant if the value of df_{fit} is anywhere between about 3 and 18. However, there is no level of smoothing at which the dip at age $= 95$ is significant. Thus, we conclude that it is "not really there". As a further example, a SiZer map for the age–income data does not find the mid-40s dip to be statistically significant.

6.10   Bibliographical Notes

The literature on inference in smoothing is large and varied. Much of it is in the local polynomial or kernel smoothing context, where theoretical properties are more tractable. The books of Fan and Gijbels (1996), Bowman and Azzalini (1997), Hart (1997), Eubank (1999), and Loader (1999) summarize and provide references to some of this literature.

Inference for spline-based smoothing is less studied. Wahba (1990) summarizes some of the earlier work. More recently, Lin and Zhang (1999) have explored the use of score tests in spline-based smoothing with mixed model representations.

6.11 Summary of Formulas

Pointwise variability band

$$\hat{f}(x) = \boldsymbol{\ell}_x^{\mathsf{T}} \mathbf{y}$$

$$\hat{f}(x) \pm 2 \times \widehat{\text{st.dev.}}\{\hat{f}(x)\}, \quad \widehat{\text{st.dev.}}\{\hat{f}(x)\} = \hat{\sigma}_\varepsilon \|\boldsymbol{\ell}_x\|$$

Pointwise confidence band

$$\hat{f}(x) \pm \begin{cases} t\left(1 - \frac{\alpha}{2}; df_{\text{res}}\right)\hat{\sigma}_\varepsilon \|\boldsymbol{\ell}_x\| & \text{for small } n \\[2mm] z\left(1 - \frac{\alpha}{2}\right)\hat{\sigma}_\varepsilon \|\boldsymbol{\ell}_x\| & \text{for large } n \end{cases}$$

Pointwise prediction band

$$\hat{f}(x) \pm \begin{cases} t\left(1 - \frac{\alpha}{2}; df_{\text{res}}\right)\hat{\sigma}_\varepsilon \sqrt{1 + \|\boldsymbol{\ell}_x\|^2} & \text{for small } n \\[2mm] z\left(1 - \frac{\alpha}{2}\right)\hat{\sigma}_\varepsilon \sqrt{1 + \|\boldsymbol{\ell}_x\|^2} & \text{for large } n \end{cases}$$

Pointwise confidence band with bias allowance

$$\hat{f}(x) \pm \begin{cases} t\left(1 - \frac{\alpha}{2}; df_{\text{res}}\right)\hat{\sigma}_\varepsilon \sqrt{\mathbf{C}_x\left(\mathbf{C}^{\mathsf{T}}\mathbf{C} + \frac{\sigma_\varepsilon^2}{\sigma_u^2}\mathbf{D}\right)^{-1}\mathbf{C}_x^{\mathsf{T}}} & \text{for small } n \\[4mm] z\left(1 - \frac{\alpha}{2}\right)\hat{\sigma}_\varepsilon \sqrt{\mathbf{C}_x\left(\mathbf{C}^{\mathsf{T}}\mathbf{C} + \frac{\hat{\sigma}_\varepsilon^2}{\hat{\sigma}_u^2}\mathbf{D}\right)^{-1}\mathbf{C}_x^{\mathsf{T}}} & \text{for large } n \end{cases}$$

Simultaneous confidence band with bias allowance

$$\hat{f}(x) \pm m_{1-\alpha}\hat{\sigma}_\varepsilon \sqrt{\mathbf{C}_x\left(\mathbf{C}^{\mathsf{T}}\mathbf{C} + \frac{\sigma_\varepsilon^2}{\sigma_u^2}\mathbf{D}\right)^{-1}}$$

for large n, where $m_{1-\alpha}$ is the $(1 - \alpha)$ quantile of

$$\sup_{x \in \mathcal{X}} \left| \frac{\hat{f}(x) - f(x)}{\widehat{\text{st.dev.}}\{\hat{f}(x) - f(x)\}} \right|$$

Restricted likelihood ratio test for linearity
Model is:

$$f(x) = \beta_0 + \beta_1 x + \sum_{k=1}^{K} u_k (x - \kappa_k)_+, \quad u_k \overset{\text{ind.}}{\sim} N(0, \sigma_u^2)$$

Restricted log-likelihood is:

$$\ell_R(\sigma_u^2, \sigma_\varepsilon^2; \mathbf{y})$$
$$= \tfrac{1}{2}\{n \log(2\pi) + \log|\mathbf{V}| + (\mathbf{y} - \mathbf{X}\boldsymbol{\beta})^{\mathsf{T}}\mathbf{V}^{-1}(\mathbf{y} - \mathbf{X}\boldsymbol{\beta}) + \log|\mathbf{X}^{\mathsf{T}}\mathbf{V}^{-1}\mathbf{X}|\}$$
$$\mathbf{V} = \sigma_u^2 \mathbf{Z}\mathbf{Z}^{\mathsf{T}} + \sigma_\varepsilon^2 \mathbf{I}$$

Test statistic is:

$$-2 \log \mathrm{LR}_R(\mathbf{y}) = -2\{\ell_R(0, \hat{\sigma}^2_{\varepsilon,0}; \mathbf{y}) - \ell_R(\hat{\sigma}^2_u, \hat{\sigma}^2_\varepsilon; \mathbf{y})\}$$

where $\hat{\sigma}^2_{\varepsilon,0}$ minimizes $-2\ell_R(0, \sigma^2_\varepsilon; \mathbf{y})$ and $(\hat{\sigma}^2_u, \hat{\sigma}^2_\varepsilon)$ minimizes $-2\ell_R(\sigma^2_u, \sigma^2_\varepsilon; \mathbf{y})$.

Likelihood ratio test for no effect
Model is:

$$f(x) = \beta_0 + \beta_1 x + \sum_{k=1}^{K} u_k (x - \kappa_k)_+, \quad u_k \overset{\text{ind.}}{\sim} N(0, \sigma^2_u)$$

Log-likelihood is:

$$\ell(\beta_0, \beta_1, \sigma^2_u, \sigma^2_\varepsilon; \mathbf{y}) = -\tfrac{1}{2}\{n \log(2\pi) + \log|\mathbf{V}| + (\mathbf{y} - \mathbf{X}\boldsymbol{\beta})^{\mathsf{T}} \mathbf{V}^{-1}(\mathbf{y} - \mathbf{X}\boldsymbol{\beta})\}$$

$$\mathbf{V} = \sigma^2_u \mathbf{Z}\mathbf{Z}^{\mathsf{T}} + \sigma^2_\varepsilon \mathbf{I}$$

Test statistic is:

$$-2 \log \mathrm{LR}(\mathbf{y}) = -2\{\ell(\hat{\beta}_0, 0, 0, \hat{\sigma}^2_{\varepsilon,0}; \mathbf{y}) - \ell(\hat{\beta}_0, \hat{\beta}_1, \hat{\sigma}^2_u, \hat{\sigma}^2_\varepsilon; \mathbf{y})\}$$

where $(\hat{\beta}_0, \hat{\sigma}^2_{\varepsilon,0})$ minimizes $-2\ell(\beta_0, 0, 0, \sigma^2_\varepsilon; \mathbf{y})$ and $(\hat{\beta}_0, \hat{\beta}_1, \hat{\sigma}^2_u, \hat{\sigma}^2_\varepsilon)$ minimizes $-2\ell(\beta_0, \beta_1, \sigma^2_u, \sigma^2_\varepsilon; \mathbf{y})$.

F-test for linearity and no effect

$$F = \frac{R^2_{\text{larger}} - R^2_{\text{smaller}}}{(1 - R^2_{\text{larger}})(df_{\text{res, smaller}} - df_{\text{res, larger}})/df_{\text{res, larger}}}$$

$$F \overset{\text{approx.}}{\sim} F(df_{\text{res, smaller}} - df_{\text{res, larger}}, df_{\text{res, larger}})$$

First derivative estimation
Quadratic penalized splines:

$$\hat{f}(x) = \hat{\beta}_0 + \hat{\beta}_1 x + \hat{\beta}_2 x^2 + \sum_{k=1}^{K} u_k (x - \kappa_k)^2_+$$

$$\hat{f}'(x) = \hat{\beta}_1 + 2\hat{\beta}_2 x + \sum_{k=1}^{K} 2\hat{u}_k (x - \kappa_k)_+$$

Cubic radial basis functions:

$$\hat{f}(x) = \hat{\beta}_0 + \hat{\beta}_1 x + \sum_{k=1}^{K} \hat{u}_k |x - \kappa_k|^3$$

$$\hat{f}'(x) = \hat{\beta}_1 + \sum_{k=1}^{K} 3\hat{u}_k (x - \kappa_k)|x - \kappa_k|$$

7

Simple Semiparametric Models

7.1 Introduction

Until now we have confined discussion to scatterplot smoothers. This setting served well to illustrate the main concepts behind smoothing. However, there is a gap between the methodology and the needs of practitioners. As exemplified by the problems described in Chapter 1, most applications of regression involve several predictors. To begin closing the gap, this chapter introduces a class of multiple regression models that have a nonparametric component involving only a single predictor and a parametric component for the other predictors. Having both parametric and nonparametric components means the models are *semiparametric*. This class of simple semiparametric models is important in its own right but also serves as an introduction to more complex semiparametric regression models of later chapters, where the effects of several predictors are modeled nonparametrically.

7.2 Beyond Scatterplot Smoothing

The end of the previous chapter closed off quite a lengthy description of how to smooth out a scatterplot and perform corresponding inference. In Chapter 3 we described three general approaches: penalized splines, local polynomial fitting, and series approximation. For penalized splines, we presented both an algorithmic approach based on ridge regression and a mixed model approach based on maximum likelihood and best prediction. There are other approaches to scatterplot smoothing that we did not describe at all.

It is expedient to choose just one method of scatterplot smoothing to extend in this book. Which should it be? Our preference for the penalized spline scatterplot smoothing with mixed model representation is based on the following reasons.

1. It is a model-based approach to smoothing that utilizes two basic principles of statistics: maximum likelihood and best prediction. This makes it easy to extend to other models such as logistic regression. The incorporation of likelihood-based models for complications such as dependence, measurement error, and missing data is also more straightforward.

2. Software for mixed models is becoming more accessible, since it is now featured in at least two prominent statistical computing environments: SAS and

Figure 7.1 Scatterplot of the density and log.yield for the onions data. The plotting symbols indicate the two locations where the onions were cultivated. The lines correspond to the linear additive model fit to the data.

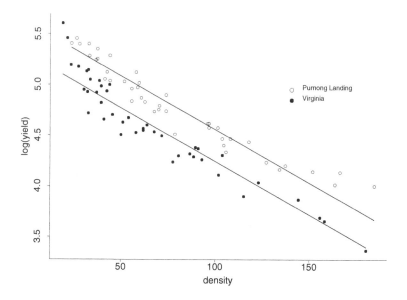

S-PLUS. This allows semiparametric modeling to be done without the dependence on specialist smoothing software and thus means that more complex models can be built.

3. Mixed model–based smoothers come equipped with an automatic smoothing parameter choice that corresponds to maximum likelihood and restricted maximum likelihood estimation of variance components. We are not able to say whether these smoothing parameter choices outperform traditional model selection choices such as GCV. But their availability in software packages makes ML or REML smoothing parameter selection quite attractive.

4. Inference can be performed within the mixed model framework. For example, many hypothesis tests of interest can be performed by appealing to the likelihood ratio principle, as illustrated in Section 6.6.1.

5. Mixed models are extendible to a full hierarchical Bayesian model, which when analyzed via Markov chain Monte Carlo allows the most satisfactory approach to inference. See Chapter 16.

7.3   Semiparametric Binary Offset Model

The onions data are taken from Ratkowsky (1983). A detailed semiparametric analysis of the data is given by Young and Bowman (1995) and in Bowman and Azzalini (1997).

Figure 7.1 contains data on yields (g/plant) of white Spanish onions in two locations: Purnong Landing and Virginia, South Australia. The horizontal axis corresponds to areal density of plants (plants/m^2). The dashed lines in Figure 7.1 correspond to fitting the linear additive model

$$\log(\texttt{yield}_i) = \beta_0 + \beta_1\text{PL}_i + \beta_2\texttt{density}_i + \varepsilon_i,$$

where

$$\text{PL}_i = \begin{cases} 0 & \text{if } i\text{th measurement is from Virginia,} \\ 1 & \text{if } i\text{th measurement is from Purnong Landing.} \end{cases}$$

The effect of Purnong Landing compared with Virginia is estimated to be

$$\hat{\beta}_1 = 0.3154 \quad \text{with} \quad \widehat{\text{st.dev.}}(\hat{\beta}_1) = 0.0311.$$

An approximate 95% confidence interval for the locational effect is then

$$(0.254, 0.376) \tag{7.1}$$

on the log-yield scale.

Close inspection of Figure 7.1 reveals some curvature apparent in the scatterplots for each location, suggesting the model

$$\log(\texttt{yield}_i) = \beta_1 \texttt{PL}_i + f(\texttt{density}_i) + \varepsilon_i. \tag{7.2}$$

We call (7.2) the *semiparametric binary offset model* for these data. The model has a nonparametric component, $f(\texttt{density})$, and a parametric component, $\beta_1 \texttt{PL}$. The binary variable PL vertically offsets the relationship between $E\{\log(\texttt{yield}_i)\}$ and $\texttt{density}$ according to location.

We can fit (7.2) using a penalized linear spline through the mixed model

$$\log(\texttt{yield}_i) = \beta_0 + \beta_1 \texttt{PL}_i + \beta_2 \texttt{density}_i + \sum_{k=1}^{K} u_k (\texttt{density}_i - \kappa_K)_+ + \varepsilon_i,$$

where

$$u_k \overset{\text{ind.}}{\sim} N(0, \sigma_u^2) \quad \text{and} \quad \varepsilon_i \overset{\text{ind.}}{\sim} N(0, \sigma_\varepsilon^2).$$

Note that this is a special case of the Gaussian linear mixed model

$$\mathbf{y} = \mathbf{X}\boldsymbol{\beta} + \mathbf{Z}\mathbf{u} + \boldsymbol{\varepsilon}$$

with $\mathbf{y}$ containing the $\log(\texttt{yield}_i)$ values

$$\mathbf{X} = \begin{bmatrix} 1 & \texttt{PL}_1 & \texttt{density}_1 \\ \vdots & \vdots & \vdots \\ 1 & \texttt{PL}_{84} & \texttt{density}_{84} \end{bmatrix}, \qquad \boldsymbol{\beta} = \begin{bmatrix} \beta_0 \\ \beta_1 \\ \beta_2 \end{bmatrix},$$

$$\mathbf{Z} = \begin{bmatrix} (\texttt{density}_1 - \kappa_1)_+ & \cdots & (\texttt{density}_1 - \kappa_K)_+ \\ \vdots & \ddots & \vdots \\ (\texttt{density}_{84} - \kappa_1)_+ & \cdots & (\texttt{density}_{84} - \kappa_K)_+ \end{bmatrix}, \qquad \mathbf{u} = \begin{bmatrix} u_1 \\ \vdots \\ u_K \end{bmatrix},$$

where $\text{Cov}(\mathbf{u}) = \sigma_u^2 \mathbf{I}$ and $\text{Cov}(\boldsymbol{\varepsilon}) = \sigma_\varepsilon^2 \mathbf{I}$.

Figure 7.2 shows the resulting fit based on REML estimation of σ_u^2 and σ_ε^2. The estimated locational effect from this model is

$$\hat{\beta}_1 = 0.3331 \quad \text{with} \quad \widehat{\text{st.dev.}}(\hat{\beta}_1) = 0.0239,$$

and an approximate 95% confidence interval for the locational effect is

$$(0.286, 0.380).$$

This interval is 77% of the length of that obtained from the model with density entering linearly, given at (7.1). It is also less biased because density is modeled in a less parametric manner.

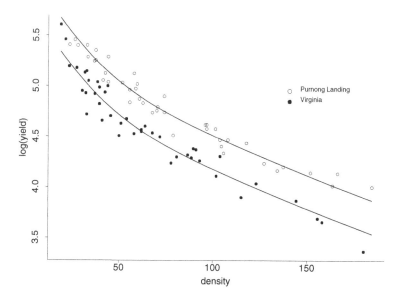

Figure 7.2 Onion data: fit to the additive model (7.2). The response is log.yield. The effect of density is fit by a penalized quadratic spline using REML to select the penalty parameter.

7.4 Additivity and Interactions

Model (7.2) is said to be *additive,* since the effects of location and density on E(log.yield) are simply added together to obtain their joint effect. The effect on E(log.yield) at Purnong Landing versus Virginia is β_1, regardless of the fixed value of density. Likewise, if density changes from a to b then the change in E(log.yield) is $f(a) - f(b)$, regardless of the fixed location.

Additivity is an *assumption,* often a reasonable one, but like all assumptions it should be checked. If the assumption seems seriously incorrect, then alternative models should be considered. When two predictors do not act additively on the mean response then there is an *interaction* between them. Chapters 12 and 13 describe models for interactions.

For the onions data, the most general interaction model is

$$E\{\log(\texttt{yield}_i)\} = \begin{cases} f_{\text{PL}}(\texttt{density}_i) & \text{if Purnong Landing,} \\ f_{\text{VA}}(\texttt{density}_i) & \text{if Virginia.} \end{cases} \tag{7.3}$$

This model can be fit by smoothing the log(yield) versus density scatterplots for each location separately. The fits based on penalized linear splines with a REML smoothing parameter choice are shown in Figure 7.3. The fits look approximately the same, suggesting that the additivity assumption is reasonable in this case. Formal tests for interaction are given in Section 12.2.1.

7.5 General Parametric Component

A natural extension of the semiparametric binary offset model is to the model with the offset term replaced by a general linear component $\boldsymbol{\beta}_{\mathbf{x}}^{\mathsf{T}}\mathbf{x}_i$, where $\mathbf{x}_i$ is a vector of covariates that enter the model linearly and the subscript $\mathbf{x}$ indicates the

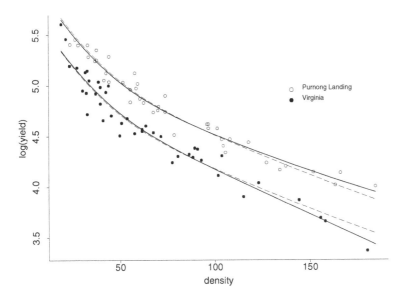

Figure 7.3 Additive and interaction fits to onion data. The solid curves are REML-based fits of the general interaction model (7.3). The dashed curves are REML-based fits of the additive model (7.2).

component of $\boldsymbol{\beta}$ consisting of coefficients of $\mathbf{x}_i$. We will illustrate this extension through the analysis of some pollen count data.

The variable of central interest is the level of ragweed pollen. The data were recorded during the 1993 ragweed season in Kalamazoo, Michigan. Since avoidance plays a large role in the treatment of pollen-related allergies, a major objective in aerobiology is the development of accurate forecasting models for daily pollen levels. Stark et al. (1997) developed some parametric regression models geared toward this aim. The data set at hand consists of:

$$\begin{aligned}
\texttt{ragweed} &= \text{ragweed level for that day (grains/m}^3\text{);} \\
\texttt{temperature} &= \text{temperature of following day (°F);} \\
\texttt{rain} &= \text{indicator of significant rain for following day} \\
&\quad (1 = \text{at least 3 hours of steady or brief but intense rain,} \\
&\quad\ 0 = \text{otherwise);} \\
\texttt{wind} &= \text{wind speed forecast for following day (knots);} \\
\texttt{day.in.seas} &= \text{day number in the current ragweed pollen season.}
\end{aligned}$$

The ragweed pollen data and corresponding meteorological measurements were provided to us by Professor Harriet Burge and Dr. Paul Stark of the Harvard School of Public Health.

Stark et al. (1997) also used the variable `temp.resid`,

$$\begin{aligned}
\texttt{temp.resid} &= \text{the difference between } \texttt{temperature} \text{ and an estimate} \\
&\quad \text{of the time trend of temperature,}
\end{aligned}$$

which corresponds to deviations from the average temperature for a particular day in the ragweed season. Since `day.in.seas` takes care of seasonal temperature variation, we will use `temp.resid` rather than `temperature`. The response, `ragweed`, is quite skewed, so we work with its square root. The marginal relationships between $\sqrt{\texttt{ragweed}}$ and the other variables are shown in Figure 7.4.

Since `day.in.seas` has a pronounced nonlinear relationship with the mean pollen level, a useful semiparametric regression model for these data is

Figure 7.4
Relationships between
$\sqrt{\texttt{ragweed}}$ and each
possible predictor.

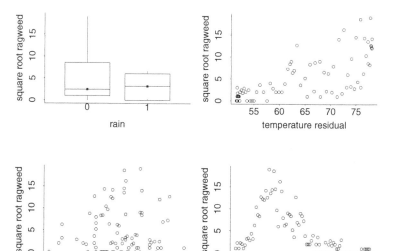

$$\sqrt{\texttt{ragweed}_i}$$

$$= \beta_0 + \beta_1 \texttt{rain}_i + \beta_2 \texttt{temp.resid}_i + \beta_3 \texttt{wind}_i + f(\texttt{day.in.seas}_i) + \varepsilon_i.$$

This can be achieved with penalized splines through the linear spline model

$$\sqrt{\texttt{ragweed}_i}$$

$$= \beta_0 + \beta_1 \texttt{rain}_i + \beta_2 \texttt{temp.resid}_i + \beta_3 \texttt{wind}_i + \beta_4 \texttt{day.in.seas}_i$$

$$+ \sum_{k=1}^{K} u_k (\texttt{day.in.seas}_i - \kappa_k)_+ + \varepsilon_i, \tag{7.4}$$

where $\kappa_1, \ldots, \kappa_K$ are knots over the range of $\texttt{day.in.seas}$ values and $u_1, \ldots, u_K$ are taken as independent $N(0, \sigma_u^2)$ variates. The model can then be written as

$$\mathbf{y} = \mathbf{X}\boldsymbol{\beta} + \mathbf{Z}\mathbf{u} + \boldsymbol{\varepsilon},$$

where

$$\mathbf{X} = \begin{bmatrix} 1 & \texttt{rain}_1 & \texttt{temp.resid}_1 & \texttt{wind}_1 & \texttt{day.in.seas}_1 \\ \vdots & \vdots & \vdots & \vdots & \vdots \\ 1 & \texttt{rain}_{87} & \texttt{temp.resid}_{87} & \texttt{wind}_{87} & \texttt{day.in.seas}_{87} \end{bmatrix},$$

$$\boldsymbol{\beta} = \begin{bmatrix} \beta_0 \\ \beta_1 \\ \beta_2 \\ \beta_3 \\ \beta_4 \end{bmatrix},$$

$$\mathbf{Z} = \begin{bmatrix} (\texttt{day.in.seas}_1 - \kappa_1)_+ & \cdots & (\texttt{day.in.seas}_1 - \kappa_K)_+ \\ \vdots & \ddots & \vdots \\ (\texttt{day.in.seas}_{87} - \kappa_1)_+ & \cdots & (\texttt{day.in.seas}_{87} - \kappa_K)_+ \end{bmatrix},$$

and $\mathbf{u} = [u_1, \ldots, u_k]^{\mathsf{T}}$, with $\mathrm{Cov}(\mathbf{u}) = \sigma_u^2 \mathbf{I}$ and $\mathrm{Cov}(\boldsymbol{\varepsilon}) = \sigma_\varepsilon^2 \mathbf{I}$.

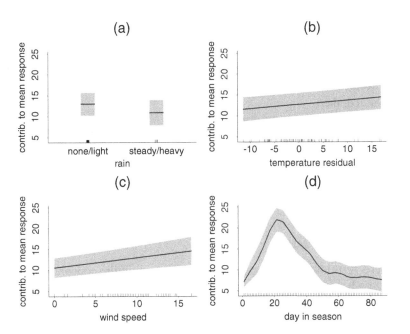

Figure 7.5
Components of fits to the pollen data. The shaded regions correspond to approximate 95% pointwise confidence intervals for the expected response at the average value of the other components.

A summary of the fit based on REML and BLUP is shown in Figure 7.5. The variability bars are calculated by extending the calculations given in Section 2.5.1.2 and, discounting bias, can be interpreted as approximate 95% pointwise confidence intervals.

Although the ragweed pollen data provides a nice illustration of a semiparametric model with a parametric component, model (7.4) has some deficiencies. Firstly, the response is really a count variable and so a discrete distribution such as the Poisson might be more appropriate; parametric Poisson regression (Chapter 10) was used by Stark et al. (1997). Secondly, the effect of `temp.resid` and `wind` may also be nonlinear. The extension to models that permit multiple nonlinear predictors is made in the next chapter. Additionally, data are available on the years 1991–1994 but with a different `day.in.seas` effect for each year. This means that interactions between `day.in.seas` and year number should be considered for the full data set. We give a full explanation and analysis in Section 12.3.2.

7.6   Inference

Inference for the parametric components in $\boldsymbol{\beta}$ can be based on the result

$$\mathrm{Cov}(\hat{\boldsymbol{\beta}}) = (\mathbf{X}^{\mathsf{T}}\mathbf{V}^{-1}\mathbf{X})^{-1}.$$

One would then need to replace $\mathbf{V}$ by $\hat{\mathbf{V}}$, in which the variance components in $\mathbf{V}$ are replaced by their REML estimates. For example, in the onions example the significance of location (Purnong Landing versus Virginia) can be made through the Z-statistic

$$Z_1 = \hat{\beta}_1 / \sqrt{\text{second diagonal entry of } (\mathbf{X}^T \hat{\mathbf{V}}^{-1} \mathbf{X})^{-1}}.$$

Such statistics are reported in mixed model packages such as `lme()` in S-PLUS. However, its interpretation as an approximate standard normal variate under $H_0 : \beta_1 = 0$ needs to be made with caution. If normality of the errors can be reasonably assumed, then

$$Z_1 \overset{\text{approx.}}{\sim} N(0, 1). \tag{7.5}$$

Otherwise, the asymptotic distribution theory is complicated by dependence inherent in the y_i. Full theoretical justification for use of Z_1 for penalized spline mixed models is an area of current research, although cursory justification (via e.g. Heckman 1986) is possible.

7.6.1 *Hypothesis Tests*

Hypothesis tests about the overall effect and linearity of the nonparametric component can be achieved by extending the tests described in Section 6.6. For example, the test for linearity of $f(\texttt{density})$ in (7.2) reduces to

$$H_0 : \sigma_u^2 = 0 \quad \text{versus} \quad H_1 : \sigma_u^2 > 0, \tag{7.6}$$

and that for overall effect of density reduces to

$$H_0 : \beta_2 = \sigma_u^2 = 0 \quad \text{versus} \quad H_1 : \beta_2 \neq 0 \text{ or } \sigma_u^2 > 0. \tag{7.7}$$

For testing (7.6), minus twice the log of the restricted likelihood ratio was 35.90; for testing (7.7), minus twice the log of the likelihood ratio was 229.49.

Monte Carlo simulation was used to compute p-values for both (7.6) and (7.7). In each case, 1,000,000 Monte Carlo samples were drawn from the null hypothesis, and the likelihood ratio statistic calculated from the data exceeded the test statistic calculated from all 1,000,000 simulated data sets. Thus, the p-values are both less than 10^{-6}.

The null distribution for testing (7.6) is well approximated by the chi-squared mixture

$$0.66 \chi_0^2 + 0.34 \chi_1^2$$

(Crainiceanu and Ruppert 2002). Using this approximation, we obtain a p-value of 7×10^{-10}.

The null distribution for testing (7.7) is well approximated by a χ_1^2 distribution (Crainiceanu and Ruppert 2002), and this approximation gives a p-value of essentially zero.

7.7 Bibliographical Notes

The simple semiparametric regression model exemplified by (7.4) goes by a number of names in the literature: partial linear model, partially linear model, partly linear model and partial spline model. Theory for the parametric components has been treated in several papers and contexts, including Heckman (1986), Chen (1988), and Speckman (1988). Härdle, Liang, and Gao (2000) devote a book to simple semiparametric regression models.

Engle and colleagues (1986) developed simple semiparametric regression models for electricity sales and weather. A customer's income, price, and monthly indicators entered the model linearly, while weather variables were modeled nonlinearly with smoothing splines. Young and Bowman (1995) applied similar models to the onion data.

8

Additive Models

8.1 Introduction

The previous chapter showed how to construct flexible regression models for a single continuous predictor modeled as a smooth function, with all the other predictors entering the model linearly. However, many regression problems involve several continuous covariates that may have nonlinear relationships with the response. An example is illustrated in Figure 8.1, where birthweight is plotted against four maternal variables for all 1990 births in the Upper Cape Cod region of Massachusetts.

The plots in Figure 8.1 show only the marginal effects of each maternal variable on birthweight, but they nonetheless suggest that some degree of nonlinearity will be present when all the predictor variables are considered together. The extension to models that allow multiple smooth functions is relatively straightforward.

Figure 8.1
Scatterplots of birthweight versus four maternal variables: parity, cigarettes per day, years of education, and number of prenatal visits; for all 1990 births in Upper Cape Cod. Smooths of each scatterplot also shown.

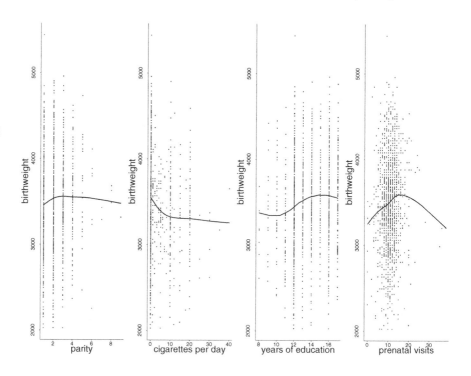

Since the only assumption made is that of additivity, they are referred to as *additive models* (Ezekiel 1924; Friedman and Stuetzle 1981).

8.2   Fitting an Additive Model

In Section 2.5 we considered data on

$y \equiv$ minimum temperature,
$s \equiv$ degrees latitude, and
$t \equiv$ degrees longitude

for 23 cities in the United States and fitted the *parametric* additive model

$$y_i = \beta_0 + \beta_s s_i + \beta_{t1} t_i + \beta_{t2} t_i^2 + \beta_{t3} t_i^3 + \varepsilon_i.$$

The *nonparametric* additive model for these data is

$$y_i = \beta_0 + f(s_i) + g(t_i) + \varepsilon_i, \tag{8.1}$$

where f and g are smooth (but otherwise unspecified) functions of latitude and longitude, respectively. Penalized splines are easily extended to handle (8.1): to model f and g by linear splines, fit

$$y_i = \beta_0 + \beta_s s_i + \sum_{k=1}^{K_s} u_k^s (s_i - \kappa_k^s)_+ + \beta_t t_i + \sum_{k=1}^{K_t} u_k^t (t_i - \kappa_k^t)_+ + \varepsilon_i \tag{8.2}$$

using least squares, but penalize the knot coefficients u_k^s and u_k^t. Here $\kappa_1^s, \ldots, \kappa_{K_s}^s$ and $\kappa_1^t, \ldots, \kappa_{K_t}^t$ are knots in the s and t directions that can be chosen using rules such as those given in Section 5.5.3. The vector of fitted values is given by

$$\hat{\mathbf{y}} = \mathbf{C}(\mathbf{C}^{\mathsf{T}}\mathbf{C} + \boldsymbol{\Lambda})^{-1}\mathbf{C}^{\mathsf{T}}\mathbf{y},$$

where

$$\mathbf{C} = \left[1, s_i, t_i, (s_i - \kappa_k^s)_+, (t_i - \kappa_k^t)_+ \right]_{\substack{1 \le k \le K^s \\ 1 \le k \le K^t}} \right]_{1 \le i \le n}$$

and

$$\boldsymbol{\Lambda} = \text{diag}(0, 0, 0, \lambda_s^2 \mathbf{1}_{K_s \times 1}, \lambda_t^2 \mathbf{1}_{K_t \times 1}).$$

The smoothing parameters λ_s and λ_t induce smoothing in the s and t directions (respectively) by penalizing the knot coefficients u_k^s and u_k^t.

As we have seen for scatterplot smoothing, penalization of the u_k^s and u_k^t is equivalent to treating them as random effects in a mixed model. Specifically, if we define

$$\boldsymbol{\beta} = [\beta_0, \beta_s, \beta_t], \qquad \mathbf{u} = [u_1^s, \ldots, u_{K_s}^s, u_1^t, \ldots, u_{K_t}^t]^{\mathsf{T}},$$

$$\mathbf{X} = [1 \ s_i \ t_i]_{1 \le i \le n}, \qquad \mathbf{Z} = \left[(s_i - \kappa_k^s)_+, (t_i - \kappa_k^t)_+ \right]_{\substack{1 \le k \le K^s \\ 1 \le k \le K^t}} \right]_{1 \le i \le n},$$

then penalized least squares is equivalent to best linear unbiased prediction in the mixed model

Figure 8.2
Components of fit
of (2.20) to the U.S.
temperature data.

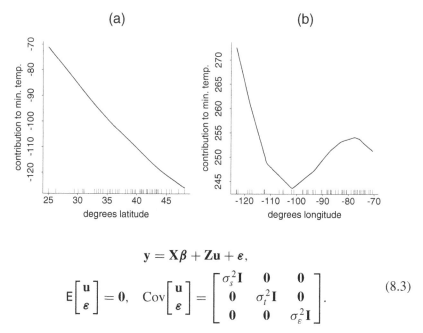

$$y = X\boldsymbol{\beta} + Z\mathbf{u} + \boldsymbol{\varepsilon},$$

$$\mathrm{E}\begin{bmatrix}\mathbf{u}\\\boldsymbol{\varepsilon}\end{bmatrix} = \mathbf{0}, \quad \mathrm{Cov}\begin{bmatrix}\mathbf{u}\\\boldsymbol{\varepsilon}\end{bmatrix} = \begin{bmatrix}\sigma_s^2\mathbf{I} & \mathbf{0} & \mathbf{0}\\\mathbf{0} & \sigma_t^2\mathbf{I} & \mathbf{0}\\\mathbf{0} & \mathbf{0} & \sigma_\varepsilon^2\mathbf{I}\end{bmatrix}. \tag{8.3}$$

Once again, the mixed model representation can be used to facilitate fitting, inference, and model selection. Its extension to higher numbers of smooth functions and other bases is straightforward. Linear components can be incorporated as fixed effects in the $X\boldsymbol{\beta}$ term.

Apart from the presence of random effects, there is no inherent difference between the parametric additive model and the nonparametric one. Therefore, the material in Section 2.5 extends rather simply to the nonparametric case. Note that we are using the U.S. temperature data only to illustrate the operational details of nonparametric additive models. For these data it appears that the parametric additive model (2.20) is adequate.

With this mixed model representation, the number of degrees of freedom used to estimate $f(\cdot)$ and $g(\cdot)$ can be chosen via REML (we will give a precise definition of degrees of freedom in Section 8.3). However, as explained in Section 8.4, we may also want to fix the degrees-of-freedom values in advance. Figure 8.2 shows latitude and longitude components of the fit to (8.2) with 3 degrees of freedom used for degrees latitude and 6 degrees of freedom used for degrees longitude for each function. As we pointed out for the parametric model in Section 2.5, the vertical positioning of the curves is somewhat arbitrary. An improvement is that shown in Figure 8.3. The curve for latitude is a slice of the fitted surface at the average longitude value, as described in Section 2.5.1.1. The curve for longitude is the slice at the average latitude value. This means that the vertical axes correspond to the minimum temperature values, rather than only having a relative interpretation as in Figure 8.2.

The next embellishment involves variability bands, as shown in Figure 8.4. The arithmetic for their construction is analogous to that described in Section 2.5.1.2 but is based on the covariance matrix

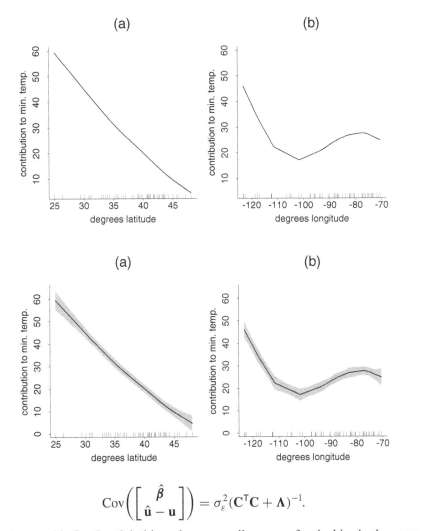

Figure 8.3
Components of fit of
(8.2) with vertical
alignment as described
in Section 2.5.1.1.

Figure 8.4 Fits
of Figure 8.3 with
variability bands
added.

$$\mathrm{Cov}\left(\begin{bmatrix} \hat{\beta} \\ \hat{\mathbf{u}} - \mathbf{u} \end{bmatrix}\right) = \sigma_\varepsilon^2 (\mathbf{C}^\mathsf{T}\mathbf{C} + \boldsymbol{\Lambda})^{-1}.$$

As discussed in Section 6.4, this makes some allowances for the bias in the curve
estimates.

Figure 8.5 adds on partial residuals, as described in Section 2.5.1.3. The apparent random scatter about the curves indicates a good fit in this case.

Estimates of the derivatives – based on quadratic spline fits with the same degrees of freedom as used in Figure 8.5 – are plotted in Figure 8.6 along with
corresponding variability bands. These reveal that the effect of latitude is significantly negative across the range of values of that variable. For longitude, there is
a significant negative effect between the west coast and Rocky Mountain region
(approximately −105 degrees longitude) and then a significant positive effect
across the prairies east of the Rocky Mountains. The effect of longitude from
about −85 degrees eastward is not significant.

The reader may wonder why we used a linear spline for estimating the curve
when a quadratic spline was used to estimate the derivative. A quadratic fit is
needed to have a continuous first derivative, but why not use a quadratic fit for

Figure 8.5 Fits of
Figure 8.4 with partial
residuals added.

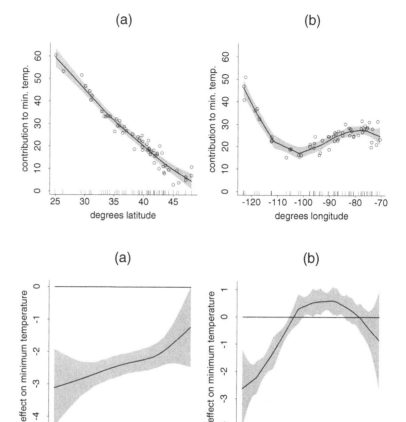

Figure 8.6
Derivatives of
components of
additive model fit to
the U.S. temperature
data, along with
variability bands.

both purposes? In fact, the quadratic and linear spline estimates of the curve are
similar, so it didn't really matter which we used.

8.3   Degrees of Freedom

The amount of smoothing used for each component is a defining characteristic of
an additive model fit. The degrees-of-freedom notion, as defined for scatterplot
smoothing in Section 3.13, is a natural and attractive way of quantifying this. Linear terms have 1 degree of freedom; whereas nonlinear terms have some number
greater than 1, depending on the curviness of the function.

Let

$$y_i = \beta_0 + \sum_{j=1}^{d} f_j(x_{ji}) + \varepsilon_i \tag{8.4}$$

be a general penalized spline additive model with mixed model representation

$$\mathbf{y} = \mathbf{X}\boldsymbol{\beta} + \mathbf{Z}\mathbf{u} + \boldsymbol{\varepsilon}, \quad \text{Cov}(\boldsymbol{\varepsilon}) = \sigma_\varepsilon^2 \mathbf{I},$$

and with centering possibly incorporated into the design matrices $\mathbf{X}$ and $\mathbf{Z}$. The functions f_j, $1 \leq j \leq d$, may be parametric or nonparametric. For given variance components, the fitted values are

$$\hat{\mathbf{y}} = \mathbf{X}\hat{\boldsymbol{\beta}} + \mathbf{Z}\hat{\mathbf{u}} = \mathbf{C}(\mathbf{C}^{\mathsf{T}}\mathbf{C} + \boldsymbol{\Lambda})^{-1}\mathbf{C}^{\mathsf{T}}\mathbf{y},$$

where

$$\mathbf{C} \equiv [\mathbf{X}|\mathbf{Z}] \quad \text{and} \quad \boldsymbol{\Lambda} \equiv \begin{bmatrix} \mathbf{0} & \mathbf{0} \\ \mathbf{0} & \sigma_\varepsilon^2 \, \mathrm{Cov}(\mathbf{u})^{-1} \end{bmatrix}. \tag{8.5}$$

The total degrees of freedom of the fit is

$$df_{\mathrm{fit}} = \mathrm{tr}\{\mathbf{C}(\mathbf{C}^{\mathsf{T}}\mathbf{C} + \boldsymbol{\Lambda})^{-1}\mathbf{C}^{\mathsf{T}}\} = \mathrm{tr}\{(\mathbf{C}^{\mathsf{T}}\mathbf{C} + \boldsymbol{\Lambda})^{-1}\mathbf{C}^{\mathsf{T}}\mathbf{C}\}.$$

However, we can also compute the degrees of freedom for each component. Let P denote the number of columns in $\mathbf{C}$, and let

$$\{\mathcal{I}_0, \mathcal{I}_1, \dots, \mathcal{I}_d\}$$

be a partition of the column indices $\{1, \dots, P\}$ such that $\mathcal{I}_0$ corresponds to the intercept β_0 and $\mathcal{I}_j$ corresponds to $f_j(\cdot)$ for each $1 \leq j \leq d$. For example, in the additive model described by (8.1) and (8.3) with $K_s = K_t = 20$, we would have $P = 43$ and

$$\mathcal{I}_0 = \{1\}, \quad \mathcal{I}_1 = \{2, 4, \dots, 23\}, \quad \mathcal{I}_2 = \{3, 24, \dots, 43\}.$$

For a general matrix $\mathbf{A}$ having P columns, define

$$\mathbf{A}_{\mathcal{I}} \equiv \text{submatrix of } \mathbf{A} \text{ consisting of columns with indices in } \mathcal{I}.$$

According to this notation,

$$\{\mathbf{C}_{\mathcal{I}_0}, \mathbf{C}_{\mathcal{I}_1}, \dots, \mathbf{C}_{\mathcal{I}_d}\}$$

represents a partition of the columns of $\mathbf{C}$ corresponding to the terms of the additive model (8.4).

Define $\mathbf{E}_j$ to be the $P \times P$ diagonal matrix with ones in the diagonal positions with indices in $\mathcal{I}_j$ and zeros elsewhere on the diagonal. The fitted values for the jth term are

$$\begin{bmatrix} \hat{f}_j(x_{j1}) \\ \vdots \\ \hat{f}_j(x_{jn}) \end{bmatrix} = \{\mathbf{C}\mathbf{E}_j(\mathbf{C}^{\mathsf{T}}\mathbf{C} + \boldsymbol{\Lambda})^{-1}\mathbf{C}^{\mathsf{T}}\}\mathbf{y},$$

so the corresponding degrees of freedom may be computed as

$$df_j = \mathrm{tr}\{\mathbf{C}\mathbf{E}_j(\mathbf{C}^{\mathsf{T}}\mathbf{C} + \boldsymbol{\Lambda})^{-1}\mathbf{C}^{\mathsf{T}}\} = \mathrm{tr}\{\mathbf{E}_j(\mathbf{C}^{\mathsf{T}}\mathbf{C} + \boldsymbol{\Lambda})^{-1}\mathbf{C}^{\mathsf{T}}\mathbf{C}\},$$

which is the sum over the indices of $\mathcal{I}_j$ of the diagonal elements of the matrix $(\mathbf{C}^{\mathsf{T}}\mathbf{C} + \boldsymbol{\Lambda})^{-1}\mathbf{C}^{\mathsf{T}}\mathbf{C}$. This last fact helps us appreciate that $df_{\mathrm{fit}} = df_0 + \cdots + df_d$.

We thus have a readily computable formula for the degrees of freedom of the components of a penalized spline additive model after the model has been fit. However, sometimes it is desirable to specify one or more of the df_j values *before* a model is fit. For example, it is common to ask that a particular component have 3 degrees of freedom; this is the default used by the function gam()

Table 8.1
Approximate and exact degrees of freedom for each nonparametric component of the additive model analysis in Section 8.4.1 of the Cape Cod birthweight data.

Component	adf	df
parity	2.906	2.895
cig's per day	2.372	2.363
years of education	3.687	3.665
prenatal visits	2.359	2.341

in the S-PLUS package. For additive models with more than one or two components, solving for these prespecified df_j values is computationally expensive because several additive model fits and multiparameter root-finding algorithms are required. A much simpler alternative is to approximate df_j by the degrees of freedom in the nonparametric regression model involving only x_j:

$$y_i = f_j(x_{ji}) + \varepsilon_i.$$

Then an approximation to df_j is

$$adf_j = \text{tr}\{(\mathbf{C}_j^\mathsf{T}\mathbf{C}_j + \mathbf{\Lambda}_j)^{-1}\mathbf{C}_j^\mathsf{T}\mathbf{C}_j\} - 1,$$

where

$$\mathbf{C}_j \equiv \mathbf{C}_{\mathcal{I}_j \cup \mathcal{I}_0}, \qquad \mathbf{\Lambda}_j = \text{diag}\{0, 0, (\sigma_\varepsilon^2/\sigma_{u_j}^2)\mathbf{1}_{K_j \times 1}\}$$

corresponds to a linear penalized spline scatterplot smooth, and $\sigma_{u_j}^2$ is the variance component corresponding to the spline coefficients for estimation of f_j. (The extension to higher-degree smooths is trivial.) It is relatively simple to solve for the $\lambda_j \equiv \sigma_\varepsilon/\sigma_{u_j}$ that give rise to a prespecified adf_j value. An algorithm for efficient computation of adf_j values is given in Appendix B. Asymptotics where $\lambda_j \to 0$ lead to the result $adf_j/df_j \to 1$ (Aerts, Claeskens, and Wand 2002).

Table 8.1 shows the values of df_j and adf_j for the Cape Cod birthweight data corresponding to the analysis given in Section 8.4.1. In this case there is practically no difference between the approximate and exact degrees-of-freedom values. This is the case for all examples that we have observed.

8.4 Smoothing Parameter Selection

In theory, the methods described in Chapter 5 are extendible to the additive model setting for automatic smoothing parameter selection. For example, in the additive model (8.1) with mixed model representation (8.3), the smoothing parameters

$$\lambda_s = \sigma_\varepsilon/\sigma_s \quad \text{and} \quad \lambda_t = \sigma_\varepsilon/\sigma_t$$

can be selected via REML estimation of the variance components σ_s^2, σ_t^2, and σ_ε^2. Mixed model software such as lme() in S-PLUS and PROC MIXED in SAS can help overcome the computational burden, since multiple variance component estimation via REML is now standard. Classical model selection criteria such as GCV and AIC can also be devised for choice of λ_s and λ_t; see Wood (2000).

Owing to the computational challenges arising from having to minimize a function in several variables, early implementations of additive models avoided

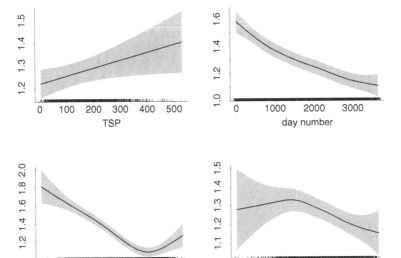

Figure 8.7 Additive
model fit of Milan
mortality data with
smoothing parameters
chosen by "3 degrees
of freedom per
smooth" rule. The
seasonal effect of
day.num is lost due to
gross oversmoothing.

automatic smoothing parameter selection. In particular, the S-PLUS additive
model function gam() defaults to the rule: 3 degrees of freedom for each additive component. This is a reasonable starting point but should not be the only
amount of smoothing used. Consider, for example, the Milan mortality data introduced in Section 1.6. Figure 8.7 shows the fits to the model

$$\sqrt{\texttt{mortality}_t}$$
$$= \beta_0 + \beta\texttt{TSP}_t + f_1(t) + f_2(\texttt{temperature}_t) + f_3(\texttt{humidity}_t) + \varepsilon_t$$

using the default of 3 degrees of freedom per smooth. The fits for temperature
and humidity look plausible, but the one for day number (t) does not. For several
years of data one would expect an oscillatory relationship corresponding to the
seasons. This pattern is apparent in Figure 8.8, where the amount of smoothing
is chosen via REML.

This particular example makes REML look good, but automatic smoothing
parameter selectors are, unfortunately, somewhat erratic. In Figure 8.8, 35 knots
are used for mean temperature and for relative humidity, while 60 knots are used
for day number. If instead 30 knots are used for mean temperature and relative
humidity, then REML leads to fits depicted in Figure 8.9, where the fit to day number shows none of the seasonal variation that exists in the data. When 30 knots
are used, REML chooses a zero variance component for day number, leading to
a gross oversmooth of this variable. We cannot explain this apparent sensitivity
of REML to the number of knots. However, it is in keeping with previous work
(by e.g. Härdle, Hall, and Marron 1988) that raises concerns about the instability of automatic smoothing parameter selection even for single predictor models.
Although we are attracted by the automatic nature of the mixed model–REML
approach to fitting additive models, we discourage blind acceptance of whatever
answer it provides and recommend looking at other amounts of smoothing.

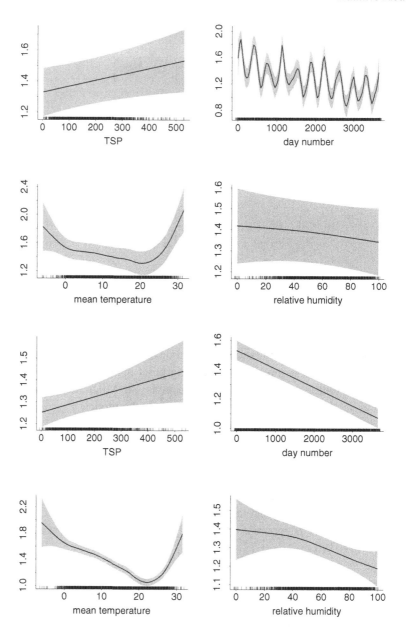

Figure 8.8 Additive model fit of Milan mortality data with smoothing parameters chosen by REML, with 35 knots for mean temperature and relative humidity and 60 knots for day number. In this case REML chooses an approximate amount of smoothing for all variables. Similar fits are achieved by Schwartz's rule of 3 degrees of freedom per year.

Figure 8.9 Additive model fit of Milan mortality data with smoothing parameters chosen by REML, but with 30 knots rather than 35 used for mean temperature and relative humidity. The seasonal effect of day number is lost due to gross oversmoothing.

In Section 6.9 we described maps (Chaudhuri and Marron 1999) that, for scatterplot smoothing, present a range of smoothing parameter choices. The extension to additive models is still in the developmental stages but would be useful for dealing with the smoothing parameter selection problem.

Finally, we mention that prior experience with the type of functional relationship between response and predictor can aid smoothing parameter selection in practice. For example, in Figures 8.7 and 8.9 the conditional mortality–temperature and mortality–humidity relationships both seem plausible. However, as mentioned previously, such is not the case for day number. An environmental epidemiologist with whom we work, Joel Schwartz, routinely uses 3 degrees of freedom per year when controlling for day number in an additive model analysis. Indeed, the REML answer depicted in Figure 8.8 chooses 39 degrees of freedom for the ten-year series, which matches Schwartz's rule reasonably well.

8.4.1 *Upper Cape Cod Birthweight Data*

A number of environmental health studies have taken place in the Upper Cape Cod region of Massachusetts after elevated cancer rates were observed there in the mid-1980s. Several possible sources of health risks have been identified and include fuel dumping at a large military reservation, pesticide use in cranberry bogs, and poly-chlorinated biphenyl in water pipes. However, the studies have been largely inconclusive.

In the late 1990s the Department of Public Health for the Commonwealth of Massachusetts commissioned a new study into geographical variation of health outcomes in Upper Cape Cod. In the latest phase, reproductive outcomes, birthweight, and gestational age have been considered. Birthweight is measured on nearly all newborns and is sensitive to recent exposures, thus facilitating the determination of exposures of biological importance. For example, a 170–200-gram decrease in mean birthweight may be seen in babies whose mothers smoke over 16 cigarettes per day during pregnancy compared with those who do not smoke.

The Upper Cape Cod reproductive data correspond to all 1630 births in 1990 across five towns; Barnstable, Bourne, Falmouth, Mashpee, and Sandwich. Apart from geographical location (longitude and latitude) and the response variables birthweight (grams) and gestational age (weeks), there are 39 covariates. A preliminary analysis showed that many have no significant association with birthweight. Those that are significantly associated with either outcome (birthweight or gestational age) include maternal age, years of education, number of cigarettes per day, and number of drinks per week. Table 8.2 lists all other variables that exhibited some association with birthweight, together with the abbreviated names that are used in Tables 8.3 and 8.4.

While the goal of the Upper Cape Cod study is the assessment of geographical variation in birthweight, the effect of these infant and maternal covariates is also of scientific interest. So, with a view to illustrating the usefulness of additive models in practice, we will ignore the geographical aspect in this section. Kammann and Wand (2003) analyze these data with the incorporation of geographical information, as described in Section 13.6.

A penalized spline additive model, as described in Section 8.2, was fit to the variables in Table 8.2 with smooth functions for the four variables depicted in Figure 8.1. The linear components were chosen according to the approximate Z-value given by lme() in S-PLUS (see Section 7.6). Table 8.3 provides a numerical summary of the final model. We see that fourteen linear predictors have a significant effect on birthweight. The four nonlinear predictors involve smooths with degrees of freedom ranging from 2.4 to 3.7.

Table 8.2 Covariates that had some association with birthweight and gestational age according to a preliminary analysis of the Cape Cod data.

Abbreviation	Description
Infant covariates	
male	indicator for infant being male
black	indicator for infant being black
asian	indicator for infant being Asian
plurality	1 = single, 2 = twin, etc.
Maternal covariates	
parity	number of live births from mother
prenatal visits	number of prenatal care visits
preg. hyperten.	pregnancy-related hypertension
incomp. cervix	indicator for incomplete cervix
eclampsia	indicator for eclampsia
light prev. birth	previous pre-term infant
heavy prev. birth	previous infant $\geq$ 4000 grams
psychiatric	indicator for psychiatric disorder
renal disease	indicator for renal disease
uterine bleeding	indicator for uterine bleeding

Table 8.3 Summary of REML-based fit of geoadditive model for Upper Cape Cod birthweight data.

	Coeff.	St. dev.	Z-Ratio	p-Value
male	161.12	23.6310	6.8182	0.0000
maternal age	−7.4278	2.6925	−2.7587	0.0058
preg. hyperten.	−187.82	95.1320	−1.9743	0.0484
light prev. birth	−448.42	132.67	−3.38	0.0007
heavy prev. birth	310.11	97.61	3.1770	0.0015
renal disease	−636.40	332.69	−1.9129	0.0558
black	−142.95	66.0910	−2.1629	0.0306
asian	−221.05	112.34	−1.9677	0.0491
drinks per week	−39.5970	16.2740	−2.4331	0.0150
plurality	−843.02	89.5930	−9.4094	0.0000
uterine bleeding	−410.31	158.90	−2.5822	0.0098
psychiatric	−387.12	210.29	−1.8409	0.0657
incomp. cervix	−931.09	470.35	−1.9796	0.0478
eclampsia	−1080.70	469.24	−2.3031	0.0213

	df
parity	2.9
cig's per day	2.4
years of education	3.7
prenatal visits	2.3

Figure 8.10 displays all nonlinear covariate effects. Though our primary concern in this study is geographical effects on reproductive outcomes, the nonlinear covariate effects depicted here are quite interesting in their own right. The corresponding derivative estimates are plotted in Figure 8.11. From this we see, for example, that the effect of cigarettes is significant only when the number is

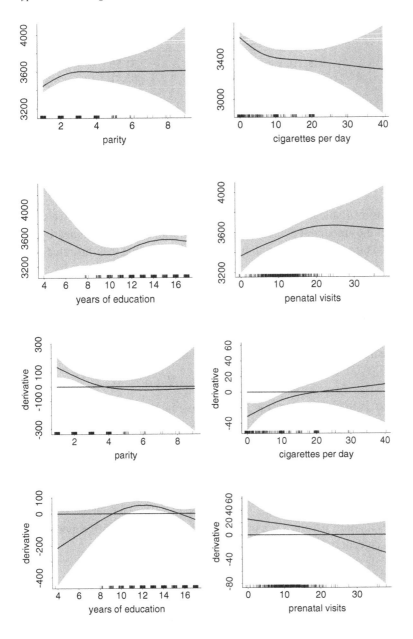

Figure 8.10
Components of full
additive model fit
to Upper Cape Cod
reproductive data.

Figure 8.11
Derivatives of
additive model fit
to Upper Cape Cod
reproductive data.

between none and about ten per day. The cigarette effect "plateaus" from about
10 per day onward.

8.5 Hypothesis Testing

The overall effect of a continuous predictor – and whether or not it has a nonlinear
effect – is of fundamental interest when building an additive model. Research into
corresponding hypothesis tests is somewhat scant, and has been mainly confined

Table 8.4 Likelihood
ratio statistics for
nonlinear terms.

Null hypothesis	$-2\log\{LR(\mathbf{y})\}$
effect of parity	31.55
effect of cig's per day	82.71
effect of years of education	36.82
effect of prenatal visits	62.99
linearity of parity	6.29
linearity of cig's per day	3.39
linearity of years of education	3.03
linearity of prenatal visits	0.74

to F-tests available in the S-PLUS gam() function. These tests give some indication of the amount of evidence regarding overall effect or nonlinearity, but they are based on an ad hoc smoothing parameter choice that (as illustrated by Figure 8.7) could be inappropriate.

8.5.1   *Likelihood Ratio Tests*

If a linear spline model is used, then linearity of the effect of a general predictor s can be assessed through a test of the hypotheses

$$H_0 : \sigma_s^2 = 0 \quad \text{versus} \quad H_1 : \sigma_s^2 > 0,$$

where σ_s^2 is the variance of the spline basis function coefficients for estimating the effect of s. We can also use the likelihood ratio statistic to test for the overall effect of each predictor. For linear splines, this corresponds to the hypotheses

$$H_0 : \beta_s = \sigma_s^2 = 0 \quad \text{versus} \quad H_1 : \beta_s \neq 0 \text{ or } \sigma_s^2 > 0. \tag{8.6}$$

Table 8.4 shows likelihood ratio statistics for the additive model fit to the Upper Cape Cod birthweight data. As described in Section 6.6.1.1, the determination of corresponding p-values is a delicate problem. In the additive model situation, this is a current research problem and is not yet settled.

8.5.2   *F-tests*

The F-test paradigm (Section 2.4.7) can also be adapted to the additive model setting. Consider, for example, the model

$$y_i = \beta_0 + f(s_i) + g(t_i) + \varepsilon_i. \tag{8.7}$$

Linearity of f can be assessed by comparing the quality of the fit of (8.7) with that of

$$y_i = \beta_0 + \beta_s s_i + g(t_i) + \varepsilon_i. \tag{8.8}$$

Let $\hat{\mathbf{y}}_{\text{smaller}}$ be the fitted values under the smaller model (8.8) and let $\hat{\mathbf{y}}_{\text{larger}}$ be the same for the larger model (8.7). Then the degree to which (8.7) improves upon (8.8) can be measured by the difference between

$$R_{\text{smaller}}^2 = \text{square of correlation coefficient between } \mathbf{y} \text{ and } \hat{\mathbf{y}}_{\text{smaller}}$$

Table 8.5 Results
from approximate
F-tests for nonlinear
terms.

Null hypothesis	F	Approx. p-value
effect of parity	7.98	0.0000
effect of cig's per day	7.18	0.0000
effect of years of education	4.48	0.0005
effect of prenatal visits	2.79	0.0111
linearity of parity	4.39	0.0055
linearity of cig's per day	3.60	0.0273
linearity of years of education	3.16	0.0185
linearity of prenatal visits	1.98	0.1378

and

$$R^2_{\text{larger}} = \text{square of correlation coefficient between } \mathbf{y} \text{ and } \hat{\mathbf{y}}_{\text{larger}}.$$

Arguments similar to those given in Section 2.4.7 lead to the F-statistic

$$F = \frac{R^2_{\text{larger}} - R^2_{\text{smaller}}}{(1 - R^2_{\text{larger}})(df_{\text{res, smaller}} - df_{\text{res, larger}})/df_{\text{res, larger}}},$$

where $df_{\text{res, larger}}$ and $df_{\text{res, smaller}}$ are the residual degrees of freedom for the respective models, as defined in Section 8.3. Under the null hypothesis that the larger model is true, we have

$$F \overset{\text{approx.}}{\sim} F(df_{\text{res, smaller}} - df_{\text{res, larger}}, df_{\text{res, larger}})$$

(Hastie and Tibshirani 1990). Such tests are used by the S-PLUS function gam() to check for nonlinearities. However, as mentioned in Section 8.4, the default used there is 3 degrees of freedom per smooth function. For example, if model (8.7) is compared with (8.8), then gam() defaults to

$$df_{\text{fit, larger}} = 7 \quad \text{and} \quad df_{\text{fit, smaller}} = 5,$$

which corresponds to

$$df_{\text{res, smaller}} - df_{\text{res, larger}} \simeq 2 \quad \text{and} \quad df_{\text{res, larger}} \simeq n - 7$$

regardless of the shape of f. Thus, there is the potential for undersmoothing and, especially, oversmoothing, which could affect the power of the test. This limitation in the choice of smoothing parameter needs to be kept in mind when using such tests.

An improvement is to use automatic smoothing parameter selection methods for choosing $df_{\text{fit, larger}}$ and $df_{\text{fit, smaller}}$.

Table 8.5 contains F-statistics and approximate p-values for the same null hypotheses as in Table 8.4.

8.6 Model Selection

Model selection is a vital component of parametric regression, particularly when there are several candidate predictors. The extension of selection techniques to additive models has only recently started. In this section we will briefly survey these developments.

8.6.1   *All-Subsets Algorithms*

If there are d candidate predictors then the total number of subset models is 2^d. Hence, if d is not too large then all such models can be fit and then compared through a model selection criterion. Simonoff and Tsai (1999) take such an approach and use an additive model extension of the AIC_C criterion defined in Section 5.3.4. This approach also requires that the degrees of freedom for each term be chosen by some objective criterion, and Simonoff and Tsai (1999) achieve this using AIC_C as well. The same general approach could be used with other automatic degrees-of-freedom choices, such as those based on REML (Section 8.4) and other model selection criteria for choosing among the subset models. We caution that this method of model selection can sometimes lead to spurious results; see Burnham and Anderson (2002) for discussion. It is likely that no automatic model selector will ever be found that is foolproof. Good scientific knowledge together with model selection criteria will provide better model selection than automatic methods alone.

Alternatively, the degrees of freedom can be chosen subjectively and in multitude. For example, the user could nominate that a particular predictor, s, enter the model with either 4, 7, or 10 degrees of freedom. Another predictor, t, could enter the model with 1 (i.e. linearly) or 4 degrees of freedom. This will obviously increase the number of subsets, but it overcomes the need for automatic degrees-of-freedom choice.

8.6.2   *Stepwise Algorithms*

All-subset approaches can be time-consuming, even for moderate numbers of candidate predictors. In parametric regression this is circumvented through algorithms that step through a smaller number of subset models, usually called *stepwise* algorithms. The same principle can be applied to additive models. That said, there is little published work on stepwise algorithms for additive models. However, Chambers and Hastie (1993, pp. 280–5) describe one algorithm based on *ordered regimens*. This is implemented in the S-PLUS package via the function `step.gam()`.

8.6.3   *MCMC Model Selection Algorithms*

Markov chain Monte Carlo (MCMC) refers to a growing body of methods for fitting complex models when traditional approaches become infeasible. A special case of MCMC with relatively simple structure is *Gibbs sampling*. Some references on MCMC are Gilks, Richardson, and Spiegelhalter (1996) and Robert and Casella (1999). Chapter 16 describes such an approach for the penalized spline models used throughout this book.

Smith and Kohn (1996) used Gibbs sampling with a regression spline approach, as briefly described for scatterplot smoothing in Section 3.4, to perform variable selection for additive models. Shively, Kohn, and Wood (1999) proposed an alternative MCMC-based algorithm with smoothing splines used for function estimation.

8.7 Bibliographical Notes

Early developments of the additive model are due to Ezekiel (1924) and Friedman and Stuetzle (1981). However, the catalyst for widespread use of additive models is the monograph of Hastie and Tibshirani (1990) and its corresponding implementation in the S-PLUS language as the function gam(). Most early references, and many of the later ones, use the *backfitting algorithm* to fit additive models. This algorithm reduces the fit to a sequence of scatterplot smooths. Buja, Hastie, and Tibshirani (1989) describe the convergence properties of backfitting; Opsomer and Ruppert (1997) derive statistical properties for backfitting with local polynomial smoothers. See also Härdle and Hall (1993).

Alternatives to backfitting have been proposed more recently. These include marginal integration (Newey 1994; Tjøstheim and Auestad 1994; Linton and Nielsen 1995) and direct fitting based on low-rank smoothers (Hastie 1996; Marx and Eilers 1998). The approach described in this chapter falls into this last category.

Inference and model selection in additive models has also been treated by Härdle and Korostelev (1996), Smith, Wong, and Kohn (1998), Lin and Zhang (1999), and Cantoni and Hastie (2002).

Finally, we mention that many disciplines have already embraced additive models. Examples include economics (e.g. Linton and Härdle 1996), environmental epidemiology (Schwartz 1994), political science (Beck and Jackman 1998), environmental science (Davis and Speckman 1999), public health (Engels, Rosenberg, and Biggar 1999), and ecology (Roland, Keyghobadi, and Fownes 2000).

9

Semiparametric Mixed Models

9.1 Introduction

As we have seen in the last few chapters, semiparametric regression models based on penalized splines can be couched in the mixed model framework, allowing for mixed model estimation and for inferential and computational tools to be used. This synergy is similar in spirit to the mixed model approach to analyzing longitudinal data that commenced with the paper of Laird and Ware (1982). Most of the work that has been undertaken to model longitudinal data has been *parametric,* in the sense that the effects of continuous covariates have been modeled linearly or by using some parametric nonlinear model (Lindstrom and Bates 1990; Davidian and Giltinan 1995; Pinheiro and Bates 2000). An alternative to nonlinear mixed modeling is to incorporate smoothing methods. The mixed model representation of penalized splines allows for a seamless fusion between parametric mixed models and smoothing, which we call *semiparametric* mixed models.

9.2 Additive Mixed Models

In Chapter 4 we explained how the longitudinal pig weight data set could be fit using the mixed model

$$\texttt{weight}_{ij} = \beta_0 + U_i + \beta_1 \texttt{week}_{ij} + \varepsilon_{ij}, \quad U_i \text{ i.i.d. N}(0, \sigma_u^2).$$

Figure 9.1 shows a data set, similar in nature to the pig weight data, that corresponds to spinal bone mineral density (SBMD) measurements of 230 female subjects aged between 8 and 27. Each subject is measured either one, two, three, or four times. Hastie and Tibshirani (2000) provide a thorough semiparametric analysis of these data for the subjects with two or more measurements. Other analyses are given in James, Hastie, and Sugar (2000) and James and Hastie (2001).

We are grateful to Professors Trevor Hastie and Gareth James for making the spinal bone mineral density data available for this analysis.

The main difference between Figure 9.1 and Figure 4.1(b) (p. 92) for the pig weight data is that the linearity assumption is not reasonable here. Therefore, an appropriate model is

$$\text{SBMD}_{ij} = U_i + f(\texttt{age}_{ij}) + \varepsilon_{ij}, \quad 1 \le j \le n_i, \ 1 \le i \le m, \quad (9.1)$$

where f is some smooth function. For the spinal bone mineral density data, $m = 230$ and $n_i \in \{1, 2, 3, 4\}$.

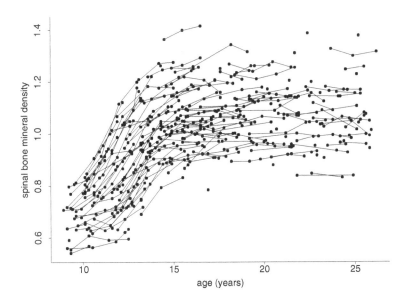

Figure 9.1 Spinal bone mineral density of 230 female subjects. Lines connect measurements on the same subject.

Fitting (9.1) via penalized splines is relatively straightforward. If we define

$$
\mathbf{X} = \begin{bmatrix}
1 & \text{age}_{11} \\
\vdots & \vdots \\
1 & \text{age}_{1n_1} \\
\vdots & \vdots \\
1 & \text{age}_{m1} \\
\vdots & \vdots \\
1 & \text{age}_{mn_m}
\end{bmatrix}, \tag{9.2}
$$

$$
\mathbf{Z} = \begin{bmatrix}
1 & \cdots & 0 & (\text{age}_{11} - \kappa_1)_+ & \cdots & (\text{age}_{11} - \kappa_K)_+ \\
\vdots & \ddots & \vdots & \vdots & \ddots & \vdots \\
1 & \cdots & 0 & (\text{age}_{1n_1} - \kappa_1)_+ & \cdots & (\text{age}_{1n_1} - \kappa_K)_+ \\
\vdots & \vdots & \vdots & \vdots & \ddots & \vdots \\
0 & \cdots & 1 & (\text{age}_{m1} - \kappa_1)_+ & \cdots & (\text{age}_{m1} - \kappa_K)_+ \\
\vdots & \ddots & \vdots & \vdots & \ddots & \vdots \\
0 & \cdots & 1 & (\text{age}_{mn_m} - \kappa_1)_+ & \cdots & (\text{age}_{mn_m} - \kappa_K)_+
\end{bmatrix},
$$

and

$$
\mathbf{u} = \begin{bmatrix}
U_1 \\
\vdots \\
U_m \\
u_1 \\
\vdots \\
u_K
\end{bmatrix}
$$

then we can simultaneously estimate variance components for the random intercept and the amount of smoothing for f by using the mixed model

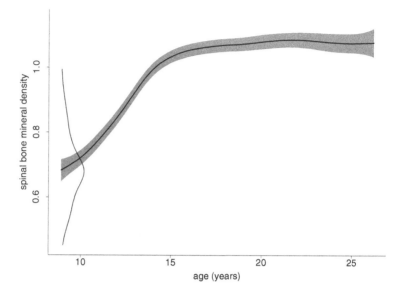

$$\mathbf{y} = \mathbf{X}\boldsymbol{\beta} + \mathbf{Z}\mathbf{u} + \boldsymbol{\varepsilon}, \quad \mathrm{Cov}\begin{bmatrix} \mathbf{u} \\ \boldsymbol{\varepsilon} \end{bmatrix} = \begin{bmatrix} \sigma_U^2\mathbf{I} & \mathbf{0} & \mathbf{0} \\ \mathbf{0} & \sigma_u^2\mathbf{I} & \mathbf{0} \\ \mathbf{0} & \mathbf{0} & \sigma_\varepsilon^2\mathbf{I} \end{bmatrix}. \quad (9.3)$$

Here σ_U^2 measures the between-subject variation, σ_ε^2 measures within-subject variation, and σ_u^2 controls the amount of smoothing done to estimate f.

Figure 9.2 shows the fit based on REML smoothing parameter selection. The estimate of f involves 9.56 degrees of freedom while the random intercept involves 219.5 degrees of freedom, so the entire fit uses about 229 degrees of freedom.

9.2.1 *Additive Model Extension*

The study for which the spinal bone mineral density data was collected is actually concerned with effects of ethnicity on bone density. As illustrated in Figure 9.3, the data can be categorized according to four ethnic groups: Asian, Black, Hispanic, and White.

To address the main question of interest, define indicator variables `black`$_i$, `hispanic`$_i$, and `white`$_i$ to correspond to membership of each of these three ethnic groups, and consider the model

$$\mathrm{SBMD}_{ij} = U_i + f(\mathtt{age}_{ij}) + \beta_1\mathtt{black}_i + \beta_2\mathtt{hispanic}_i + \beta_3\mathtt{white}_i + \varepsilon_{ij},$$
$$1 \le j \le n_i, \ 1 \le i \le m. \quad (9.4)$$

With this formulation, the Asian subjects comprise the reference group and $\beta_1, \beta_2, \beta_3$ represent mean differences in spinal bone mineral density between the other ethnic groups.

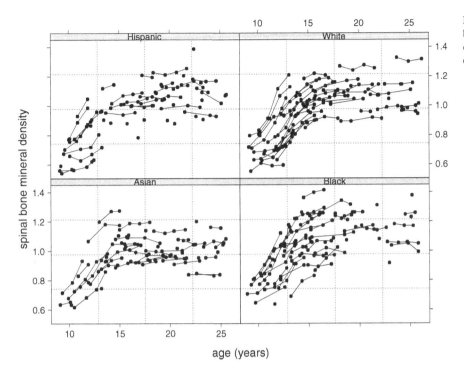

Figure 9.3 Spinal bone mineral density data broken down by ethnic group.

Model (9.4) is easily accommodated by adding the columns

$$\begin{bmatrix} \text{black}_1 & \text{hispanic}_1 & \text{white}_1 \\ \vdots & \vdots & \vdots \\ \text{black}_1 & \text{hispanic}_1 & \text{white}_1 \\ \vdots & \vdots & \vdots \\ \text{black}_m & \text{hispanic}_m & \text{white}_m \\ \vdots & \vdots & \vdots \\ \text{black}_m & \text{hispanic}_m & \text{white}_m \end{bmatrix}$$

to the matrix $\mathbf{X}$ given at (9.2) to form an

$$\left(\sum_{i=1}^m n_i \right) \times 5 = 547 \times 5$$

$\mathbf{X}$-matrix and then fitting a mixed model of the same form as (9.3). The mixed model output corresponding to the fixed effects (from the lme() function in S-PLUS) is summarized in Table 9.1.

A graphical summary of these contrasts is given in Figure 9.4, which shows the difference in mean bone mineral density between Black and Asian subjects to be highly significant. There is some suggestion of a difference between Hispanic and Asian subjects. Contrasts not involving Asian subjects can be done by taking other ethnic groups as the reference group, or through contrast coding.

Table 9.1 Fixed effects component of mixed model output for the linear mixed model fit of (9.4).

Variable	Value	Approx. st. dev.	Z-Ratio
black	0.1062	0.02066	5.141
hispanic	0.0260	0.02164	1.203
white	0.0131	0.02165	0.6069

Figure 9.4 Estimates (horizontal lines) and approximate 95% confidence intervals for the differences in mean spinal bone mineral density between female subjects with Black, Hispanic, and White ethnic origin contrasted with those of Asian origin.

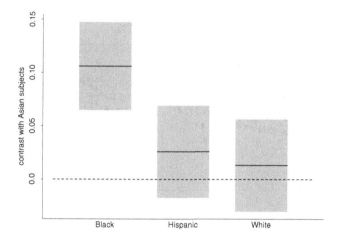

Extensions to more elaborate additive mixed models (e.g., several smooth functions) are quite straightforward by way of combining material presented here and in Chapter 8.

9.2.2 Serially Correlated Errors

The model (9.4) assumes that the errors ε_{ij} are uncorrelated. Since the data are collected over time, it is likely (see Diggle et al. 2002) that some serial correlation would exist within each subject. The spinal bone mineral density measurements for each subject are not equally spaced in time, so such serial correlation is difficult to model by (say) an autoregressive process. However, equation (9.7) discussed in the next section models serial correlation with a random function g_i that is specific to the ith individual.

An example where serial correlation is easily formulated was given in Coull, Schwartz, and Wand (2001) with respect to daily data collected by Pope et al. (1991) on respiratory health and air pollution. In this study, 41 Utah schoolchildren had peak expiratory flow (PEF) measurements recorded for 109 consecutive days. Coull et al. (2001) fit the model

$$\text{PEF}_{ij} = \beta_0 + U_i + (\beta_1 + V_i)\text{PM}_{10j} + f(\texttt{low.temp}_j) + g(j) + \varepsilon_{ij}, \quad (9.5)$$

where PEF_{ij} is the PEF measurement for child i on day j, PM_{10j} is the amount of particulate matter of aerodynamic diameter less than $10\,\mu\text{m}$ on day j, $\texttt{low.temp}_j$

	Variable	Value	Approx. st. dev.	Z-Ratio
Independent errors	PM_{10}	−0.086	0.032	−2.7
AR(1) errors	PM_{10}	−0.070	0.025	−2.8

Table 9.2 Analysis of the peak expiratory flow data for the fits of (9.5) with and without AR(1) errors.

is the lowest temperature on day j, and U_i and V_i are (respectively) random intercept and slope effects for the effect of PM_{10}. However, the errors for each child were taken to be serially correlated:

$$\varepsilon_{ij} = \rho \varepsilon_{i,\,j-1} + \xi_{ij}, \tag{9.6}$$

where the ξ_{ij} are independent and $|\rho| < 1$. Correlation of this type is also referred to as *first-order autoregressive,* abbreviated as AR(1). AR(1) serial correlation in the errors can be accommodated by standard mixed model software packages.

Coull et al. (2001) fit (9.5) and (9.6) using PROC MIXED in SAS and estimated ρ as $\hat{\rho} = 0.510$ with a standard error of 0.014. This represents a significant amount of serial correlation. Table 9.2 compares the regression coefficients and their estimated standard deviations for models with and without AR(1) errors. Note the decrease in standard deviation when the autocorrelation is taken into account.

9.3 Subject-Specific Curves

Models such as (9.1), (9.4), and (9.5) have the feature that the nonparametric function $f(\cdot)$ is a *global* function and does not depend on the individual. Consider, for example, model (9.1). Another way of thinking about this model is that individuals have their own functions but that the functions differ from one another only in their intercept.

In many contexts, it seems sensible to allow the functions to vary with the individuals by more than just the intercept. Thus, for example, instead of (9.1) we might have the model

$$\text{SBMD}_{ij} = f(\text{age}_{ij}) + g_i(\text{age}_{ij}) + \varepsilon_{ij}, \tag{9.7}$$

where now the $g_i(\cdot)$ are random functions with mean zero. We can model these random functions as regression splines and place them into the mixed model framework.

Recall that the ordinary linear regression spline has two components: a fixed component for the linear part of the model, and a random component describing the deviations from linearity; see (9.3). Each g_i has two components, but now we must remember that the functions $g_i(\cdot)$ are random with mean zero. This means that the linear part of the regression spline is also random, rather than a fixed effect.

We now describe a strategy for fitting (9.7). Define the linear predictors $\mathbf{x}_{ij}^{\mathsf{T}} = [1, \text{age}_{ij}]$ and the nonlinear components

$$\mathbf{z}_{ij}^{\mathsf{T}} = [(\text{age}_{ij} - \kappa_1)_+, \ldots, (\text{age}_{ij} - \kappa_K)_+].$$

Then (as before) we will write

$$f(\text{age}_{ij}) = \mathbf{x}_{ij}^{\mathsf{T}} \boldsymbol{\beta} + \mathbf{z}_{ij}^{\mathsf{T}} \mathbf{U}, \tag{9.8}$$

where $\text{Cov}(\mathbf{U}) = \sigma_U^2 \mathbf{I}$. The random functions follow the same basic framework,

$$g_i(\text{age}_{ij}) = \mathbf{x}_{ij}^{\mathsf{T}} \mathbf{u}_{i1} + \mathbf{z}_{ij}^{\mathsf{T}} \mathbf{u}_{i2}. \tag{9.9}$$

However, these are random functions with mean zero, so there are no fixed effects. We set $\text{Cov}(\mathbf{u}_{i1}) = \boldsymbol{\Sigma}$, an unstructured 2×2 covariance matrix, and thus allow for complex departures from the common linear component. Finally, we set $\text{Cov}(\mathbf{u}_{i2}) = \sigma_2^2 \mathbf{I}$. We can now place (9.7) in a mixed model framework. Define

$$\mathbf{Z} = \begin{bmatrix}
\mathbf{z}_{11}^{\mathsf{T}} & \mathbf{x}_{11}^{\mathsf{T}} & 0 & \cdots & 0 & \mathbf{z}_{11}^{\mathsf{T}} & 0 & \cdots & 0 \\
\vdots & \vdots & \vdots & \ddots & \vdots & \vdots & \vdots & \ddots & \vdots \\
\mathbf{z}_{1n_1}^{\mathsf{T}} & \mathbf{x}_{1n_1}^{\mathsf{T}} & 0 & \cdots & 0 & \mathbf{z}_{1n_1}^{\mathsf{T}} & 0 & \cdots & 0 \\
\mathbf{z}_{21}^{\mathsf{T}} & 0 & \mathbf{x}_{21}^{\mathsf{T}} & \cdots & 0 & 0 & \mathbf{z}_{21}^{\mathsf{T}} & \cdots & 0 \\
\vdots & \vdots & \vdots & \ddots & \vdots & \vdots & \vdots & \ddots & \vdots \\
\mathbf{z}_{2n_2}^{\mathsf{T}} & 0 & \mathbf{x}_{2n_2}^{\mathsf{T}} & \cdots & 0 & 0 & \mathbf{z}_{2n_2}^{\mathsf{T}} & \cdots & 0 \\
\vdots & \vdots & \vdots & \ddots & \vdots & \vdots & \vdots & \ddots & \vdots \\
\mathbf{z}_{m1}^{\mathsf{T}} & 0 & 0 & \cdots & \mathbf{x}_{m1}^{\mathsf{T}} & 0 & 0 & \cdots & \mathbf{z}_{m1}^{\mathsf{T}} \\
\vdots & \vdots & \vdots & \ddots & \vdots & \vdots & \vdots & \ddots & \vdots \\
\mathbf{z}_{mn_m}^{\mathsf{T}} & 0 & 0 & \cdots & \mathbf{x}_{mn_m}^{\mathsf{T}} & 0 & 0 & \cdots & \mathbf{z}_{mn_m}^{\mathsf{T}}
\end{bmatrix}$$

and

$$\mathbf{u} = [\mathbf{U}^{\mathsf{T}}, \mathbf{u}_{11}^{\mathsf{T}}, \mathbf{u}_{21}^{\mathsf{T}}, \ldots, \mathbf{u}_{m1}^{\mathsf{T}}, \mathbf{u}_{12}^{\mathsf{T}}, \mathbf{u}_{22}^{\mathsf{T}}, \ldots, \mathbf{u}_{m2}^{\mathsf{T}}]^{\mathsf{T}},$$

so that

$$\mathbf{y} = \mathbf{X}\boldsymbol{\beta} + \mathbf{Z}\mathbf{u} + \boldsymbol{\varepsilon}, \quad \text{Cov}(\mathbf{u}) = \begin{bmatrix} \sigma_U^2 \mathbf{I} & \mathbf{0} & \mathbf{0} \\ \mathbf{0} & \underset{1 \le i \le m}{\text{blockdiag}(\boldsymbol{\Sigma})} & \mathbf{0} \\ \mathbf{0} & \mathbf{0} & \sigma_2^2 \mathbf{I} \end{bmatrix}. \tag{9.10}$$

Although this approach to fitting subject-specific curves seems straightforward, more work needs to be done on implementation. At this stage we have not explored the use of standard software for fitting (9.10). There may also be some advantages in centering $\mathbf{X}$ so that $\boldsymbol{\Sigma}$ becomes effectively diagonal. This would be in keeping with advice regarding Markov chain Monte Carlo fitting of mixed models, where some form of centering is advocated even for very simple mixed models (see e.g. Gelfand, Sahu, and Carlin 1995).

9.4   Bibliographical Notes

The main ideas of Section 9.3 have been discussed in a number of contexts in the smoothing spline literature. See, for example, Donnelly, Laird, and Ware

(1995), Brumback and Rice (1998), Zhang et al. (1998), Wang (1998), Verbyla et al. (1999), and Guo (2002). There are nontrivial practical differences among these papers.

Without going into the details of the smoothing spline approach, for the equivalent formulation of our (9.9) Brumback and Rice (1998) treat the linear terms as fixed rather than random effects, thus leading to the number of fixed effects being at least twice as large as the number of sampled individuals. As we do in Section 9.3, Guo (2002) lowers the dimension of the fixed effects by allowing these linear terms to be random; in his example, Σ is assumed to be a diagonal matrix. Zhang et al. (1998) use a smoothing spline parameterization for (9.8) but not for (9.9), assuming that the latter is a realization of a nonhomogeneous Ornstein–Uhlenbeck process with 1–2 fixed parameters.

Rice and Wu (2001) use B-spline basis functions with a fixed number K of knots. In their formulation, $f(x) = \sum_{k=1}^{K} \beta_k B_k(x)$, where the β_k are fixed effects. In addition, $g_i(x) = \sum_{k=1}^{K} \gamma_{ik} B_k(x)$, where the γ_{ik} have a $K \times K$ covariance matrix Σ. Their covariance matrix Σ is thus potentially of much higher dimension than ours, depending on the number of knots.

10

Generalized Parametric Regression

10.1 Introduction

The data that we have dealt with in the preceding chapters has the feature that the response variable is *continuous.* This usually means that, possibly with the help of a transformation, the data can be modeled to be normal and that linear regression techniques (such as least squares and best linear unbiased prediction) can be used for fitting. However, it is often the case that the response variable is not continuous but rather categorical or perhaps a count. Examples include: tumor present or absent; customer prefers green, pink, orange, or yellow packaging; number of emergency asthma admissions on a given day. Such response variables cannot be handled through the normal regression framework. In many fields (e.g., medicine and marketing), categorical response variables are more the rule than the exception. Some continuous response data cannot be handled satisfactorily within the normal errors framework – for example, if they are heavily skewed. Skewed data often can be transformed to near symmetry, but an alternative is to apply a Gamma model (Section 10.4.3) to the untransformed data.

Regression models that aim to handle non-Gaussian response variables such as these are usually termed *generalized linear models* (GLMs). The first part of this chapter gives a brief overview of GLMs; a reader with plenty of experience on this topic could skip this part of the chapter. The second part of the chapter deals with generalized linear mixed models (GLMMs) and is recommended for all readers.

10.2 Binary Response Data

The source of the BPD data is *Principles of Biostatistics* (Pagano and Gauvreau 1993). We are grateful to Professor Kim Gauvreau for sharing the data.

The data depicted in Figure 10.1 correspond to 223 birthweight measurements (in grams) and occurrence of *bronchopulmonary dysplasia* (BPD) for a set of babies. The BPD data are coded as

$$\text{BPD}_i = \begin{cases} 1 & \text{if } i\text{th baby has BPD,} \\ 0 & \text{otherwise.} \end{cases}$$

Such 0–1 data are usually referred to as *binary.* It is of interest to measure the effect of birthweight on the occurrence of BPD. Consider, for the moment, the model

$$\text{BPD}_i = \beta_0 + \beta_1 \text{birthweight}_i + \varepsilon_i, \quad \varepsilon_i \sim \text{N}(0, \sigma^2). \tag{10.1}$$

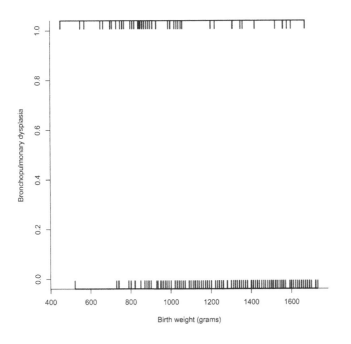

$F(x)$	Distribution
$H(x) = e^x/(1 + e^x)$	logistic
$\Phi(x)$	normal
$1 - \exp(-e^x)$	complementary log-log

Table 10.1 Some cumulative distribution functions used with a binary response.

This is easily seen to be inappropriate. First of all, the right-hand side of (10.1) is not guaranteed to be binary. Second, it implies that the BPD_i are normally distributed and with homoscedastic errors; such an assumption is easily refuted for binary data. Finally, the expected values of the fit need not be probabilities, and there is a danger that, when fit, the model will report estimated probabilities of the occurrence of BPD for certain values of birthweight that are negative, or exceed 1!

10.3 Logistic Regression

A remedy for the problems just raised is to change the model to

$$P(\text{BPD}_i = 1|\text{birthweight}) = F(\beta_0 + \beta_1\text{birthweight}_i), \qquad (10.2)$$

where F is a function that maps any real number to a number between 0 and 1. To retain interpretability of the coefficients, F should also be strictly increasing. There are many functions that have these properties. In fact, the cumulative distribution function of any continuous distribution with an everywhere positive density (e.g., the standard normal distribution) must meet these requirements. Some examples are shown in Table 10.1. They are each plotted in Figure 10.2,

Figure 10.2 Plots of
the three cumulative
distribution functions
listed in Table 10.1;
they have been shifted
and scaled to have the
same value and slope
at the origin.

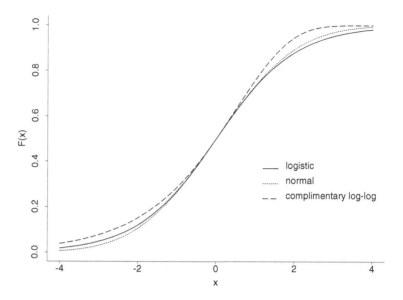

where it is seen that they have the properties mentioned previously. The functions are each shifted and scaled to have the same value and slope at the origin in order to allow easier comparison.

Logistic regression uses the logistic probability distribution function, whereas probit regression uses the normal probability distribution function. Paradoxically, although the normal distribution is used in almost all branches of statistics, for binary data the logistic distribution is used in most applications. The reason for this choice goes back many years and is both philosophical and computational. Unlike the probit model, the logistic model possesses nontrivial sufficient statistics, thus allowing data compression and exact (finite-sample) inference. In addition, although this is no longer a major issue, logistic regression requires only the exponential function – something hard-coded into computers – whereas probit regression requires the calculation of the normal distribution function. Finally, the logistic regression model leads to particularly simple expressions when it comes to fitting the model. The complementary log-log distribution function is more often used for ordered categorical data. It should be mentioned, however, that Bayesians are abandoning the logistic model in favor of the probit model because the probit is much simpler to implement with MCMC than is the logistic.

Since $H^{-1}(y) = \log\{y/(1-y)\}$, the logistic regression model can be rewritten as

$$\log\left\{\frac{P(\text{BPD}_i = 1|\text{birthweight})}{1 - P(\text{BPD}_i = 1|\text{birthweight})}\right\} = \beta_0 + \beta_1\text{birthweight}_i. \quad (10.3)$$

The left-hand side is the logarithm of the *odds* of BPD for a given birthweight, sometimes called the *log odds* for short. A convenient shorthand is to define the logit function as H^{-1} so that

$$\text{logit}(u) = \log\left(\frac{u}{1-u}\right),$$

which leads to

$$\text{logit}\{P(\text{BPD}_i = 1|\texttt{birthweight})\} = \beta_0 + \beta_1\texttt{birthweight}_i.$$

Notice that logit maps numbers in $(0, 1)$ to the real line. The logit transformation of the probability of BPD is an example of what is commonly called a *link* transformation. In particular, the logit transformation – the one that results in the simplest likelihood – is called the *canonical* link. A more precise definition of canonical link will be given in the next section.

10.4   Other Generalized Linear Models

Although the logistic regression model

$$y_i \sim \text{Bernoulli}\left(\frac{\exp\{(\mathbf{X}\boldsymbol{\beta})_i\}}{1 + \exp\{(\mathbf{X}\boldsymbol{\beta})_i\}}\right)$$

is the most common generalized linear model, there are others that are often used in practice. These include the Poisson regression model

$$y_i \sim \text{Poisson}[\exp\{(\mathbf{X}\boldsymbol{\beta})_i\}],$$

which is appropriate for count data, and the Gamma regression models such as

$$y_i \sim \text{Gamma}[\{1/(\mathbf{X}\boldsymbol{\beta})_i\}, \phi] \tag{10.4}$$

and

$$y_i \sim \text{Gamma}[\exp\{(\mathbf{X}\boldsymbol{\beta})_i\}, \phi], \tag{10.5}$$

which are appropriate for right-skewed continuous data. Although (10.4) has the canonical link, (10.5) is more commonly used because the log link guarantees a positive mean. Here $\text{Gamma}(\mu, \phi)$ means a gamma distribution with mean μ and coefficient of variation $\sqrt{\phi}$. The gamma and the Gaussian families are examples of GLMs with dispersion parameters, the standard deviation for the Gaussian family, and coefficient of variation for the gamma family.

A GLM begins with a 1-parameter exponential family of distributions for the response with density of the form

The coefficient of variation of a distribution is the ratio of the standard deviation to the mean.

$$f(y; \eta) = \exp\left(\frac{y\eta - b(\eta)}{\phi} + c(y, \phi)\right) \tag{10.6}$$

for some functions $b(\eta)$ and $c(y, \phi)$; see Table 10.2. Here ϕ is a dispersion parameter; the Bernoulli and Poisson distributions have no dispersion parameters, so for these distributions we take $\phi \equiv 1$. The parameter η is called the natural parameter. It can be shown that $\mathsf{E}(y) = b'(\eta)$ and $\text{var}(y) = \phi b''(\eta)$, where $b'(\eta)$ and $b''(\eta)$ are the first and second derivatives of b. In a GLM, it is assumed that the natural parameter for y_i, η_i, depends on a vector of predictor variables, $\mathbf{x}_i$. More

Table 10.2 Functions
b and c for some
common 1-parameter
exponential families.

Name	$b(\eta)$	$b'(\eta)$	Canonical link	$c(y, \phi)$
Bernoulli	$\log(1 + e^\eta)$	$e^\eta/(1 + e^\eta)$	$\text{logit}(\mu)$	0
Poisson	e^η	e^η	$\log(\mu)$	$-\log(y!)$
Gamma	$\log(\eta)$	$1/\eta$	$1/\mu$	see text
Gaussian	$\eta^2/2$	η	μ	see text

explicitly, it is assumed that for some function ψ, $\eta_i = \psi(\mathbf{x}_i^\mathsf{T}\boldsymbol{\beta})$. The canonical link occurs when ψ is the identity function and hence $\eta_i = \mathbf{x}_i^\mathsf{T}\boldsymbol{\beta}$.

More generally, the link function $\mathcal{L}$ is defined by the equation $\mathcal{L}\{\mathsf{E}(y_i)\} = \mathbf{x}_i^\mathsf{T}\boldsymbol{\beta}$. Later we will need the notation $\mu(\cdot) = \mathcal{L}(\cdot)^{-1}$. Note that the inverse link, $\mu(\cdot)$, converts the linear predictor $\mathbf{x}_i^\mathsf{T}\boldsymbol{\beta}$ to the expectation of y_i: $\mu(\mathbf{x}_i^\mathsf{T}\boldsymbol{\beta}) = \mathsf{E}(y_i) = \mu_i$. For logistic regression, $\mathcal{L}(u) = \text{logit}(u)$ and $\mu(u) = H(u)$, where H is the logistic function.

The dispersion parameter ϕ is assumed not to depend on i; this is a generalization of the constant variance assumption of the linear model. With these assumptions, the density of $\mathbf{y}$ is

$$f(\mathbf{y}; \boldsymbol{\beta}) = \exp\left(\frac{\mathbf{y}^\mathsf{T}\psi(\mathbf{X}\boldsymbol{\beta}) - \mathbf{1}^\mathsf{T}b\{\psi(\mathbf{X}\boldsymbol{\beta})\}}{\phi} + \mathbf{1}^\mathsf{T}c(\mathbf{y}, \phi)\right). \qquad (10.7)$$

10.4.1   *Poisson Regression and Overdispersion*

The Poisson GLM, often called *Poisson regression,* uses the Poisson density

$$P(Y = y) = \frac{\mu^y e^{-\mu}}{y!}, \quad y = 0, 1, \dots.$$

The logarithm of this density is $y\log(\mu) - \mu - \log(y!)$. Letting $\eta = \log(\mu)$, the log density is $y\eta - e^\eta - \log(y!)$. Therefore, $b'(\eta) = e^\eta$ and $c(y, \phi) = -\log(y!)$.

Poisson regression
is often appropriate
for count data, but
overdispersion can
easily occur and so
affect inference.
The Poisson regression model is often used when the response is a count. However, the assumption that a count is Poisson distributed should not be taken lightly, since the variance of a Poisson regression equals its mean. Often, a response that is a count has a variance that is larger than its mean, sometimes much larger. This is "overdispersion" relative to the Poisson model (McCullagh and Nelder 1989). In such a case, the Poisson assumption can lead to a underestimation of variability. This would, for example, cause confidence intervals to be too small, with coverage probability smaller than the nominal value. With serious overdispersion, which occurs fairly frequently, the size of the undercoverage could be substantial. See Section 10.7 for discussion of overdispersion and other models for the variance.

There are various models for overdispersion, the most well-known of which is the negative binomial distribution. Another way to achieve overdispersion is to introduce random effects (see Section 10.8 and following for details). Let u be a normal random variable with mean $-\sigma_u^2/2$ and variance σ_u^2; these parameters are chosen so that $\mathsf{E}\{\exp(u)\} = 1$. Given u and the covariate $\mathbf{x}$, suppose that y is Poisson with mean $\exp(\mathbf{x}^\mathsf{T}\boldsymbol{\beta} + u)$. Then, unconditionally, y has mean $\exp(\mathbf{x}^\mathsf{T}\boldsymbol{\beta})$ but its variance is

$$\exp(\mathbf{x}^\mathsf{T}\boldsymbol{\beta}) + \exp(2\mathbf{x}^\mathsf{T}\boldsymbol{\beta})\{\exp(\sigma_u^2) - 1\} \geq \exp(\mathbf{x}^\mathsf{T}\boldsymbol{\beta}).$$

Thus, the variance exceeds the mean (overdispersion) unless the random effects are all zero. With this mixed model formulation, overdispersion is a natural consequence of nonzero random effects.

10.4.2   The Gaussian GLM: A Model for Symmetrically Distributed and Homoscedastic Responses

The GLM is just the ordinary linear model, which shows that generalized linear models are, in fact, a generalization of linear models. The Gaussian family of densities, when parameterized by the mean μ and variance ϕ, is

Ordinary multiple linear regression is a Gaussian GLM.

$$\frac{1}{\sqrt{2\pi\phi}} \exp\left\{-\frac{1}{2\phi}(y - \mu)^2\right\}.$$

The log density is

$$\frac{y\mu - \mu^2/2}{\phi} - \frac{1}{2}\left\{\frac{y^2}{\phi} + \log(2\pi\phi)\right\}.$$

Therefore, $\eta = \mu$, $b(\eta) = \eta^2/2$, and $c(y, \phi) = -\frac{1}{2}\{y^2/\phi + \log(2\pi\phi)\}$.

10.4.3   The Gamma GLM: A Model with a Constant Coefficient of Variation

There are many ways to parameterize the gamma family. Following McCullagh and Nelder (1989), we will use the mean, μ, and squared coefficient of variation (variance over squared mean), denoted by ϕ, as the parameters. Then the gamma density with parameters (μ, ϕ) is

$$\frac{1}{y\Gamma(\phi^{-1})}\left(\frac{y}{\phi\mu}\right)^{\phi^{-1}} \exp\left(-\frac{y}{\phi\mu}\right). \tag{10.8}$$

Define $\eta = -1/\mu$. Then the log of the density in (10.8) is

$$\frac{y\eta + \log(-\eta)}{\phi} - \log\{y\Gamma(\phi^{-1})\} + \phi^{-1}\log\left(\frac{y}{\eta}\right),$$

which is in exponential family form with $b(\eta) = -\log(-\eta)$ and with $c(y, \phi) = -\log\{y\Gamma(\phi^{-1})\} + \phi^{-1}\log(y/\eta)$.

   The gamma model can be used when the responses have a right-skewed distribution and are heteroscedastic. However, only a special type of heteroscedasticity can be modeled by the gamma family: $\mathrm{var}(y_i|\mathbf{x}_i)$ must be proportional to $\{\mathsf{E}(y_i|\mathbf{x}_i)\}^2$. Other types of heteroscedasticity should be modeled by variance function models or transformation models; see Carroll and Ruppert (1988), which also has an extensive discussion of how to detect the presence and functional form of heteroscedascity. Ruppert, Carroll, and Cressie (1989, 1991) compare the gamma–GLM approach to modeling skewness and heteroscedasticity with the more flexible variance function–transformation approach. A brief discussion of variance function estimation in nonparametric regression is given in Chapter 14. Section 10.7 contains details of estimation in overdispersion and variance function models.

10.5 Iteratively Reweighted Least Squares

In the GLM family, we have seen that if the mean $E(y|\mathbf{x}) = \mu(\mathbf{x}^T\boldsymbol{\beta})$ then the variance is $\text{var}(y|\mathbf{x}) = \phi V(\mathbf{x}^T\boldsymbol{\beta})$ for some function V. In canonical exponential families, it can be shown that the first derivative of μ is V, that is, $\mu' = V$.

Computing parameter estimates in GLMs is particularly simple and uses a method called iteratively reweighted least squares. This method also turns out to be equivalent to *Fisher's method of scoring*, which is simply the Newton–Raphson method with the Hessian replaced by its expected value.

Suppose now that $y_1, \ldots, y_n$ denote the response variables and $\mathbf{x}_1, \ldots, \mathbf{x}_n$ are vectors of predictor variables. For the BPD example,

Iteratively reweighted least squares is the main computational engine for GLMs and also for some implementations of generalized linear mixed models.

$$y_i = \text{BPD}_i \quad \text{and} \quad \mathbf{x}_i = \begin{bmatrix} 1 \\ \text{birthweight}_i \end{bmatrix}.$$

Define

$$\mathbf{y} = \begin{bmatrix} y_1 \\ \vdots \\ y_n \end{bmatrix} \quad \text{and} \quad \mathbf{X} = \begin{bmatrix} \mathbf{x}_1^T \\ \vdots \\ \mathbf{x}_n^T \end{bmatrix}.$$

In the case of the birthweight data,

$$\mathbf{X} = \begin{bmatrix} 1 & \text{birthweight}_1 \\ \vdots & \vdots \\ 1 & \text{birthweight}_{223} \end{bmatrix}.$$

Notice that $\mathbf{X}$ is precisely the model matrix in a linear regression model.

In iteratively reweighted least squares, the basic idea is as follows. Suppose the current estimate is $\boldsymbol{\beta}^{(t)}$. Form the weights $w = 1/V(\mathbf{x}^T\boldsymbol{\beta}^{(t)})$. Then, in iteratively reweighted least squares, we update the current estimate by performing *one step* of Fisher's method of scoring for weighted least squares. When implemented, the algorithm takes the following form. Let

$$\mathbf{W}_{1,\boldsymbol{\beta}} \equiv \text{diag}\{\mu'(\mathbf{x}_i^T\boldsymbol{\beta})\},$$

$$\mathbf{W}_{2,\boldsymbol{\beta}} \equiv \text{diag}\{V(\mathbf{x}_i^T\boldsymbol{\beta})\}.$$

Then the updating step is

$$\hat{\boldsymbol{\beta}} \leftarrow \hat{\boldsymbol{\beta}} + (\mathbf{X}^T\mathbf{W}_{1,\hat{\boldsymbol{\beta}}}\mathbf{W}_{2,\hat{\boldsymbol{\beta}}}^{-1}\mathbf{W}_{1,\hat{\boldsymbol{\beta}}}\mathbf{X})^{-1}\mathbf{X}^T\mathbf{W}_{1,\hat{\boldsymbol{\beta}}}\mathbf{W}_{2,\hat{\boldsymbol{\beta}}}^{-1}\{\mathbf{y} - \mu(\mathbf{X}\hat{\boldsymbol{\beta}})\}. \tag{10.9}$$

For canonical links, $\mathbf{W}_{1,\boldsymbol{\beta}} = \mathbf{W}_{2,\boldsymbol{\beta}}$ and the algorithm takes the usual generalized least-squares form

$$\hat{\beta} \leftarrow \hat{\beta} + (\mathbf{X}^{\mathsf{T}} \mathbf{W}_{2,\hat{\beta}}^{-1} \mathbf{X})^{-1} \mathbf{X}^{\mathsf{T}} \{\mathbf{y} - \mu(\mathbf{X}\hat{\beta})\}. \tag{10.10}$$

In the logistic regression case, the algorithm takes the simple form

$$\hat{\beta} \leftarrow \hat{\beta} + \left(\mathbf{X}^{\mathsf{T}} \operatorname{diag}\left\{\frac{\exp[\mathbf{X}\hat{\beta}]}{(1 + \exp[\mathbf{X}\hat{\beta}])^2}\right\}\mathbf{X}\right)^{-1} \mathbf{X}^{\mathsf{T}}\left(\mathbf{y} - \frac{\exp[\mathbf{X}\hat{\beta}]}{1 + \exp[\mathbf{X}\hat{\beta}]}\right). \tag{10.11}$$

10.6 Hat Matrix, Degrees of Freedom, and Standard Errors

It is relatively easy to define analogues to the hat matrix and degrees of freedom as well as to obtain standard errors for the parameter estimates. The calculations necessary are outlined for the logistic case at the end of this chapter in Section 10.10.

Using those type of calculations, the hat matrix is defined as

$$\mathbf{H}_{\beta} = \mathbf{W}_{1,\beta} \mathbf{X} (\mathbf{X}^{\mathsf{T}} \mathbf{W}_{1,\beta} \mathbf{W}_{2,\beta}^{-1} \mathbf{W}_{1,\beta} \mathbf{X})^{-1} \mathbf{X}^{\mathsf{T}} \mathbf{W}_{1,\beta} \mathbf{W}_{2,\beta}^{-1}, \tag{10.12}$$

These quantities are the natural generalizations from ordinary linear regression to GLMs.

and the degrees of freedom of the fit is $\operatorname{tr}(\mathbf{H}_{\beta})$, which equals

$$\operatorname{tr}\{(\mathbf{X}^{\mathsf{T}} \mathbf{W}_{1,\beta} \mathbf{W}_{2,\beta}^{-1} \mathbf{W}_{1,\beta} \mathbf{X})^{-1} \mathbf{X}^{\mathsf{T}} \mathbf{W}_{1,\beta} \mathbf{W}_{2,\beta}^{-1} \mathbf{W}_{1,\beta} \mathbf{X}\} = p. \tag{10.13}$$

Moreover, the estimated variance matrix of $\hat{\beta}$ is

$$\widehat{\operatorname{Cov}(\hat{\beta})} = (\mathbf{X}^{\mathsf{T}} \mathbf{W}_{1,\hat{\beta}} \mathbf{W}_{2,\hat{\beta}}^{-1} \mathbf{W}_{1,\hat{\beta}} \mathbf{X})^{-1}. \tag{10.14}$$

Standard errors for each component of the estimate of β are formed by taking the square root of the diagonal of the matrix in (10.14).

The fit to the BPD data based on this estimation strategy is shown in Figure 10.3. One can see that the estimated probability of BPD decreases from approximately 0.80 to approximately 0.05, indicating a strong dependence of BPD on birth weight. Of course, to assess this dependence statistically we need inference procedures, and for this we use standard errors.

Application of the standard error formulas to the BPD example leads to the standard errors and t-values in Table 10.3. Note that, as expected, birth weight is a statistically significant predictor of BPD, with higher birth weights associated with lower risk of BPD.

10.7 Overdispersion and Variance Functions: Pseudolikelihood

In general regression problems, suppose that the mean is $\mu(\mathbf{x}^{\mathsf{T}}\beta)$ and the variance is $V(\mathbf{x}^{\mathsf{T}}\beta, \theta)$, where θ is an unknown parameter. For example, consider Poisson data. We discussed in Section 10.4.1 a model for count data in which

$$\mu(\mathbf{x}^{\mathsf{T}}\beta) = \exp(\mathbf{x}^{\mathsf{T}}\beta),$$

$$V(\mathbf{x}^{\mathsf{T}}\beta, \theta) = \mu(\mathbf{x}^{\mathsf{T}}\beta) + \theta\mu^2(\mathbf{x}^{\mathsf{T}}\beta).$$

Other common models include the power of the mean model, so that with $\theta = (\theta_0, \theta_1)$ we have

Figure 10.3 Plot of the occurrence of the coded bronchopulmonary dysplasia data against birthweight for 223 babies.

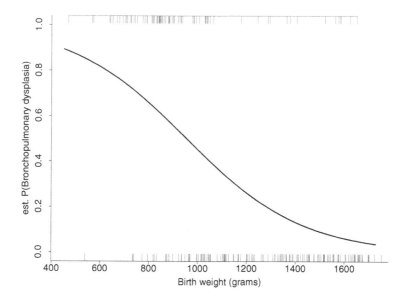

Table 10.3 Results of the logistic regression analysis of the BPD data.

Coefficient	Value	St. dev.	t-Value
intercept	1.21	1.06×10^{-1}	11.5
birth.weight	-7.46×10^{-4}	8.70×10^{-5}	-8.58

$$V(\mathbf{x}^\mathsf{T}\boldsymbol{\beta}, \boldsymbol{\theta}) = \theta_0 \mu(\mathbf{x}^\mathsf{T}\boldsymbol{\beta})^{\theta_1}.$$

Pseudolikelihood estimation is the process of iteratively alternating between estimation of one set of parameters, say $\boldsymbol{\beta}$, by maximizing the likelihood in those parameters with another set of parameters, say $\boldsymbol{\theta}$, fixed; then maximization of the likelihood in $\boldsymbol{\theta}$ with $\boldsymbol{\beta}$ fixed; and so on. Pseudolikelihood should not be confused with quasilikelihood, where the likelihood function is replaced by a so-called quasilikelihood function with some (but not all) of the properties of the likelihood function. In fact, pseudolikelihood and quasilikelihood can be combined by alternating between maximization of a quasilikelihood function over $\boldsymbol{\beta}$ and over $\boldsymbol{\theta}$.

Variance function estimation and the pseudolikelihood algorithm form a general approach to the problem of nonconstant variances in regression and are applicable to nonlinear least-squares problems as well.

Carroll and Ruppert (1988) describe the pseudolikelihood method for estimating $(\boldsymbol{\beta}, \boldsymbol{\theta})$. The algorithm is given as follows. The method refers to iteratively reweighted least squares (see Section 10.5 for details). We assume a sample of size n.

(1) Set $t = 0$ and estimate $\boldsymbol{\beta}$ by (unweighted) iteratively reweighted least squares with a constant variance function. Call the estimate $\boldsymbol{\beta}^{(t)}$.
(2) Estimate $\boldsymbol{\theta}$ by maximizing in $\boldsymbol{\theta}$ only the pseudolikelihood

$$\frac{-\sum_{i=1}^{n} \log\{V(\mathbf{x}_i^\mathsf{T}\boldsymbol{\beta}^{(t)}, \boldsymbol{\theta})\} - \sum_{i=1}^{n}\{y_i - \mu(\mathbf{x}_i^\mathsf{T}\boldsymbol{\beta}^{(t)})\}^2}{V(\mathbf{x}_i^\mathsf{T}\boldsymbol{\beta}^{(t)}, \boldsymbol{\theta})}.$$

Call the estimate $\boldsymbol{\theta}^{(t)}$.

(3) Update $\boldsymbol{\beta}^{(t)}$ to $\boldsymbol{\beta}^{(t+1)}$ using iteratively reweighted least squares with the variance function $V(\mathbf{x}^{\mathsf{T}}\boldsymbol{\beta}, \boldsymbol{\theta}^{(t)})$.

(4) Set $t = t + 1$ and return to step (2).

(5) Iterate until convergence.

10.7.1 *Quasilikelihood and Overdispersion Parameters*

One common approach to quasilikelihood starts with a full parametric model but then relaxes the assumptions by specifying that the variance function is ϕ times the variance function specified by that model, where ϕ is an unknown overdispersion parameter. For example, one might specify that the variance of count data is ϕ times the mean. Quasilikelihood estimation is used in the GLIMMIX macro of SAS that is discussed in Section 10.8.3. GLIMMIX will provide an estimate of ϕ. If $\hat{\phi} > 1$, then there is an indication of overdispersion.

10.8 Generalized Linear Mixed Models

As in the case of linear models, it is sometimes useful to incorporate random effects into a generalized linear model. The resultant models are known as *generalized linear mixed models* (GLMMs). However, their fitting presents some computational challenges. This has led to a large amount of recent research aimed at overcoming these challenges.

Generalized linear mixed models include hierarchical models, longitudinal models, and cluster variation models. This area is one of the most rapidly expanding and vigorously researched fields in statistics.

Each of the generalized linear models in Section 10.4 can be extended to allow for some effects to be random. We will denote such random effects by $\mathbf{u}$, and we will assume that they are normally distributed with mean zero and a covariance matrix $\mathbf{G}_{\boldsymbol{\theta}}$, where $\mathbf{G}_{\boldsymbol{\theta}}$ is a positive definite matrix that depends on a parameter vector $\boldsymbol{\theta}$, usually called the variance component.

The most common such models are the logistic–normal mixed model

$$y_i|\mathbf{u} \sim \text{Bernoulli}\left(\frac{\exp\{(\mathbf{X}\boldsymbol{\beta} + \mathbf{Z}\mathbf{u})_i\}}{1 + \exp\{(\mathbf{X}\boldsymbol{\beta} + \mathbf{Z}\mathbf{u})_i\}}\right), \quad \mathbf{u} \sim \text{N}(\mathbf{0}, \mathbf{G}_{\boldsymbol{\theta}}),$$

and the general Poisson–normal mixed model

$$y_i|\mathbf{u} \sim \text{Poisson}[\exp\{(\mathbf{X}\boldsymbol{\beta} + \mathbf{Z}\mathbf{u})_i\}], \quad \mathbf{u} \sim \text{N}(\mathbf{0}, \mathbf{G}_{\boldsymbol{\theta}}).$$

In what follows, for purposes of explication we will assume that the dispersion parameter ϕ is known and equal to 1 – for example, logistic or Poisson regression. We will also work entirely within the context of the canonical exponential family. We can treat both the logistic–normal and Poisson–normal models with the 1-parameter exponential family notation:

$$f(\mathbf{y}|\mathbf{u}) = \exp\{\mathbf{y}^{\mathsf{T}}(\mathbf{X}\boldsymbol{\beta} + \mathbf{Z}\mathbf{u}) - \mathbf{1}^{\mathsf{T}}b(\mathbf{X}\boldsymbol{\beta} + \mathbf{Z}\mathbf{u}) + \mathbf{1}^{\mathsf{T}}c(\mathbf{y})\},$$
$$f(\mathbf{u}) = (2\pi)^{-q/2}|\mathbf{G}_{\boldsymbol{\theta}}|^{-1/2}\exp\left(-\tfrac{1}{2}\mathbf{u}^{\mathsf{T}}\mathbf{G}_{\boldsymbol{\theta}}^{-1}\mathbf{u}\right),$$

$$(10.15)$$

where q is the dimension of $\mathbf{u}$. The second equation is the probability density function of the random effects.

10.8.1 *Estimation of Model Parameters*

The parameters in the model are $(\boldsymbol{\beta}, \boldsymbol{\theta})$, and the corresponding likelihood is

$$\mathcal{L}(\boldsymbol{\beta}, \boldsymbol{\theta}) = f(\mathbf{y}; \boldsymbol{\beta}, \boldsymbol{\theta})$$

$$= \int_{\mathbb{R}^q} f(\mathbf{y}|\mathbf{u}) f(\mathbf{u}) \, d\mathbf{u}$$

$$= (2\pi)^{-q/2} |\mathbf{G}_{\boldsymbol{\theta}}|^{-1/2} \exp\{\mathbf{1}^{\mathsf{T}} c(\mathbf{y})\} J(\boldsymbol{\beta}, \boldsymbol{\theta}),$$

where

$$J(\boldsymbol{\beta}, \boldsymbol{\theta}) = \int_{\mathbb{R}^q} \exp\{\mathbf{y}^{\mathsf{T}}(\mathbf{X}\boldsymbol{\beta} + \mathbf{Z}\mathbf{u}) - \mathbf{1}^{\mathsf{T}} b(\mathbf{X}\boldsymbol{\beta} + \mathbf{Z}\mathbf{u}) - \tfrac{1}{2}\mathbf{u}^{\mathsf{T}}\mathbf{G}_{\boldsymbol{\theta}}^{-1}\mathbf{u}\} \, d\mathbf{u}. \quad (10.16)$$

Maximum likelihood estimation of $\ell(\boldsymbol{\beta}, \boldsymbol{\theta})$ is hindered by the presence of this q-dimensional integral. As in Section 4.9, if we want to use GLMMs to fit penalized splines then q is the number of knots; even for five knots, the integral becomes essentially intractable to direct calculation.

The computational issues in GLMMs are nontrivial and require special tools. There is considerable research interest in the computational methods themselves as well as in the modeling.

There has been a great deal of research, accelerating in the 1990s, on remedies to the computational problem. There are also a variety of software options (see e.g. the website ⟨multilevel.ioe.ac.uk⟩ for various links). These remedies may be divided into four distinct categories.

(1) *Laplace approximation of* (10.16) *via PQL.* Laplace's method is a classical approximation technique for handling intractable multivariate integrals. Application to (10.16) reduces the problem to one that is akin to fitting a generalized linear model (among many others, see McGilchrist and Aisbett 1991; Schall 1991; Breslow and Clayton 1993; Wolfinger and O'Connell 1993). The only difference is that the coefficients are subject to a penalty, and nowadays the name *penalized quasilikelihood* (PQL) is usually associated with the method. Interestingly, PQL is essentially the same as maximizing the joint likelihood of the observed data and random effects simultaneously (Harville and Mee 1984; Gilmour, Anderson, and Rae 1985; Schall 1991); see below for more details. This corresponds to Henderson's (1950) justification for Gaussian mixed models; see equation (4.10). Improved Laplace approximation through higher-order expansion has been investigated by Shun and McCullagh (1995), Shun (1997), and Raudenbush, Yang, and Yosef (2000).

(2) *Bias corrections to PQL.* The approximations used by PQL induce bias in the estimates. This has resulted in a stream of research (Breslow and Lin 1995; Goldstein and Rasbash 1996; Lin and Breslow 1996) that uses asymptotic arguments to devise bias corrections to the PQL estimates.

(3) *Fitting via expectation maximization.* The expectation maximization (EM) algorithm (Dempster, Laird, and Rubin 1977) can be used to fit mixed models by treating the random effects as missing. However, the E-step involves intractable integrals, so Laplace integration (Steele 1996) or Monte Carlo methods (McCulloch, 1997; Booth and Hobert 1999) need to be employed.

(4) *Bayesian fitting via Markov chain Monte Carlo.* This involves a Bayesian formulation of the generalized linear mixed model in which $(\boldsymbol{\beta}, \boldsymbol{\theta})$ is treated as

randomly distributed according to some prior distribution. The posterior distribution of $(\boldsymbol{\beta}, \boldsymbol{\theta})$ is intractable, so Markov chain Monte Carlo algorithms (see e.g. Robert and Casella 1999) are used to generate samples from this distribution and allow estimation and inference for these parameters (Zeger and Karim 1991; Clayton 1996; Diggle, Tawn, and Moyeed 1998). See Section 17.4 for a discussion of Bayesian fitting by MCMC.

Generalized estimating equations (GEE) (see Gourieroux, Monfort, and Trognon 1984; Liang and Zeger 1986) are also, in some sense, a remedy to the maximum likelihood problem conveyed by (10.16). However, this remedy is specific to the longitudinal data setting rather than the type of mixed models that arise in smoothing. Hence, we will forego outlining details of the GEE methodology.

We now discuss each of (1)–(4).

10.8.2   *Penalized Quasilikelihood (PQL)*

Penalized quasilikelihood is a relatively simple method for fitting generalized linear mixed models. The fits from PQL also serve as useful starting values for the other fitting approaches. As we will see in Section 11.2, its application to penalized spline fitting is equivalent to the penalized likelihood approach traditionally used there. In this subsection, we state the necessary formulas required to implement PQL. Derivation of these equations is given in Section 10.10.4.

Observe that, in what follows, for purposes of explication we will assume that the data come from a canonical exponential family and that the dispersion parameter ϕ is known and equal to 1. Write $\mu = b'$ and $V = b''$ as the mean and variance functions. Also write $\boldsymbol{\mu} = \mu(\mathbf{X}\boldsymbol{\beta} + \mathbf{Z}\mathbf{u}) = b'(\mathbf{X}\boldsymbol{\beta} + \mathbf{Z}\mathbf{u}) = \mathsf{E}(\mathbf{y}|\mathbf{X}, \mathbf{Z}, \mathbf{u})$ and $\mathbf{W} = \mathrm{diag}\{b''(\mathbf{X}\boldsymbol{\beta} + \mathbf{Z}\mathbf{u})\} = \mathrm{var}(\mathbf{y}|\mathbf{X}, \mathbf{Z}, \mathbf{u})$.

Recall from (10.15) that $f(\mathbf{y}|\mathbf{u})$ is the notation we use for the likelihood of the data given the random effects $\mathbf{u}$. PQL estimates of $(\boldsymbol{\beta}, \mathbf{u})$ are obtained by treating the random effects $\mathbf{u}$ as fixed parameters, but the likelihood is penalized according to the distribution of $\mathbf{u}$. Thus, for given $\boldsymbol{\theta}$, $(\boldsymbol{\beta}, \mathbf{u})$ is obtained by maximizing the penalized log-likelihood

$$\log\{f(\mathbf{y}|\mathbf{u})\} - \tfrac{1}{2}\mathbf{u}^{\mathsf{T}}\mathbf{G}_{\theta}^{-1}\mathbf{u}.$$

The notion that the likelihood is penalized leads to the name *penalized likelihood*. Penalized quasilikelihood involves the technical extension of likelihood that we have described as quasilikelihood.

Given $\boldsymbol{\theta}$, direct differentiation of the penalized likelihood leads to the score equations for $(\boldsymbol{\beta}, \mathbf{u})$:

$$\begin{bmatrix} \mathbf{X}^{\mathsf{T}}(\mathbf{y} - \boldsymbol{\mu}) \\ \mathbf{Z}^{\mathsf{T}}(\mathbf{y} - \boldsymbol{\mu}) - \mathbf{G}_{\theta}^{-1}\mathbf{u} \end{bmatrix} = \mathbf{0}. \tag{10.17}$$

The Hessian of (10.17) is independent of $\mathbf{y}$ and is given by

$$-\begin{bmatrix} \mathbf{X}^{\mathsf{T}}\mathbf{W}\mathbf{X} & \mathbf{X}^{\mathsf{T}}\mathbf{W}\mathbf{Z} \\ \mathbf{Z}^{\mathsf{T}}\mathbf{W}\mathbf{X} & \mathbf{Z}^{\mathsf{T}}\mathbf{W}\mathbf{Z} + \mathbf{G}_{\theta}^{-1} \end{bmatrix}.$$

Here, the Newton–Raphson method and Fisher's method of scoring are identical.

PQL is only an approximation to a full likelihood analysis, except in the Gaussian GLMM (i.e., an ordinary LMM), where it is exact. Sometimes the approximation works remarkably well, but in some problems (e.g. logistic regression) the variance components may not be well estimated.

There is an identical formulation of Fisher's method of scoring that leads to the PQL estimates of $\boldsymbol{\theta}$. Consider the pseudodata

$$\mathbf{y}_{\text{pseudo}} = \mathbf{X}\boldsymbol{\beta} + \mathbf{Z}\mathbf{u} + \mathbf{W}^{-1}(\mathbf{y} - \boldsymbol{\mu}) = \mathbf{X}\boldsymbol{\beta} + \mathbf{Z}\mathbf{u} + \boldsymbol{\varepsilon}_{\text{pseudo}}.$$

This is in the form of a *linear* mixed model, where (in the notation of Section 4.4) the covariance matrix of the pseudoerrors $\boldsymbol{\varepsilon}_{\text{pseudo}}$ is $\mathbf{R} = \mathbf{W}^{-1}$. Fisher's method of scoring turns out to be nothing more than iterative updating of the LMM formula (4.9) and (4.11), using the pseudodata $\mathbf{y}_{\text{pseudo}}$ as the response.

10.8.2.1   *Estimation of Variance Components via Mixed Models*

This formulation of the GLMM as an interactively updated form of the LMM led Breslow and Clayton (1993) to suggest a PQL method for estimating $\boldsymbol{\theta}$. Specifically, fixing $\boldsymbol{\beta}$ and $\mathbf{u}$ at their current values, they suggest updating $\boldsymbol{\theta}$ at each stage of the iteration by using the ML or REML estimates of Section 4.5.4 applied to the pseudodata and with $\mathbf{R} = \mathbf{W}^{-1}$.

10.8.2.2   *Estimation of Variance Components via Cross-Validation*

An alternative method for estimating the variance components appropriate for smoothing is cross-validation; see Chapter 11.

10.8.3   *GLIMMIX*

The SAS macro GLIMMIX implements a refinement of PQL due to Wolfinger and O'Connell (1993), which they call pseudolikelihood (PL). Pseudolikelihood incorporates an overdispersion parameter, ϕ. The GLIMMIX macro makes PL available on a standard statistical package, and values of $\hat{\phi}$ substantially larger than 1 are an indication of overdispersion.

10.8.4   *Bias Correction to PQL*

PQL is based on a likelihood that is only approximate; hence estimates of the variance component $\boldsymbol{\theta}$ are asymptotically biased, as are estimates of $\boldsymbol{\beta}$. This has led Breslow and Lin (1995) and Lin and Breslow (1996) to derive corrections to the PQL estimates based on small-$\boldsymbol{\theta}$ asymptotics.

10.8.5   *Fitting via Expectation Maximization*

The EM algorithm in GLMMs often requires simulation, and it is sometimes referred to as *Monte Carlo EM.*

The expectation maximization algorithm (Dempster et al. 1977) is a general-purpose method for maximum likelihood estimation in the presence of missing data (see e.g. McLachlan and Krishnan 1997). It can be used for fitting mixed models by treating the random effects as missing data. For the Gaussian mixed model it provides an alternative to BLUP/REML for estimation of the model parameters (Laird and Ware 1982). It can also be used to guide the choice of the parameters in the generalized context. For the generalized linear mixed model

$$f(\mathbf{y}|\mathbf{u}) = \exp\{\mathbf{y}^{\mathsf{T}}(\mathbf{X}\boldsymbol{\beta} + \mathbf{Z}\mathbf{u}) - \mathbf{1}^{\mathsf{T}}b(\mathbf{X}\boldsymbol{\beta} + \mathbf{Z}\mathbf{u}) + \mathbf{1}^{\mathsf{T}}c(\mathbf{y})\},$$

$$f(\mathbf{u}) = (2\pi)^{-q/2}|\mathbf{G}_{\boldsymbol{\theta}}|^{-1/2}\exp\left(-\tfrac{1}{2}\mathbf{u}^{\mathsf{T}}\mathbf{G}_{\boldsymbol{\theta}}^{-1}\mathbf{u}\right),$$

let $\boldsymbol{\psi} \equiv (\boldsymbol{\beta}, \boldsymbol{\theta})$ be the parameter vector. The EM algorithm iterates between the *E-step* (expectation) and the *M-step* (maximization) until convergence. The E-step requires computation of an expectation:

$$Q(\boldsymbol{\psi}'|\boldsymbol{\psi}) \equiv E_{\mathbf{u}|\mathbf{y};\,\boldsymbol{\psi}}\{\log f(\mathbf{y}, \mathbf{u};\,\boldsymbol{\psi}')\},$$

while the M-step involves an update of parameter estimates through maximization,

$$\boldsymbol{\psi}_{\text{new}} = \operatorname*{argmax}_{\boldsymbol{\psi}} Q(\boldsymbol{\psi}|\boldsymbol{\psi}_{\text{old}}). \tag{10.18}$$

Because (from Bayes' rule)

$$f(\mathbf{u}|\mathbf{y};\,\boldsymbol{\psi}) = \frac{f(\mathbf{y}|\mathbf{u};\,\boldsymbol{\psi})f(\mathbf{u})}{\int_{\mathbb{R}^q} f(\mathbf{y}|\mathbf{u};\,\boldsymbol{\psi})f(\mathbf{u};\,\boldsymbol{\psi})\,d\mathbf{u}}, \tag{10.19}$$

we have the representation

$$Q(\boldsymbol{\psi}'|\boldsymbol{\psi}) = \frac{\int_{\mathbb{R}^q} \log f(\mathbf{y}, \mathbf{u};\,\boldsymbol{\psi}')f(\mathbf{y}, \mathbf{u};\,\boldsymbol{\psi})\,d\mathbf{u}}{\int_{\mathbb{R}^q} f(\mathbf{y}, \mathbf{u};\,\boldsymbol{\psi})\,d\mathbf{u}}. \tag{10.20}$$

However, computation of (10.20) is at least as difficult as computation of the log-likelihood $\ell(\boldsymbol{\psi})$.

One solution is to use Laplace's approximation to handle the integrals in (10.20) (Steele 1996). Techniques designed for approximating ratios of integrals (e.g. Tierney, Kass, and Kadane 1989) are appropriate in this case. Alternatively, one can use a *Monte Carlo EM*:

$$\hat{Q}(\boldsymbol{\psi}'|\boldsymbol{\psi}) = \frac{1}{m}\sum_{i=1}^{m}\log f(\mathbf{y}, \mathbf{u}_i;\,\boldsymbol{\psi}'), \tag{10.21}$$

where $\mathbf{u}_1, \ldots, \mathbf{u}_m$ is a Monte Carlo–generated sample from $[\mathbf{u}|\mathbf{y};\,\boldsymbol{\psi}]$ (Wei and Tanner 1990). Inspection of (10.19) shows $[\mathbf{u}|\mathbf{y};\,\boldsymbol{\psi}]$ to have a complicated distribution from which sampling is difficult. One remedy is to use the Metropolis–Hastings (MH) algorithm (McCulloch 1997). Another is to replace (10.21) by

$$\frac{1}{m}\sum_{i=1}^{m}\frac{\log f(\mathbf{y}, \mathbf{u}_i^h;\,\boldsymbol{\psi}')f(\mathbf{y}|\mathbf{u}_i^h)f(\mathbf{u}_i^h)}{h(\mathbf{u}_i^h)}, \tag{10.22}$$

where h is a standard density (such as that of the multivariate t-distribution) and $\mathbf{u}_1^h, \ldots, \mathbf{u}_m^h$ is a random sample from h. This is known as *importance sampling* (Rubinstein 1981; Booth and Hobert 1999). Note that (10.22) estimates the numerator of (10.20) rather than $Q(\boldsymbol{\psi}'|\boldsymbol{\psi})$ itself. However, since the denominator does not involve $\boldsymbol{\psi}'$, the M-step (10.18) is unaffected.

10.8.6 *Bayesian Fitting via Markov Chain Monte Carlo*

Another approach to fitting a GLMM is to put priors on all parameters and to simulate from the posterior by Markov chain Monte Carlo. This, in effect, integrates out the unobserved random effects; it also imputes the values of the random effects so that we get the equivalent of their BLUPs. The parameters are estimated by their posterior means, which are approximated by their sample averages from the MCMC output. This method is introduced in Chapter 16 (see Section 16.5).

10.8.7   *Prediction of Random Effects*

For the generalized extension of semiparametric models such as those described in Chapters 8 and 9, prediction of $\mathbf{u}$ is required. For known $(\boldsymbol{\beta}, \boldsymbol{\theta})$, the best predictor of $\mathbf{u}$ is

$$\tilde{\mathbf{u}} = E_{(\boldsymbol{\beta}, \boldsymbol{\theta})}(\mathbf{u} | \mathbf{y}),$$

which suggests the predictor

$$\hat{\mathbf{u}} = E_{(\hat{\boldsymbol{\beta}}, \hat{\boldsymbol{\theta}})}(\mathbf{u} | \mathbf{y}).$$

Note that

$$\tilde{\mathbf{u}} = \frac{\int_{\mathbb{R}^q} \mathbf{u} \exp\left\{\mathbf{y}^\mathsf{T}(\mathbf{X}\boldsymbol{\beta} + \mathbf{Z}\mathbf{u}) - \mathbf{1}^\mathsf{T} b(\mathbf{X}\boldsymbol{\beta} + \mathbf{Z}\mathbf{u}) - \frac{1}{2}\mathbf{u}^\mathsf{T} \mathbf{G}_\theta^{-1}\mathbf{u}\right\} d\mathbf{u}}{\int_{\mathbb{R}^q} \exp\left\{\mathbf{y}^\mathsf{T}(\mathbf{X}\boldsymbol{\beta} + \mathbf{Z}\mathbf{u}) - \mathbf{1}^\mathsf{T} b(\mathbf{X}\boldsymbol{\beta} + \mathbf{Z}\mathbf{u}) - \frac{1}{2}\mathbf{u}^\mathsf{T} \mathbf{G}_\theta^{-1}\mathbf{u}\right\} d\mathbf{u}},$$

so computation is hindered by the presence of higher-dimensional integrals.

Of course, PQL directly estimates $\mathbf{u}$. Methods that provide corrected estimates of $\boldsymbol{\theta}$ can then provide an estimate of $\mathbf{u}$ through the solving of (10.17). Monte Carlo EM produces an estimate of $\boldsymbol{\beta}$, and since one generates a sample of the $\mathbf{u}$s from the distribution of $\mathbf{u}$ given $\mathbf{y}$, the mean of these samples provides an estimate of $\mathbf{u}$. Similarly, in the Bayesian formulation, MCMC also provides a sample of the $\mathbf{u}$s, and the mean of this sample yields an estimate of $\mathbf{u}$.

10.8.8   *Standard Error Estimation*

Standard error estimates for $(\boldsymbol{\beta}, \mathbf{u})$ can be constructed using either the EM algorithm or MCMC. For the EM algorithm, consult Louis (1982). Bayesian methods yield standard error estimates and posterior confidence intervals as part of the MCMC calculations. Both methods account for the estimation of the variance component $\boldsymbol{\theta}$. On the other hand, known standard error estimates for PQL do not account for estimation of $\boldsymbol{\theta}$. However, using the identification of PQL as an iterative updating of the linear mixed model, we have a result similar to (4.17) for the LMM:

$$\mathrm{Cov}\left(\begin{bmatrix} \hat{\boldsymbol{\beta}} \\ \hat{\mathbf{u}} \end{bmatrix} \bigg| \mathbf{u}\right) \simeq (\mathbf{C}^\mathsf{T}\mathbf{W}\mathbf{C} + \mathbf{B})^{-1}\mathbf{C}^\mathsf{T}\mathbf{W}\mathbf{C}(\mathbf{C}^\mathsf{T}\mathbf{W}\mathbf{C} + \mathbf{B})^{-1}, \qquad (10.23)$$

where $\mathbf{C} = [\mathbf{X} \ \mathbf{Z}]$,

$$\mathbf{W} = \mathrm{diag}\{b''(\mathbf{X}\boldsymbol{\beta} + \mathbf{Z}\mathbf{u})\} = \mathrm{var}(\mathbf{y} | \mathbf{X}, \mathbf{Z}, \mathbf{u}), \qquad (10.24)$$

and

$$\mathbf{B} = \begin{bmatrix} \mathbf{0} & \mathbf{0} \\ \mathbf{0} & \mathbf{G}_\theta^{-1} \end{bmatrix}.$$

Evaluation of (10.23) allows for standard error bars for the fitted function and its linear part. Suppose that interest focuses on estimating at a given $(\mathbf{x}, \mathbf{z})$, so that the linear predictor is $\mathbf{x}^\mathsf{T}\hat{\boldsymbol{\beta}} + \mathbf{z}^\mathsf{T}\hat{\mathbf{u}}$. The estimated standard error of this prediction is

$$\sqrt{\begin{bmatrix} \mathbf{x} \\ \mathbf{z} \end{bmatrix}^\mathsf{T} \mathrm{Cov}\left(\begin{bmatrix} \hat{\boldsymbol{\beta}} \\ \hat{\mathbf{u}} \end{bmatrix} \Big| \mathbf{u}\right) \begin{bmatrix} \mathbf{x} \\ \mathbf{z} \end{bmatrix}},$$

These formulas for standard errors are useful for smoothing and semiparametric modeling.

where the $\boldsymbol{\beta}$ and $\mathbf{u}$ appearing in $\mathbf{W}$ are replaced by their estimates. Often more interesting is a standard error estimate for the estimated mean $\mu(\mathbf{x}^\mathsf{T}\hat{\boldsymbol{\beta}} + \mathbf{z}^\mathsf{T}\hat{\mathbf{u}})$. This estimate is

$$\sqrt{\{\mu'(\mathbf{x}^\mathsf{T}\hat{\boldsymbol{\beta}} + \mathbf{z}^\mathsf{T}\hat{\mathbf{u}})\}^2 \begin{bmatrix} \mathbf{x} \\ \mathbf{z} \end{bmatrix}^\mathsf{T} \mathrm{Cov}\left(\begin{bmatrix} \hat{\boldsymbol{\beta}} \\ \hat{\mathbf{u}} \end{bmatrix} \Big| \mathbf{u}\right) \begin{bmatrix} \mathbf{x} \\ \mathbf{z} \end{bmatrix}}.$$

Since $\mu' = b'' = V$, there are several ways to re-express this standard error.

10.8.9   *Bias Adjustment*

In Section 6.4 we argued in favor of confidence bands based on the unconditional covariance matrix

$$\mathrm{Cov}\left(\begin{bmatrix} \hat{\boldsymbol{\beta}} \\ \hat{\mathbf{u}} - \mathbf{u} \end{bmatrix}\right)$$

because they adjust for the additional uncertainty in the fitted curve due to bias, whereas confidence regions based on the conditional covariance

$$\mathrm{Cov}\left(\begin{bmatrix} \hat{\boldsymbol{\beta}} \\ \hat{\mathbf{u}} \end{bmatrix} \Big| \mathbf{u}\right)$$

make no such adjustment. Such confidence bands have not yet been studied for generalized regression, but we believe this would be a useful area for further inquiry.

10.9   Deviance

In Chapter 2 we saw that sums of squared deviations of y_i from its fitted value $\hat{y}_i$ played an important role in inference for Gaussian regression models. In a GLM, *deviance* is the analogue of the residual sum of squares. The deviance of a model compares the fit for that model with the fit for the so-called *saturated* model, where there is a separate parameter for each observation. More specifically, the deviance of any model is twice the difference in log-likelihoods between the saturated model and the given model.

Let $\hat{y}_i$ be the fitted value for a given model. This means that $\hat{y}_i$ is the expected value of y_i, given the covariates for the ith case, evaluated at $\hat{\boldsymbol{\beta}}$. We will assume that $\hat{y}_i = y_i$ for the saturated model, which is true for any of the GLMs we are considering. For logistic regression, the deviance is

$$D(\mathbf{y}; \hat{\mathbf{y}}) = 2\sum_{i=1}^{n}\left\{y_i \log\left(\frac{y_i}{\hat{y}_i}\right) + (1 - y_i)\log\left(\frac{1 - y_i}{1 - \hat{y}_i}\right)\right\};$$

for Poisson regression, the deviance is

$$D(\mathbf{y}; \hat{\mathbf{y}}) = 2 \sum_{i=1}^{n} \left\{ y_i \log\left(\frac{y_i}{\hat{y}_i}\right) - (y_i - \hat{y}_i) \right\},$$

where $0 \log(0) = 0$; see McCullagh and Nelder (1989), who give the deviance for other GLMs as well.

For Gaussian linear models, the deviance is just the residual sum of squares. As discussed in detail in McCullagh and Nelder (1989), the analysis of variance for Gaussian linear models can be generalized to the analysis of deviance for GLMs.

10.10   Technical Details

In this section we collect some technical details that may be of use to those who want to understand the methods of this chapter algebraically. Those readers whose interests are mainly in applications and an intuitive understanding of the theory should skip this section.

10.10.1   *Fitting a Logistic Regression*

The general logistic regression model can be written as

$$\text{logit}\{P(y_i = 1|\mathbf{x}_i)\} = \boldsymbol{\beta}^{\mathsf{T}}\mathbf{x}_i, \quad i = 1, \ldots, n. \tag{10.25}$$

The log-likelihood for this problem is

$$\ell(\boldsymbol{\beta}) = \sum_{i=1}^{n} \{y_i(\boldsymbol{\beta}^{\mathsf{T}}\mathbf{x}_i) - \log(1 + \exp[\boldsymbol{\beta}^{\mathsf{T}}\mathbf{x}_i])\}$$

$$= \mathbf{y}^{\mathsf{T}}\mathbf{X}\boldsymbol{\beta} - \mathbf{1}^{\mathsf{T}} \log(\mathbf{1} + \exp[\mathbf{X}\boldsymbol{\beta}]). \tag{10.26}$$

Differentiation with respect to $\boldsymbol{\beta}$ leads to the *score equations*

$$\mathbf{S}(\boldsymbol{\beta}) \equiv \mathbf{X}^{\mathsf{T}}\left(\mathbf{y} - \frac{\exp[\mathbf{X}\boldsymbol{\beta}]}{1 + \exp[\mathbf{X}\boldsymbol{\beta}]}\right) = \mathbf{0}.$$

The prototype of tools for solving a vector equation of the form

$$\mathbf{S}(\boldsymbol{\beta}) = \mathbf{0} \tag{10.27}$$

is the *Newton–Raphson* technique. It involves the updating step

$$\hat{\boldsymbol{\beta}} \leftarrow \hat{\boldsymbol{\beta}} - \{\mathbf{DS}(\hat{\boldsymbol{\beta}})\}^{-1}\mathbf{S}(\hat{\boldsymbol{\beta}}), \tag{10.28}$$

where $\mathbf{DS}(\boldsymbol{\beta})$ is called the Hessian: the square matrix with (i, j) entry equal to

$$\frac{\partial}{\partial \beta_j} \mathbf{S}(\boldsymbol{\beta})_i.$$

Provided $\mathbf{S}(\hat{\boldsymbol{\beta}})$ is well-behaved, the iteration defined by (10.28) leads to rapid convergence to the solution of (10.27)

Fisher's method of scoring uses the same basic algorithm as in (10.28), except that if the Hessian depends on $\mathbf{y}$ then it is replaced by its expected value. We call this the *scoring Hessian*.

For logistic regression, the Hessian and the scoring Hessian are the same and equal

$$DS(\boldsymbol{\beta}) = D\mathbf{X}^{\mathsf{T}}\left(\mathbf{y} - \frac{\exp[\mathbf{X}\boldsymbol{\beta}]}{1 + \exp[\mathbf{X}\boldsymbol{\beta}]}\right)$$

$$= -\mathbf{X}^{\mathsf{T}}\operatorname{diag}\left(\frac{\exp[\mathbf{X}\boldsymbol{\beta}]}{(1 + \exp[\mathbf{X}\boldsymbol{\beta}])^2}\right)\mathbf{X} = \mathbf{X}^{\mathsf{T}}\mathbf{W}_{\boldsymbol{\beta}}\mathbf{X},$$

so the updating step is as given in (10.11). Convergence is usually very rapid, with 5–10 iterations being sufficient in most circumstances.

In the logistic regression model, and more generally for canonical exponential models, the Newton–Raphson algorithm is identical to Fisher's method of scoring and to iteratively reweighted least squares (see Section 10.5). For other GLMs, the Newton–Raphson differs from the other two algorithms and is generally not used.

10.10.2   *Standard Error Estimation in Logistic Regression*

There are two ways to make inference about the regression parameter $\boldsymbol{\beta}$: (i) likelihood ratio tests and confidence intervals; and (ii) using standard t-test/interval methods after having obtained standard error estimates. In this subsection, we discuss how to obtain standard error estimates for logistic regression.

The maximum likelihood estimate of $\boldsymbol{\beta}$ satisfies

$$S(\hat{\boldsymbol{\beta}}) = \mathbf{0}.$$

Thus we can make the informal Taylor's theorem argument

$$\begin{aligned}\mathbf{0} &= S(\hat{\boldsymbol{\beta}}) \\ &= S(\boldsymbol{\beta} + \hat{\boldsymbol{\beta}} - \boldsymbol{\beta}) \\ &\simeq S(\boldsymbol{\beta}) + DS(\boldsymbol{\beta})(\hat{\boldsymbol{\beta}} - \boldsymbol{\beta}).\end{aligned}$$

Rearranging yields

$$\hat{\boldsymbol{\beta}} - \boldsymbol{\beta} \simeq -\{DS(\boldsymbol{\beta})\}^{-1}\mathbf{X}^{\mathsf{T}}\left(\mathbf{y} - \frac{\exp[\mathbf{X}\boldsymbol{\beta}]}{1 + \exp[\mathbf{X}\boldsymbol{\beta}]}\right).$$

This means that

$$E(\hat{\boldsymbol{\beta}} - \boldsymbol{\beta}) \simeq \mathbf{0} \quad \text{and} \quad \operatorname{Cov}(\hat{\boldsymbol{\beta}} - \boldsymbol{\beta}) \simeq (\mathbf{X}^{\mathsf{T}}\mathbf{W}_{\boldsymbol{\beta}}\mathbf{X})^{-1}.$$

From this it follows that

$$\widehat{\text{st.dev.}}(\hat{\beta}_i) = \sqrt{i\text{th diagonal entry of } (\mathbf{X}^{\mathsf{T}}\mathbf{W}_{\hat{\boldsymbol{\beta}}}\mathbf{X})^{-1}}.$$

10.10.3   *The Hat Matrix and Degrees of Freedom*

For linear regression, we defined the hat matrix $\hat{\mathbf{y}}$ in Section 2.3, noting that multiplying it by the response $\mathbf{y}$ led to the predicted values. There is an analogue of the hat matrix for GLMs, one that reflects both leverage and degrees of freedom.

In a generalized linear model, $\hat{\mathbf{y}}$ is a nonlinear function of $\mathbf{y}$; there is no matrix $\mathbf{H}$ such that $\hat{\mathbf{y}} = \mathbf{H}\mathbf{y}$ and hence the usual definition of a hat matrix does not apply. However, we can define a hat matrix by a linearization using the following analogy with a linear model. For a linear model,

$$\hat{\mathbf{y}} - \mathsf{E}(\mathbf{y}) = \mathbf{H}\{\mathbf{y} - \mathsf{E}(\mathbf{y})\} = \mathbf{H}\boldsymbol{\varepsilon};$$

the left-hand side of this equation is the error in estimating $\mathsf{E}(\mathbf{y})$ by $\hat{\mathbf{y}}$. We will define the hat matrix $\mathbf{H}_\beta$ to be the matrix such that, with μ the inverse link function,

$$\mu(\mathbf{X}\hat{\boldsymbol{\beta}}) - \mu(\mathbf{X}\boldsymbol{\beta}) \simeq \mathbf{H}_\beta\{\mathbf{y} - \mathsf{E}(\mathbf{y})\}.$$

It is a general fact that

$$\hat{\boldsymbol{\beta}} - \boldsymbol{\beta} \approx (\mathbf{X}^\mathsf{T}\mathbf{W}_\beta\mathbf{X})^{-1}\mathbf{X}^\mathsf{T}\{\mathbf{y} - \mathsf{E}(\mathbf{y})\}.$$

Then, by a Taylor approximation and using $\mathbf{W}_\beta = \mathrm{diag}\{\mathcal{L}'(\mathbf{X}\boldsymbol{\beta})\}$, we have

$$\mu(\mathbf{X}\hat{\boldsymbol{\beta}}) - \mu(\mathbf{X}\boldsymbol{\beta}) \approx \mathbf{W}_\beta\mathbf{X}(\mathbf{X}^\mathsf{T}\mathbf{W}_\beta\mathbf{X})^{-1}\mathbf{X}^\mathsf{T}\{\mathbf{y} - \mathsf{E}(\mathbf{y})\}.$$

Thus, an appropriate definition for the hat matrix is

$$\mathbf{H}_\beta \equiv \mathbf{W}_\beta\mathbf{X}(\mathbf{X}^\mathsf{T}\mathbf{W}_\beta\mathbf{X})^{-1}\mathbf{X}^\mathsf{T}.$$

Notice that

$$\mathrm{tr}(\mathbf{H}_\beta) = \mathrm{tr}\{(\mathbf{X}^\mathsf{T}\mathbf{W}_\beta\mathbf{X})^{-1}\mathbf{X}^\mathsf{T}\mathbf{W}_\beta\mathbf{X}\} = \mathrm{tr}(\mathbf{I}_p) = p,$$

where

$$p = \text{number of parameters in model}.$$

This argument shows that

$$df_{\mathrm{fit}} \equiv \mathrm{tr}(\mathbf{H}_{\hat{\beta}})$$

is a reasonable definition for degrees of freedom in generalized parametric regression models. Later we will see that the same definition can be used to quantify the effective number of parameters in generalized semiparametric models.

10.10.4   *Derivation of PQL*

Here we show how PQL can be derived as an approximation to the maximum likelihood solution. We hold $\boldsymbol{\theta}$ fixed and, as before, consider only the canonical model with $\phi = 1$ known.

Write

$$J(\boldsymbol{\beta}, \boldsymbol{\theta}) = \int_{\mathbb{R}^q} \exp\{h(\mathbf{u})\}\, d\mathbf{u}, \tag{10.29}$$

where

$$h(\mathbf{u}) = \mathbf{y}^\mathsf{T}(\mathbf{X}\boldsymbol{\beta} + \mathbf{Z}\mathbf{u}) - \mathbf{1}^\mathsf{T}b(\mathbf{X}\boldsymbol{\beta} + \mathbf{Z}\mathbf{u}) - \tfrac{1}{2}\mathbf{u}^\mathsf{T}\mathbf{G}_\theta^{-1}\mathbf{u}. \tag{10.30}$$

Recall that D refers to first-order differentiation whereas H refers to second-order differentiation. Laplace approximation of $J(\boldsymbol{\beta}, \boldsymbol{\theta})$ starts with the Taylor series approximation

$$h(\mathbf{u}) \simeq h(\mathbf{u}^0) + \mathsf{D}h(\mathbf{u}^0)(\mathbf{u} - \mathbf{u}^0) + \tfrac{1}{2}(\mathbf{u} - \mathbf{u}^0)^{\mathsf{T}} \mathsf{H}h(\mathbf{u}^0)(\mathbf{u} - \mathbf{u}^0).$$

Choose $\mathbf{u}^0$ to solve

$$\mathsf{D}h(\mathbf{u}^0) = \mathbf{0}.$$

This leads to the approximation

$$h(\mathbf{u}) \simeq h(\mathbf{u}^0) + \tfrac{1}{2}(\mathbf{u} - \mathbf{u}^0)^{\mathsf{T}} \mathsf{H}h(\mathbf{u}^0)(\mathbf{u} - \mathbf{u}^0).$$

Using the expression for the density of a $N(\boldsymbol{\mu}, \boldsymbol{\Sigma})$ random vector, we have

$$\int_{\mathbb{R}^q} (2\pi)^{-q/2} |\boldsymbol{\Sigma}|^{-1/2} \exp\{-\tfrac{1}{2}(\mathbf{x} - \boldsymbol{\mu})^{\mathsf{T}} \boldsymbol{\Sigma}^{-1}(\mathbf{x} - \boldsymbol{\mu})\} \, d\mathbf{x} = 1.$$

Combining this result with (10.29) and (10.30) results in the approximation

$$J(\boldsymbol{\beta}, \boldsymbol{\theta}) \simeq (2\pi)^{q/2} |-\mathsf{H}h(\mathbf{u}^0)|^{-1/2} \exp\{h(\mathbf{u}^0)\}. \tag{10.31}$$

Vector differential calculus (see Appendix A) shows that

$$\mathsf{D}h(\mathbf{u}) = \{\mathbf{y} - b'(\mathbf{X}\boldsymbol{\beta} + \mathbf{Z}\mathbf{u})\}^{\mathsf{T}}\mathbf{Z} - \mathbf{u}^{\mathsf{T}}\mathbf{G}_{\theta}^{-1},$$
$$\mathsf{H}h(\mathbf{u}) = -\mathbf{Z}^{\mathsf{T}} \operatorname{diag}\{b''(\mathbf{X}\boldsymbol{\beta} + \mathbf{Z}\mathbf{u})\}\mathbf{Z} - \mathbf{G}_{\theta}^{-1}.$$

The resulting loglikelihood approximation is then

$$\ell(\boldsymbol{\beta}, \boldsymbol{\theta}) \simeq \mathbf{y}^{\mathsf{T}}(\mathbf{X}\boldsymbol{\beta} + \mathbf{Z}\mathbf{u}^0) - \mathbf{1}^{\mathsf{T}} b(\mathbf{X}\boldsymbol{\beta} + \mathbf{Z}\mathbf{u}^0) + \mathbf{1}^{\mathsf{T}} c(\mathbf{y}) - \tfrac{1}{2}\mathbf{u}^{0\mathsf{T}}\mathbf{G}_{\theta}^{-1}\mathbf{u}^0$$
$$- \tfrac{1}{2} \log|\mathbf{I} + \mathbf{G}_{\theta}\mathbf{Z}^{\mathsf{T}} \operatorname{diag}\{b''(\mathbf{X}\boldsymbol{\beta} + \mathbf{Z}\mathbf{u}^0)\}\mathbf{Z}|.$$

However, for ease of fitting, PQL uses one final approximation that is based on the assumption that $b''(\mathbf{X}\boldsymbol{\beta} + \mathbf{Z}\mathbf{u}^0)$ is relatively constant as a function of $\boldsymbol{\beta}$. For the purpose of maximizing $\ell(\boldsymbol{\beta}, \boldsymbol{\theta})$ with respect to $\boldsymbol{\beta}$, this gives some justification for its omission from the log-likelihood to yield

$$\ell(\boldsymbol{\beta}, \boldsymbol{\theta}) \simeq \mathbf{y}^{\mathsf{T}}(\mathbf{X}\boldsymbol{\beta} + \mathbf{Z}\mathbf{u}^0) - \mathbf{1}^{\mathsf{T}} b(\mathbf{X}\boldsymbol{\beta} + \mathbf{Z}\mathbf{u}^0) + \mathbf{1}^{\mathsf{T}} c(\mathbf{y}) - \tfrac{1}{2}\mathbf{u}^{0\mathsf{T}}\mathbf{G}_{\theta}^{-1}\mathbf{u}^0.$$

Maximizing this expression leads to the solutions described in Section 10.8.2.

10.11   Bibliographical Notes

Chapter 9 of McCullagh and Nelder (1989) provides a thorough overview of quasi-likelihood estimation. McCulloch and Searle (2001) is an excellent introduction to generalized linear and mixed linear models. Other good sources of information on generalized linear models include Aitkin et al. (1989), Green and Silverman (1994), Lindsey (1997), Fahrmeir and Tutz (2001), Gill (2001), Hardin and Hilbe (2001), Myers, Montgomery, and Vining (2001), and Dobson (2002).

11

Generalized Additive Models

11.1 Introduction

The generalized parametric models of Chapter 10 are nonlinear because of the link function, but nonetheless they are parametric and have none of the flexibility of nonparametric models.

The wage data were taken from the Internet at the Statlib site at Carnegie Mellon University. The URL of Statlib is ⟨lib.stat.cmu.edu/⟩.

As an example of this lack of flexibility, consider data on wages (denoted as wages) and union membership (union). These data are from 1985 and appear in Berndt (1991). The variable union is binary and wages is continuous. To understand the relationship between these variables, one can examine the binary regression of union on wages. In Figure 11.1 we have the fits to these data using linear, quadratic, and cubic logistic regression. There is strong evidence from the quadratic and cubic fits that the linear logistic model is inadequate. In fact, the quadratic and cubic coefficients are both significant in the cubic logistic fit. Seeing how dramatically the fits change as the degree is increased, one could be suspicious – and rightly so – as to whether even the cubic fit is adequate.

Figure 11.1
Polynomial-logistic fits to the union and wages scatterplot. Raw data are plotted as short vertical bars but with values of 1 for union replaced by 0.5 for graphical purposes. A worker making $44.50/hour was used in the fitting but not shown on the plot to increase the detail.

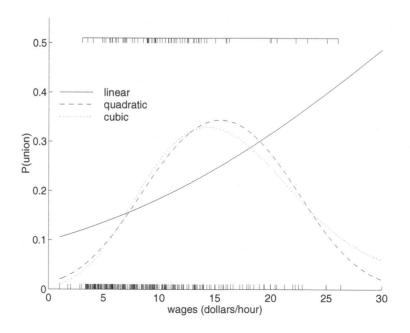

The solution to this problem of inflexibility is not hard to guess: one replaces the linear predictor in the generalized linear model by a spline. The resulting model is called a generalized nonparametric regression model.

The nonmonotonic behavior seen in the quadratic and cubic fits is interesting. The probability of union membership increases as wages increase, but only up to a point. Once wages are above approximately $15/hour, the probability of union membership decreases with increasing wages. An obvious question at this point is whether the nonmonotonic pattern will persist when we switch to a nonparametric model. We will see that, in fact, it does.

In Section 11.2 we look at the simplest case of generalized nonparametric regression, a binary response and a univariate predictor variable having a smooth but otherwise unknown effect on the response. In Section 11.3 this simple model is extended in two ways: the response distribution can be in an arbitrary exponential family, and there may be several predictors. In this case, we replace the linear predictor of the generalized linear model by an additive model predictor. Such models are called generalized additive models (GAMs). In a GAM, some variables may enter the the additive predictor linearly but the effects of others are modeled as splines.

The remaining sections cover estimation of standard errors, approximating degrees of freedom, and connections with mixed models.

11.2   Generalized Scatterplot Smoothing

Suppose that we observed pairs (x_i, y_i), where the conditional distribution of y_i given x_i is in an exponential family with density (10.6). For example, y_i might be binary. In Chapter 10 we saw that we could estimate $f(x) = \mathsf{E}(y|x)$ under a set of parametric assumptions called the generalized linear model (GLM). In this chapter we estimate f assuming only that f is a smooth function. In other words, the "linear" part of the GLM assumptions will be relaxed but the remaining structure of the GLM will be retained. Thus, we assume that y_i given x_i has density

$$f(y_i; \eta_i) = \exp\left(\frac{y\eta_i - b(\eta_i)}{\phi} + c(y, \phi)\right),$$

where η_i depends on x_i. More precisely, we will assume that $\eta_i = \eta(x_i)$ for a smooth function $\eta(\cdot)$, and we will use the notation

$$\eta = [\eta(x_1), \ldots, \eta(x_n)]^{\mathsf{T}}.$$

For simplicity, throughout this chapter we will assume a canonical link function. Assume also that $\phi \equiv 1$ – for example, binary or Poisson regression.

The most common method for smoothing such data is penalized likelihood, which is a generalization to non-Gaussian data of smoothing splines that minimize the penalized sum of squares (3.16). For example, the smoothing spline solution is

$$\hat{\mathbf{f}} = (b')^{-1}(\hat{\boldsymbol{\eta}}),$$

where

$$\hat{\boldsymbol{\eta}} = \underset{\eta(\cdot)}{\operatorname{argmax}} \{\mathbf{y}^{\mathsf{T}}\boldsymbol{\eta} - \mathbf{1}^{\mathsf{T}}b(\boldsymbol{\eta})\} - \frac{1}{2}\lambda^3 \int_{-\infty}^{\infty} \eta''(x)^2 \, dx. \tag{11.1}$$

For any space of splines, there is a matrix $\mathbf{K}$ such that $\boldsymbol{\eta}^{\mathsf{T}}\mathbf{K}\boldsymbol{\eta}$ equals $\int_{-\infty}^{\infty} \eta''(x)^2 \, dx$ for all $\eta(\cdot)$ in this space. Thus, we can find such a $\mathbf{K}$ and substitute $\boldsymbol{\eta}^{\mathsf{T}}\mathbf{K}\boldsymbol{\eta}$ for $\int_{-\infty}^{\infty} \eta''(x)^2 \, dx$ in (11.1). Note that finiteness of $\int_{-\infty}^{\infty} \eta''(x)^2 \, dx$ requires splines of quadratic degree or higher.

The penalized linear spline solution to the same problem, for the basis functions

$$\mathbf{X} = [1 \;\; x_i], \qquad \mathbf{Z} = \left[(x_i - \kappa_k)_+\right]_{\substack{1 \le k \le K \\ 1 \le i \le n}},$$

is

$$\hat{\boldsymbol{\eta}} = \mathbf{X}\hat{\boldsymbol{\beta}} + \mathbf{Z}\hat{\mathbf{u}},$$

where

$$\begin{bmatrix} \hat{\boldsymbol{\beta}} \\ \hat{\mathbf{u}} \end{bmatrix} = \underset{\boldsymbol{\beta},\mathbf{u}}{\operatorname{argmax}} \{\mathbf{y}^{\mathsf{T}}(\mathbf{X}\boldsymbol{\beta} + \mathbf{Z}\mathbf{u}) - \mathbf{1}^{\mathsf{T}}b(\mathbf{X}\boldsymbol{\beta} + \mathbf{Z}\mathbf{u})\} - \frac{1}{2}\lambda^3 \|\mathbf{u}\|^2. \tag{11.2}$$

Recall that we are assuming a canonical link; for noncanonical links, (11.2) would need to be modified appropriately. For a fixed value of λ, (11.2) is equivalent to PQL estimation in the generalized linear mixed model

$$f(\mathbf{y}|\mathbf{u}) = \exp\{\mathbf{y}^{\mathsf{T}}(\mathbf{X}\boldsymbol{\beta} + \mathbf{Z}\mathbf{u}) - \mathbf{1}^{\mathsf{T}}b(\mathbf{X}\boldsymbol{\beta} + \mathbf{Z}\mathbf{u}) + \mathbf{1}^{\mathsf{T}}c(\mathbf{y})\},$$

$$\mathbf{u} \sim N(\mathbf{0}, \sigma_u^2\mathbf{I}),$$

with $\sigma_u^2 = (1/\lambda^3)$.

This generalized linear mixed model representation comes equipped with a method for selecting the amount of smoothing via the PQL and REML algorithm given in Section 10.8.2. See Section 11.5.

11.2.1 *Application to Wage–Union Data*

Figure 11.2 shows a logistic spline fit to the `union` and `wages` scatterplot using a 20-knot linear spline. The value of λ was chosen by an extension of generalized cross-validation; Section 11.5 provides the details.

The nonmonotonic behavior seen in the previous figure is seen here as well. However, the actual shape of the fit here is different from the shape of the parametric fits, since the shape is now determined by the data rather than by the functional form of a parametric model.

Regression analysis by itself cannot be used to infer causality. Therefore, we do not know why the probability of union membership increases with increasing wages when wages are below \$15/hour. One cannot determine from the regression analysis whether union membership causes wages to increase, whether increasing wages causes a worker to consider union membership, or whether (as is likely)

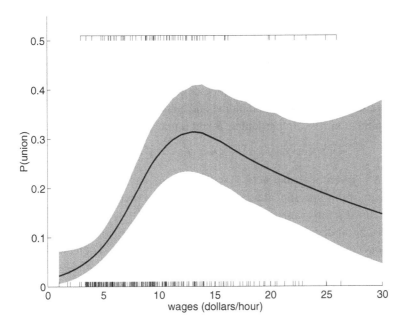

Figure 11.2 Logistic spline fit to the union and wages scatterplot (solid) with 95% confidence bands (shaded). Raw data are plotted as short vertical bars but with values of 1 for union replaced by 0.5 for graphical purposes. A worker making $44.50/hour was used in the fitting but not shown on the plot to increase the detail.

something more complex is happening such as both wages and union are being influenced by other variables. Similarly, we do not know why the probability of union membership decreases for wages above $15/hour. Regression analysis can only raise these interesting questions.

11.3 Generalized Additive Mixed Models

As shown in Section 8.2, the additive model can be represented as a mixed model with a variance component controlling the amount of smoothing for each additive component. For example, the normal additive model

$$y_i \overset{\text{ind.}}{\sim} N(\beta_0 + f(s_i) + g(t_i), \sigma^2), \tag{11.3}$$

where f and g are linear splines with knots at $\{\kappa_k^s\}_{k=1}^{K^s}$ and $\{\kappa_k^t\}_{k=1}^{K^t}$ (respectively), can be fit through the normal mixed model

$$\mathbf{y}|\mathbf{u} \sim N(\mathbf{X}\boldsymbol{\beta} + \mathbf{Z}\mathbf{u}, \sigma_\varepsilon^2 \mathbf{I}) \tag{11.4}$$

with

$$\mathbf{X} = [1 \ s_i \ t_i]_{1 \le i \le n}, \qquad \mathbf{Z} = [\mathbf{Z}_s | \mathbf{Z}_t],$$

$$\mathbf{Z}_s = \left[(s_i - \kappa_k^s)_+ \right]_{\substack{1 \le k \le K^s \\ 1 \le i \le n}}, \qquad \mathbf{Z}_t = \left[(t_i - \kappa_k^t)_+ \right]_{\substack{1 \le k \le K^t \\ 1 \le i \le n}}, \tag{11.5}$$

where

$$\mathbf{u} \sim N\left(\mathbf{0}, \begin{bmatrix} \sigma_s^2 \mathbf{I} & \mathbf{0} \\ \mathbf{0} & \sigma_t^2 \mathbf{I} \end{bmatrix} \right).$$

The extension to the generalized case can be achieved by replacement of (11.4) by

$$\log\{f(\mathbf{y}|\mathbf{u})\} = \mathbf{y}^{\mathsf{T}}(\mathbf{X}\boldsymbol{\beta} + \mathbf{Z}\mathbf{u}) - \mathbf{1}^{\mathsf{T}}b(\mathbf{X}\boldsymbol{\beta} + \mathbf{Z}\mathbf{u}) + \mathbf{1}^{\mathsf{T}}c(\mathbf{y}), \tag{11.6}$$

where (as usual in this chapter) a canonical link and $\phi \equiv 1$ are assumed. This is now a generalized linear mixed model and can be fit using any of the techniques given in Section 10.8. In particular, $\hat{\boldsymbol{\beta}}$ and $\hat{\mathbf{u}}$ maximize

$$\mathbf{y}^{\mathsf{T}}(\mathbf{X}\boldsymbol{\beta} + \mathbf{Z}\mathbf{u}) - \mathbf{1}^{\mathsf{T}}b(\mathbf{X}\boldsymbol{\beta} + \mathbf{Z}\mathbf{u}) + \mathbf{1}^{\mathsf{T}}c(\mathbf{y}) - \tfrac{1}{2}\mathbf{u}^{\mathsf{T}}\boldsymbol{\Lambda}\mathbf{u},$$

where $\boldsymbol{\Lambda}$ is given by equation (8.5) with $\sigma_\varepsilon^2 = \phi = 1$.

Instead of this specific additive model with two predictor variables, consider a generalized semiparametric additive model with $D_1 + D_2$ predictor variables. The first D_1 predictors are assumed to enter the model linearly, while the last D_2 predictors enter nonparametrically as pth-degree splines. Then the first D_1 predictors form columns of the matrix $\mathbf{X}$ representing the fixed effects. For each of the last D_2 predictors, the powers of degree 1 through p are columns of $\mathbf{X}$ while the truncated power functions form columns of $\mathbf{Z}$. Then model (11.6) still holds for the appropriate choices of $\mathbf{X}$ and $\mathbf{Z}$.

As an example, we return to the union membership data set. The response is again the indicator of union membership (union). The predictor variables are:

- wages (wages in dollars/hour);
- age (age in years);
- ed (number of years of education);
- race (indicator of white race);
- gender (indicator of female);
- South (indicator of living in southern region of the U.S.).

The effects of the first three (wages, age, and ed) were modeled nonparametrically using splines and GCV. The last three variables (race, gender, and South) are all indicators, so they were modeled with linear effects. The t-statistics for these variables were -2.23, -3.12, and -2.12, respectively. These values indicate that U.S. union membership is significantly lower for whites, for females, and for workers living in the South.

Figure 11.3 shows the effects of the variables entering the model nonparametrically. Each plot shows the estimated probability of union membership as the variable on the horizontal axis varies across its observed range while the other variables are fixed at their observed means; 95% confidence bands are also shown. The confidence bands for the variable ed are quite wide on the left. The reason is that not many workers have fewer than eight years of education, so the function is not accurately estimated in this region.

We see from Figure 11.3 that the effect of wages on union does not change qualitatively when adjusted for the effects of the other predictor variables. In particular, the effect of wages still appears to be nonmonotonic. The probability of

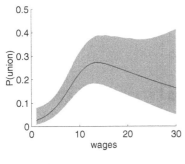

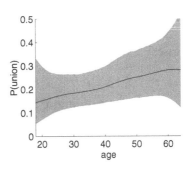

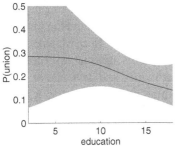

Figure 11.3
Components of the
additive model fit
(solid) and 95%
confidence bands.
Each solid curve is
the probability of
union membership
as a function of one
covariate with the
other covariates fixed
at their sample means.

union membership increases with age and decreases with ed. The effect of age is nearly linear, as is the effect of ed (at least for ed above 10).

11.4 Degrees-of-Freedom Approximations

The hat matrix and the degrees of freedom of the fit can be defined by generalizing the results given in Section 10.6 for ordinary maximum likelihood estimation to penalized maximum likelihood. Let $\mathbf{C}$ and $\mathbf{\Lambda}$ be defined by (8.5).

Let $\mu(\eta)$ and $V(\eta)$ be the conditional mean and variance of y given $\eta = \mathbf{x}^{\mathsf{T}}\boldsymbol{\beta} + \mathbf{z}^{\mathsf{T}}\mathbf{u}$. As in (10.24), define

$$\mathbf{W} = \operatorname{diag}\{\mu'(\mathbf{X}\hat{\boldsymbol{\beta}} + \mathbf{Z}\hat{\mathbf{u}})\}.$$

Analogous to (10.12), the hat matrix is

$$\mathbf{H}_{\hat{\boldsymbol{\beta}},\hat{\mathbf{u}}} = \mathbf{W}\mathbf{C}(\mathbf{C}^{\mathsf{T}}\mathbf{W}\mathbf{C} + \tfrac{1}{2}\mathbf{\Lambda})^{-1}\mathbf{C}^{\mathsf{T}} \tag{11.7}$$

and the degrees of freedom of the fit is $\operatorname{tr}(\mathbf{H}_{\hat{\boldsymbol{\beta}},\hat{\mathbf{u}}})$, which equals

$$\operatorname{tr}\{(\mathbf{C}^{\mathsf{T}}\mathbf{W}\mathbf{C} + \tfrac{1}{2}\mathbf{\Lambda})^{-1}\mathbf{C}^{\mathsf{T}}\mathbf{W}\mathbf{C}\}. \tag{11.8}$$

Moreover, the estimated conditional covariance matrix of $(\hat{\boldsymbol{\beta}}, \hat{\mathbf{u}})^{\mathsf{T}}$ given $\mathbf{u}$ is calculated by a "sandwich" formula somewhat analogous to (6.10) and (10.23):

$$(\mathbf{C}^{\mathsf{T}}\mathbf{W}\mathbf{C} + \tfrac{1}{2}\mathbf{\Lambda})^{-1}(\mathbf{C}^{\mathsf{T}}\mathbf{W}\mathbf{C})(\mathbf{C}^{\mathsf{T}}\mathbf{W}\mathbf{C} + \tfrac{1}{2}\mathbf{\Lambda})^{-1}, \tag{11.9}$$

which reduces to (10.14) when the penalty matrix $\mathbf{\Lambda}$ is 0.

Recall that this chapter assumes a canonical GLM and a ϕ that is known and equal to 1. The results in this section would need to be modified slightly in other cases.

11.5   Automatic Smoothing Parameter Selection

As was seen in Chapter 5, there are two general approaches to selecting the amount of smoothing from the data. One is to use the generalized linear mixed model framework – in which the amount of smoothing is controlled by variance components – and estimate the variance components via (approximate) maximum likelihood as described in Section 10.8. The second is to extend the various model selection criteria of Section 5.3 to the generalized case. Here we will concentrate on generalized cross-validation (GCV) and Akaike's information criterion (AIC).

Let $\mathbf{H}_{\beta,\hat{u}}(\Lambda)$ and $D(\mathbf{y};\hat{\mathbf{y}}:\Lambda)$ be the hat matrix and deviance of the model, respectively, for a fixed value of the smoothing parameter matrix Λ. The GCV deviance is

$$\text{GCV}(\Lambda) = \frac{n^{-1}D(\mathbf{y};\hat{\mathbf{y}}:\Lambda)}{[1 - n^{-1}\operatorname{tr}\{\mathbf{H}_{\hat{\beta},\hat{u}}(\Lambda)\}]^2}. \tag{11.10}$$

See Hastie and Tibshirani (1990), who also define AIC as

$$\text{AIC}(\Lambda) = n^{-1}[D(\mathbf{y};\hat{\mathbf{y}}:\Lambda) + 2\operatorname{tr}\{\mathbf{H}_{\hat{\beta},\hat{u}}(\Lambda)\}\phi], \tag{11.11}$$

where ϕ is the dispersion parameter and assumed to be equal to 1 in this chapter. One selects Λ by minimizing $\text{GCV}(\Lambda)$ or $\text{AIC}(\Lambda)$.

There are several other ways by which GCV or AIC can be extended to the generalized situation; the extensions given in (11.10) and (11.11) represent the simplest ones. Examples of other extensions may be found in Gu (1992a) and Xiang and Wahba (1996), for example.

11.6   Hypothesis Testing

Hypothesis testing can be approached in the same ways as for additive models in Section 8.5.

For example, within the GLMM framework, many hypotheses are equivalent to assuming that one or more variance components are zero and can be tested using likelihood ratios. As discussed in Section 8.5.1, currently available asymptotic theory is inadequate for likelihood ratio testing even for ordinary additive models, so the null distribution of the likelihood ratio must be approximated by simulation.

The F-tests for additive models in Section 8.5.2 treat all parameters as fixed effects, but they are estimated by penalized least squares. Hastie and Tibshirani (1990, sec. 6.8.3) describe approximate F-tests for generalized additive models based on deviances. However, we are not aware of any investigations of the accuracy of such approximation, so for formal testing we recommend that critical values be determined by Monte Carlo.

11.7 Model Selection

Model selection methodology for GAMs is still in its infancy. One can compare models using the GCV deviance or AIC as defined in the previous section. For each model, GCV or AIC would be minimized over Λ to obtain the model GCV or AIC. Then models with small GCV or AIC would be considered best. Of course one should not just blindly minimize GCV or AIC. Rather, all models with reasonably small GCV or AIC values should be considered as potentially appropriate and evaluated according to their simplicity and scientific relevance.

The function step.gam() in S-PLUS builds GAMs stepwise using AIC; see Chambers and Hastie (1993, chap. 7). Shively et al. (1999) develop a Bayesian approach to model selection for probit additive models using smoothing splines.

11.8 Density Estimation

A common statistical problem is that we observe a sample $x_1, \ldots, x_n$ from a density f that we need to estimate. There is a huge literature on density estimation that we will not even attempt to summarize, but the references in Section 11.9 will provide the reader with a place to start.

A convenient method for univariate density estimation is converting the density estimation problem to a regression problem (Eilers and Marx 1996). Specifically, one takes a histogram of $x_1, \ldots, x_n$ with many equal-width bins, say 200. After normalization, the plot of the bin heights versus the bin centers is a rough estimate of f. If we smooth these points, then we have a satisfactory density estimate. Moreover, conditional on n, the bin counts are binomially distributed. They are also approximately Poisson distributed, since a binomial(n, p) variable is approximately Poisson(np) if n is large and p is small. Therefore, generalized regression based on either a binomial or Poisson model is appropriate.

Suppose that there are B histogram bins of equal width on the interval $[a, b]$, so that the bin widths are $\Delta = (b - a)/B$. Let N_j be the number of x_i in the jth bin and let c_j be the center of the jth bin. If the height of the jth bin is $y_j = N_j/(n\Delta)$ then the area under the histogram will be 1, so that it is an estimate of f – albeit not a very smooth estimate. In order to obtain a smooth estimate, one regresses the N_j on the c_j using either a logistic or Poisson spline model. The fitted curve is then divided by $n\Delta$ to yield the smooth density estimate.

As an example, we will use Monte Carlo data. The data are from a Markov chain Monte Carlo Bayesian analysis of the Canadian age–income data. The analysis is discussed fully in Chapter 16. All that one needs to understand at present is that the data are a sample of 3000 observations from a particular density of interest – specifically, the posterior density of the residual standard deviation σ_ε. In the top panel of Figure 11.4 we have the raw "data," the bin counts divided by $n\Delta$ versus the bin centers. There are 200 bins. One can see that the data are scattered about a roughly bell-shaped curve. The bottom plot is a smooth fit to these data, using a Poisson model and a log link function.

Figure 11.4 Age and
income data: estimate
of posterior density
of σ_ε. *Top:* Plot of
normalized bin counts
versus bin centers.
Bottom: Smooth fit to
the data on top.

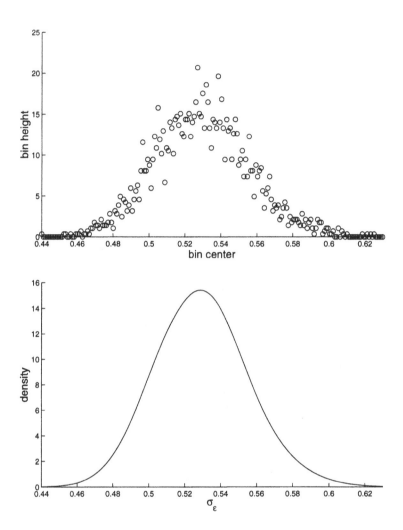

11.9   Bibliographical Notes

The definitive reference for generalized additive models is the book of Hastie and
Tibshirani (1990). Most of that book is concerned with full-rank smoothers with
implementation through the backfitting algorithm. Lin and Zhang (1999) develop
the mixed model approach to generalized additive models, including the incorporation of random effects.

  Density estimation has a very large literature. Summaries may be found, for
example, in Wand and Jones (1995), Simonoff (1996), and Loader (1999).

12

Interaction Models

12.1 Introduction

The additive models of Chapters 8 and 11 have many attractive features. The joint effect of all the predictor variables upon the response is expressed as a sum of individual effects. These individual effects show how the expected response varies as any single predictor varies with the others held fixed at arbitrary values; because of the additivity, the effect of one predictor does not depend on the values at which the others are fixed. Thus, the individual component functions can be plotted separately to visualize the effect of each predictor, and these functions – taken together – allow us to understand the joint effects of all the predictors upon the expected response. If, for example, we wish to find conditions under which the expected response is maximized, then we need only maximize separately each of the component functions of the additive model. In summary, it is extremely convenient whenever an additive model provides an accurate summary of the data.

However, there are no guarantees that an additive model will provide a satisfactory fit in any given situation. Nonadditivity means that, as one predictor is varied, the effect on the expected response *depends* on the fixed values of the other predictors. A deviation from additivity is called an *interaction*. Consider a general multiple regression model $y_i = f(\mathbf{x}_i) + \varepsilon_i$, where f is a smooth function of a vector $\mathbf{x}_i$ of predictor variables. Suppose that $\{\mathbf{x}_i\}_{i=1}^n$ are contained in a region that is small relative to the curvature in f. Then both additivity and linearity can be justified by a first-order Taylor approximation: $f(\mathbf{x}) \approx f(\mathbf{x}^*) + \mathsf{D}f(\mathbf{x}^*)(\mathbf{x} - \mathbf{x}^*)$, where $\mathbf{x}^*$ is at the center of the $\mathbf{x}_i$ (e.g., is their sample mean); see Appendix A for vector differential calculus notation. If there is enough curvature in f over the region in which $\mathbf{x}$ varies, then more terms in the Taylor approximation are needed. The next terms are quadratic. The pure quadratic terms, where components of $\mathbf{x}$ are squared, give nonlinear but additive departures from a linear model. Mixed quadratic terms are those where two different components of $\mathbf{x}$ are multiplied; they represent two-way interactions. Thus, whenever there is enough curvature that nonlinearities appear, we might expect that interactions will also appear. In fact, the popularity of additive models as an alternative to linear models is undoubtedly their simplicity – not their ability to provide highly accurate summaries of data where a linear model fails.

In this chapter we consider models with interactions. These models have two uses. The first use is checking for additivity. One can do this by testing the null hypothesis of a pure additive model (i.e., one without any interaction terms) against the alternative of the additive model plus interaction terms. If we accept the null hypothesis, then we have some assurance that using the additive model provides a reasonable fit to the data.

The second use of interaction models is as alternative models when an additive model does not fit important aspects of the data. If the hypothesis of an additive model is rejected, then one can add appropriate interaction terms to the additive model.

In Section 12.2 we introduce the basic ideas of interactions in semiparametric regression with the simple special case of one binary factor that possibly interacts with a single continuous covariate. In Section 12.3 we generalize to interactions between a discrete factor at more than two levels and a continuous covariate, and we consider adding interactions in generalized additive models. In Section 12.4 we look at varying coefficient models. These are models for the interaction between two continuous predictors, where the effect of one variable is linear but with intercept and slope depending nonparametrically on the second variable.

12.2 Binary-by-Continuous Interaction Models

In Chapter 7 we fit a model of the form

$$y_i = \gamma_0 z_i + f(x_i) + \varepsilon_i \qquad (12.1)$$

to the onion data, where

$$(x_i, y_i, z_i) = (\texttt{location}_i, \texttt{log.yield}_i, \texttt{density}_i).$$

As we mentioned there, this model assumes that `location` and `density` act *additively* on `log.yield`. A more general model is

$$y_i = f_{z_i}(x_i) + \varepsilon_i,$$

$$z_i = \begin{cases} 1 & \text{if } (x_i, y_i) \text{ is from Purnong Landing,} \\ 2 & \text{if } (x_i, y_i) \text{ is from Virginia.} \end{cases} \qquad (12.2)$$

Here f_1 and f_2 are any two smooth curves. This model dispenses with the additivity assumption and allows for the possibility that `location` and `density` *interact* with one another, meaning that the effect of `density` on `yield` can depend in a completely general way upon `location`. Model (12.2) is the simplest example of a *nonparametric interaction model.* It involves the interaction between a binary factor (`location`) and a nonparametric function of a continuous variable (`density`).

To fit (12.2) using penalized splines, we may start with the following representation of (12.1):

$$y_i = \beta_0 + \sum_{j=1}^{p} \beta_j x_i^j + \sum_{k=1}^{K} u_k (x_i - \kappa_k)_+^p + \gamma_0 z_i + \varepsilon_i. \qquad (12.3)$$

The $\gamma_0 z_i$ term represents the vertical shift between the two curves in the additive model.

Define the indicator of the ℓth location as

$$z_{i\ell} = \begin{cases} 1 & \text{if } z_i = \ell, \\ 0 & \text{otherwise.} \end{cases}$$

To cater for (12.2) we simply extend (12.3) to

$$y_i = \beta_0 + \sum_{j=1}^{p} \beta_j x_i^j + \sum_{k=1}^{K} u_k (x_i - \kappa_k)_+^p$$

$$+ \sum_{\ell=1}^{2} z_{i\ell} \left(\gamma_{0\ell} + \sum_{j=1}^{p} \gamma_{j\ell} x_i^j + \sum_{k=1}^{K} v_{k\ell} (x_i - \kappa_k)_+^p \right) + \varepsilon_i. \qquad (12.4)$$

A mixed model representation can be obtained by taking

$$u_k \text{ i.i.d. } N(0, \sigma_u^2) \quad \text{and} \quad v_{k\ell} \text{ i.i.d. } N(0, \sigma_v^2).$$

In order for the fixed effects parameters to be identified (i.e., so that they are uniquely defined), we need constraints on the $\{\gamma_{j\ell}\}$. One possible set of constraints is that $\gamma_{j2} = 0$ for all $j = 0, \ldots, p$. Another possibility is that $\sum_{\ell=1}^{2} \gamma_{j\ell} = 0$ for all $j = 0, \ldots, p$. Very similar constraints are used for certain non–full-rank analysis of variance models to identify parameters.

The fitted curves do not depend on which set of constraints is chosen, but the interpretation of the parameters in the curve does. Note that, if we adopt the constraints that $\gamma_{j2} = 0$ for $j = 0, \ldots, p$, then

$$\beta_0 + \sum_{j=1}^{p} \beta_j x_i^j + \sum_{k=1}^{K} u_k (x_i - \kappa_k)_+^p \qquad (12.5)$$

is the fitted curve for $\ell = 2$ (Virginia) and

$$\gamma_{0\ell} + \sum_{j=0}^{p} \gamma_{j\ell} x_i^j + \sum_{k=1}^{K} v_{k\ell} (x_i - \kappa_k)_+^p \qquad (12.6)$$

is the difference of the fitted curves for the ℓth location and Virginia – that is, the difference between Purnong Landing and Virginia when $\ell = 1$ and zero when $\ell = 2$. If we adopt the constraints $\sum_{\ell=1}^{2} \gamma_{j\ell} = 0$ for all $j = 0, \ldots, p$, then (12.5) is the average curve for the two sites and (12.6) is deviation of the curve for the ℓth location from the average curve.

The maximum likelihood fit of this model for $p = 2$ is shown in Figure 12.1, with the additive model fit shown for comparison. The fits are reasonably close to each other and differ mostly where the data are sparse on the right side of the

Figure 12.1
Interaction model
fit to the onion data
based on maximum
likelihood smoothing
parameter selection
(solid curve). The
dashed curves show
the additive model fit.

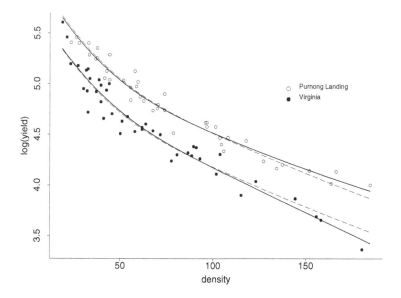

plot, suggesting adequacy of the additivity assumption in this case. The next section describes a way of testing this hypothesis.

12.2.1 *Testing for Additivity*

A test for additivity can be conducted by comparing the additive model to the interaction model and checking whether the interaction model offers a significant improvement in fit. In particular, one can compare the log-likelihoods for the additive and interaction models through the likelihood ratio statistic. In terms of the parameters in (12.4), the null hypothesis of additivity corresponds to all terms in (12.6) except the intercept being zero for all ℓ. Thus, the null hypothesis is

$$H_0 : \gamma_{j\ell} = 0, \ \ell = 1, \ldots, 2, \ j = 1, \ldots, p, \ \text{and} \ \sigma_v^2 = 0.$$

After accounting for the constraints imposed on the fixed effects to ensure identifiability, the null hypothesis restricts $(2 - 1)p$ fixed effects parameters and one variance component.

12.3 Factor-by-Curve Interactions in Additive Models

12.3.1 *Modularity of Spline Models*

One of the real advantages of spline modeling is its modularity. By this we mean that concepts like main effects, interaction effects, generalized regression, and the mixed model formulation with smoothing parameter selection by REML can be viewed as modules and put together into an almost endless variety of statistical

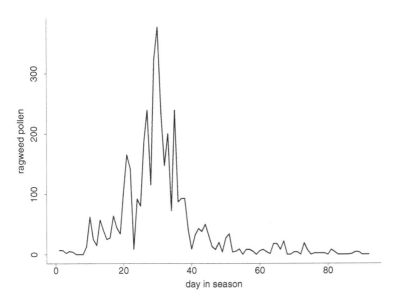

Figure 12.2
Ragweed pollen
levels in Kalamazoo,
Michigan, 1991.

models. Therefore, one can easily tailor a model to a specific application. In particular, there are many ways in which the two-way interaction between a binary and a continuous variable presented in Section 12.2 can be extended or combined with other techniques discussed in this book. For example: higher-order interactions can easily be defined; the binary-by-continuous interactions used in the onions example can be extended to discrete-by-continuous interactions; and interactions can be added to generalized additive models. In this section we will discuss some of these extensions.

12.3.2   *Example: Ragweed Pollen Revisited*

As a motivating example, we will use the case study in Section 7.5 of daily ragweed pollen counts. The interaction model for the pollen data presented here is taken from Coull, Ruppert, and Wand (2001). Figure 12.2 shows data on daily ragweed pollen counts from four consecutive pollen seasons in Kalamazoo, Michigan. Stark et al. (1997) and Brumback et al. (2000) fit generalized linear and generalized additive models, respectively, to these data to investigate the predictive power of meteorological variables on pollen level. Because each year's pollen season starts at a different time and will progress at a different rate, the effect of the seasonal trend is thought to be different for each year. This year-to-year heterogeneity led both sets of authors to fit models to each year's data separately. Later in this section, we will present a single analysis based on data from all four pollen seasons.

Some intuition for why a model like this is appropriate is shown in Figures 12.3 and 12.4, which show the marginal relationships between $\sqrt{\text{ragweed}}$ and day.in.season and wind, respectively. In Figure 12.3 we see that the relationship between $\sqrt{\text{ragweed}}$ and day.in.season is relatively strong and also

Figure 12.3
Interaction fits
for $\sqrt{\text{ragweed}}$ on
day.in.season.

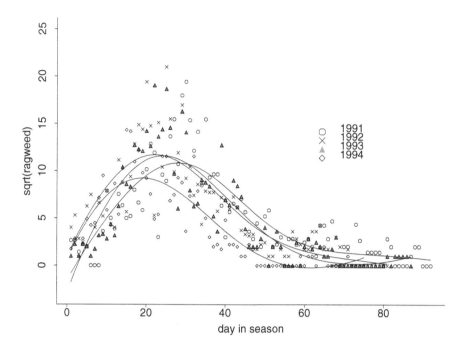

Figure 12.4
Interaction fits for
$\sqrt{\text{ragweed}}$ on wind.

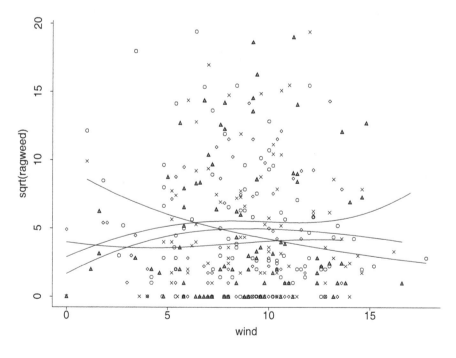

somewhat different for each season, so an interaction term seems appropriate. On the other hand, the $\sqrt{\text{ragweed}}$–wind relationship is rather weak, and there is little to be gained from going beyond using the same smooth function relationship across all years.

The pollen example requires two extensions. First, ragweed counts are discrete responses and should be modeled by the Poisson distribution – possibly with overdispersion. Second, Figure 12.3 suggest an interaction between the discrete factor, year, and the continuous variable, day of the year. (Strictly speaking, day of the year is discrete, but it makes sense for this variable to be modeled as a continuous variable because expected ragweed pollen counts are a continuous function of day of the year.) After discussing these and other extensions of the basic interactions models, we will return to the pollen data.

12.3.3 Discrete-by-Continuous Interactions

For simplicity, we first explain "discrete factor by curve" interactions for a single continuous predictor and a single categorical factor. Consider the set of triples (x_i, y_i, z_i), $1 \le i \le n$, where the x_i and y_i represent continuous predictor and response recordings, respectively, and where $z_i \in \{1, \ldots, L\}$ represents a coded factor. The type of model that we wish to fit is

$$y_i = f_{z_i}(x_i) + \varepsilon_i, \quad 1 \le i \le n, \tag{12.7}$$

where $f_1, \ldots, f_L$ are L different functions depending on the value of z_i and where the ε_i are i.i.d. N$(0, \sigma_\varepsilon^2)$.

Let $\kappa_1, \ldots, \kappa_K$ be a set of knots inside the range of the x_i. Define

$$z_{i\ell} = \begin{cases} 1 & \text{if } z_i = \ell, \\ 0 & \text{otherwise}, \end{cases} \tag{12.8}$$

for $\ell = 1, \ldots, L$. A linear penalized spline model for (12.7) is

$$y_i = \beta_0 + \beta_1 x_i + \sum_{k=1}^{K} u_k (x_i - \kappa_k)_+ + \sum_{\ell=1}^{L} z_{i\ell} (\gamma_{0\ell} + \gamma_{1\ell} x_i)$$

$$+ \sum_{\ell=1}^{L} z_{i\ell} \left(\sum_{k=1}^{K} v_k^\ell (x_i - \kappa_k)_+ \right) + \varepsilon_i, \tag{12.9}$$

where the u_k are i.i.d. N$(0, \sigma_u^2)$ and the v_k^ℓ are i.i.d. N$(0, \sigma_{v\ell}^2)$, $\ell = 1, \ldots, L$, for appropriate values of σ_u and $\sigma_{v\ell}$, $\ell = 1, \ldots, L$. Henceforth, we use this mixed model formulation of penalized spline models.

In model (12.9), $\sum_{k=1}^{K} v_k^\ell (x_i - \kappa_k)_+$ represents deviations from the overall smooth term $\sum_{k=1}^{K} u_k (x_i - \kappa_k)_+$. Note that model (12.9) is overparameterized in that (for example) we could add a constant to β_0 and subtract the same constant from each $\gamma_{0\ell}$ without changing the expectation of y_i in any way. This overparameterization is similar to that of non–full-rank analysis of variance (ANOVA) models and can be handled by the same types of constraints used in ANOVA models. For example, Coull et al. (2001) assume that $\gamma_{01} = \gamma_{11} = 0$. With this constraint, $(\gamma_{0\ell} + \gamma_{1\ell} x_i)$ models the linear deviation between f_1 and f_ℓ, $\ell = 2, \ldots, L$.

No constraint is needed on the v_k^ℓ. They are modeled as having zero means, so one cannot add an arbitrary constant to them. In effect, the assumption of a zero mean serves the same role as a constraint on the fixed effects.

We can write model (12.9) in matrix notation as

$$\mathbf{y} = \mathbf{X}\boldsymbol{\beta} + \mathbf{Z}\mathbf{u} + \boldsymbol{\varepsilon}, \tag{12.10}$$

where

$$\boldsymbol{\beta} = [\beta_0, \beta_1, \gamma_{02}, \ldots, \gamma_{0L}, \gamma_{12}, \ldots, \gamma_{1L}]^{\mathsf{T}},$$

$$\mathbf{u} = [u_1, \ldots, u_K, v_1^1, \ldots, v_K^1, \ldots, v_1^L, \ldots, v_K^L]^{\mathsf{T}},$$

$$\mathbf{X} = [1 \ \ x_i \ \ z_{i2} \ \cdots \ z_{iL} \ \ z_{i2}x_i \ \cdots \ z_{iL}x_i]_{1 \le i \le n},$$

$$\mathbf{Z} = \left[\underset{1 \le k \le K}{(x_i - \kappa_k)_+} \ \ \underset{1 \le k \le K}{z_{i1}(x_i - \kappa_k)_+} \ \cdots \ \underset{1 \le k \le K}{z_{iL}(x_i - \kappa_k)_+} \right]_{1 \le i \le n},$$

and

$$\begin{bmatrix} \mathbf{u} \\ \boldsymbol{\varepsilon} \end{bmatrix} \sim \mathrm{N}\left(\mathbf{0}, \begin{bmatrix} \mathbf{G} & \mathbf{0} \\ \mathbf{0} & \sigma_\varepsilon^2 \mathbf{I} \end{bmatrix}\right),$$

with $\mathbf{G} = \mathrm{diag}(\sigma_u^2 \mathbf{1}_K, \sigma_{v1}^2 \mathbf{1}_K, \ldots, \sigma_{vL}^2 \mathbf{1}_K)$.

Extension to models with truncated polynomials $(x_i - \kappa_k)_+^p$ for $p > 1$ is straightforward. For example, the pth-order penalized spline model for (12.7) is

$$y_i = \beta_0 + \sum_{j=1}^p \beta_j x_i^p + \sum_{k=1}^K u_k (x_i - \kappa_k)_+^p$$

$$+ \sum_{\ell=1}^L z_{i\ell} \left(\gamma_{0\ell} + \sum_{j=1}^p \gamma_{j\ell} x_i^j \right) + \sum_{\ell=1}^L z_{i\ell} \left(\sum_{k=1}^K v_k^\ell (x_i - \kappa_k)_+^p \right) + \varepsilon_i,$$

where again the u_k are i.i.d. $\mathrm{N}(0, \sigma_u^2)$ and the v_k^ℓ are i.i.d. $\mathrm{N}(0, \sigma_{v\ell}^2)$, $\ell = 1, \ldots, L$. We have assumed that $\sigma_{v\ell}^2$ depends on ℓ, meaning that the variance of the jumps in the pth derivative at the knots depends on ℓ. A simpler model uses a constant variance component, σ_v^2.

Model (12.9) specifies a different smooth function $f(x_i)$ for each subset of observations defined by the levels of z. Thus, one can effectively fit this model (apart from the assumption of homoscedastic errors across factor levels) by fitting a nonparametric regression model to each subset separately. For a multiple regression model in which some terms do not interact with z, however, the penalized spline approach holds a substantial advantage over the data subsetting approach because the latter must be nested within a backfitting algorithm.

12.3.4 *Interactions in Additive Models*

We now incorporate the additive effects of other variables into the discrete-by-continuous interaction model (12.9). To keep notation simple, consider a semi-parametric model with a single parametric term, a single nonparametric term, and a factor-by-curve interaction. (Extension to models with more than one term of each type is straightforward.) Consider the multiple regression setting with response y_i, general predictor x_i, continuous predictors (s_i, t_i), and categorical predictor z_i, $i = 1, \ldots, n$. A semiparametric model for y_i that allows the functional form of the effect of t_i on y_i to vary according to the level of z_i is

$$y_i = \alpha_0 + \alpha_1 x_i + g(s_i) + f_{z_i}(t_i) + \varepsilon_i, \quad 1 \le i \le n. \tag{12.11}$$

To generalize model (12.9), let $\kappa_1^s, \ldots, \kappa_{K_s}^s$ and $\kappa_1^t, \ldots, \kappa_{K_t}^t$ be the K_s and K_t knots corresponding to s_i and t_i, respectively. In addition, let $z_{i\ell}$ $(i = 1, \ldots, n,$ $\ell = 1, \ldots, L)$ be defined as in (12.8). Then a linear penalized spline model for (12.11) is

$$y_i = \alpha_0 + \alpha_1 x_i + \beta_1^s s_i + \sum_{k=1}^{K_s} u_k^s (s_i - \kappa_k^s)_+ + \beta_1^t t_i + \sum_{k=1}^{K_t} u_k^t (t_i - \kappa_k^t)_+$$

$$+ \sum_{\ell=2}^{L} z_{i\ell} (\gamma_{0\ell} + \gamma_{1\ell} t_i) + \sum_{\ell=1}^{L} z_{i\ell} \left(\sum_{k=1}^{K_t} v_k^\ell (t_i - \kappa_k^t)_+ \right) + \varepsilon_i, \tag{12.12}$$

where the u_k^s are i.i.d. $N(0, \sigma_{us}^2)$, the u_k^t are i.i.d. $N(0, \sigma_{ut}^2)$, and the v_k^ℓ are i.i.d. $N(0, \sigma_{v\ell}^2)$. Model (12.12) also falls within the mixed model framework (12.10), making estimation and inference no more difficult than that for the single covariate model (12.9).

For penalized spline model (12.12), smoothing parameter selection is a byproduct of model fitting with variance component estimation. The amount of smoothing for $g(\cdot)$ and $f_\ell(\cdot)$, $\ell = 1, \ldots, L$, is governed by

$$\frac{\sigma_\varepsilon^2}{\sigma_{us}^2} \quad \text{and} \quad \frac{\sigma_\varepsilon^2}{\sigma_{ut}^2 + \sigma_{v\ell}^2}, \quad \ell = 1, \ldots, L,$$

respectively. Thus, smoothing parameter selection reduces to variance component estimation in a mixed model, with a small variance component corresponding to more smoothness for a particular curve. Note that models (12.11) and (12.12) specify independent amounts of smoothing for each curve f_ℓ. One can obtain either maximum likelihood (ML) or restricted maximum likelihood (REML) estimates (Searle et al. 1992) of the variance components – and hence of the smoothing parameters – by using, for example, PROC MIXED in SAS or the S-PLUS function lme(). Alternatively, one can fit these models using a prespecified amount of smoothing for a given curve by fixing the value of the corresponding variance component. This can be accomplished, for instance, using the parms option in the SAS procedure PROC MIXED.

12.3.5 *Generalized Additive Models with Interactions*

It is very simple to add interaction terms to the generalized additive models discussed in Chapter 11. One simply takes the additive model predictor – for example, $\beta_0 + f(s) + g(t)$ in (11.3) – and adds interaction terms. Thus, if all variables are continuous then the linear predictor becomes the right-hand side of (12.12) but without ε_i.

12.3.6 *Pollen Data*

We now consider both generalized linear and generalized additive interaction models for the pollen data. Let μ_{ij} denote the mean pollen count on day i of the pollen season j, $j = 1, \ldots, 4$. We first fit the most general parametric Poisson regression model that contains terms corresponding to rain, wind, temperature, and day in season, with each of these effects varying according to year. Specifically, we fit the Poisson GLM

$$\log(\mu_{ij}) = \beta_{0j} + \beta_{1j}\texttt{rain}_{ij} + \beta_{2j}\texttt{wind}_{ij} + \beta_{3j}\texttt{temp}_{ij}$$
$$+ \beta_{4j}\texttt{temp.resid}_{ij} + \beta_4 i + \beta_{4j} \log(i + 1). \quad (12.13)$$

Here, for day i in year j, $\texttt{rain}_{ij}$ is a rain indicator, $\texttt{wind}_{ij}$ denotes wind speed, $\texttt{temp}_{ij}$ denotes the fitted values from a smooth of temperature as a function of day in season, and $\texttt{temp.resid}_{ij}$ denotes the residual from this smooth. The last two terms on the right-hand side of (12.13) aim to capture the nonlinearity in day number. This model corresponds to fitting a Poisson regression model to data from each year separately. This model does not appear to fit the data well, yielding a deviance of 4169.6 on 312 residual degrees of freedom.

We next fit generalized additive models to investigate whether this lack of fit arises from the linearity assumptions in the Poisson GLM. Consider the semi-parametric regression model

$$\log(\mu_{ij}) = \alpha_0 + \alpha_{1j}\texttt{rain}_{ij} + g_1(\texttt{wind}_{ij}) + g_2(\texttt{temp.resid}_{ij}) + f_j(i). \quad (12.14)$$

This model specifies a rain-by-year interaction and a factor-by-curve interaction representing distinct seasonal trends for the four years. The functions g_1 and g_2, however, are the same for every year j. That is, we assume that the relationships between pollen and residual temperature and wind speed do not change from year to year. The penalized quasilikelihood fit of the appropriate linear penalized spline model yields a deviance of 2577.9 on approximately 293.5 degrees of freedom, or an almost 40% decrease relative to the deviance from model (12.13). Overdispersion is still present, however, and GLIMMIX yields an overdispersion parameter estimate of $\hat{\phi} = 8.5$; see Section 10.8.3 for a discussion of GLIMMIX and overdispersion parameters. This overdispersion persists under the more general models containing heterogeneous wind and/or residual temperature effects for different seasons.

Table 12.1 shows the estimates of the rain coefficients and corresponding standard errors, adjusted for overdispersion, from the fit of model (12.14). The range of these estimates is larger than those of previous analyses, which is primarily due to the linear assumption for the residual temperature effect in earlier models. Fitting the data to all four years simultaneously allows us to test formally the plausibility of a homogeneous rain effect across the four years. In particular, we compare the fit of model (12.14) to that obtained under the constraint $\alpha_{11} = \cdots = \alpha_{14}$. Because of overdispersion, we compare the difference between the deviances, scaled by $\hat{\phi}$, to the quantiles of an F-distribution (McCullagh and

Year	$\hat{\alpha}_1$	St. dev.
1991	0.56	0.33
1992	0.72	0.21
1993	0.99	0.25
1994	0.87	0.42
Pooled	0.80	0.14

Table 12.1
Yearly and pooled
pseudolikelihood
estimates of the
rain coefficients
from semiparametric
model (12.14).

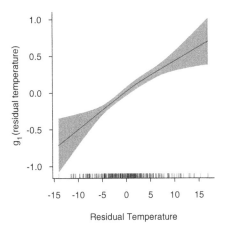 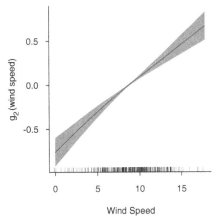

Figure 12.5 Fitted
curves for residual
temperature and wind
speed for the pollen
data.

Nelder 1989, p. 203) instead of the usual χ^2 cut-off. Wolfinger and O'Connell's
(1993) pseudolikelihood approximation of this test is $F = 0.74$ on 3 df, suggesting that a common homogeneous rain effect is plausible; see Section 10.8.3 for
a discussion of pseudolikelihood. Table 12.1 shows the pooled estimate and associated standard error adjusted for overdispersion. Note the improved precision
of the pooled estimate that results from estimating the rain effect from all of the
data.

Figure 12.5 shows plots of the estimated curves and pointwise 95% confidence
bands for the effects of residual temperature and wind speed on daily pollen
counts. The model that specifies equivalent smoothness for $\{f_j\}$ via $\sigma_{v1}^2 = \cdots =$
σ_{v4}^2 yields a scaled (by $\hat{\phi}$) difference in deviance of 1.5 on three additional degrees of freedom, suggesting that this simplification is reasonable. This model
specifies the same amount of smoothness for the different functions, but the random effects are independent from function to function and so there is no implied
similarity between the effects.

Figure 12.6 shows plots of the estimates and pointwise 95% confidence bands
for f_j, $j = 1, \ldots, 4$, from this model. The fit of the model specifying a common
rain effect and additivity between year and seasonal trends yields a deviance of
4451.5 on approximately 316.2 residual degrees of freedom, supporting the hypothesis that seasonal trend of pollen counts does indeed vary across year.

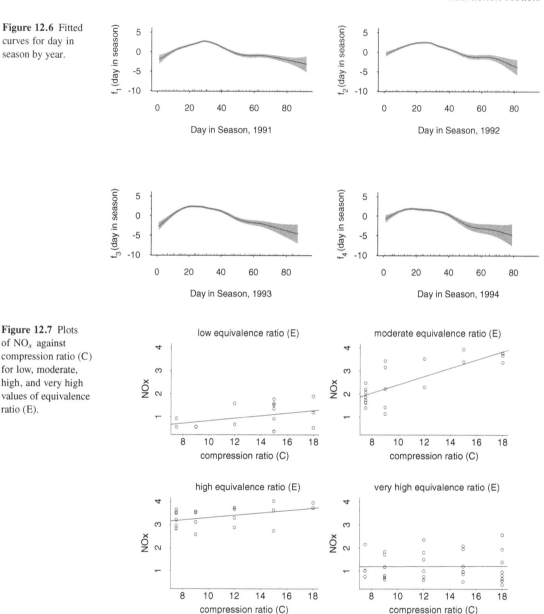

Figure 12.6 Fitted curves for day in season by year.

Figure 12.7 Plots of NO_x against compression ratio (C) for low, moderate, high, and very high values of equivalence ratio (E).

12.4   Varying Coefficient Models

The data corresponding to Figure 12.7 are part of the S-PLUS computing package, where it is referred to as ethanol.

Until now, we have developed models for two-way interactions between two predictors only in the special case where at least one of the two variables is discrete – including binary as a special case of discrete. Now we introduce a special class of continuous-by-continuous interactions.

A motivating data set is shown in Figure 12.7, corresponding to data from an experiment in which ethanol was burned in a single-cylinder automobile test engine. Each panel corresponds to a scatterplot of the concentration of NO_x (nitric oxide and nitrogen dioxide in engine exhaust, normalized by the work done by the engine) versus the compression ratio of the engine. However, the data are stratified according to a third variable: the equivalence ratio at which the engine was run – a measure of the richness of the air–ethanol mix. Notice that, for a given equivalence ratio, there appears to be a linear relationship between mean NO_x concentration and compression ratio. However, the equivalence ratio *modifies* this relationship.

An appropriate model is one where the effect of one predictor, conditional upon a fixed value of the second predictor, is modeled linearly. However, the intercept and slope parameters in this model vary nonparametrically as a function of the second predictor. Models of this type are called *varying coefficient models* (Hastie and Tibshirani 1993).

Let x be a predictor variable that, for given values of a *modifying* predictor s, has a linear relationship with the mean of the response variable y. If (x_i, s_i, y_i), $1 \le i \le n$, are measurements on each, then a *varying coefficient model* for these data is

$$y_i = \alpha(s_i) + \beta(s_i)x_i + \varepsilon_i. \tag{12.15}$$

The model allows the intercept and slope coefficients to be arbitrary smooth functions of s. The penalized linear spline version of this model is

$$y_i = \alpha_0 + \alpha_1 s_i + \sum_{k=1}^{K} u_k^\alpha (s_i - \kappa_k)_+ + \left(\beta_0 + \beta_1 s_i + \sum_{k=1}^{K} u_k^\beta (s_i - \kappa_k)_+ \right) x_i + \varepsilon_i,$$

where $\kappa_1, \ldots, \kappa_K$ are knots over the range of the s_i values. A mixed model representation $\mathbf{y} = \mathbf{X}\boldsymbol{\beta} + \mathbf{Z}\mathbf{u} + \boldsymbol{\varepsilon}$ is obtained by setting

$$\mathbf{X} = [1 \ \ s_i \ \ x_i \ \ s_i x_i]_{1 \le i \le n}, \qquad \mathbf{Z} = \left[(s_i - \kappa_k)_+ \ \ x_i (s_i - \kappa_k)_+ \atop 1 \le k \le K \qquad 1 \le k \le K \right]_{1 \le i \le n},$$

$\mathbf{u} = [u_1^\alpha, \ldots, u_K^\alpha, u_1^\beta, \ldots, u_K^\beta]^\mathsf{T}$, and $\mathrm{Cov}(\mathbf{u}) = \mathrm{diag}\{\sigma_\alpha^2 \mathbf{1}_{K \times 1}, \sigma_\beta^2 \mathbf{1}_{K \times 1}\}$.

Figure 12.8 shows the fitted varying coefficient model for the ethanol data. Notice, for example, that the positive linear relationship peaks when the equivalence ratio is about 0.85 but then declines to zero for higher values of the equivalence ratio.

Local polynomial regression methods have been developed for varying coefficient models. However, we find penalized spline models to be at least as simple and easier to implement than local regression; furthermore, penalized splines retain all of the flexibility of local regression.

12.5 Continuous-by-Continuous Interactions

So far we have seen several special cases of two-way interactions. In Sections 12.2 and 12.3 we studied interactions when one of the covariates was discrete. The

Figure 12.8 Fit of
varying coefficient
model to ethanol data.

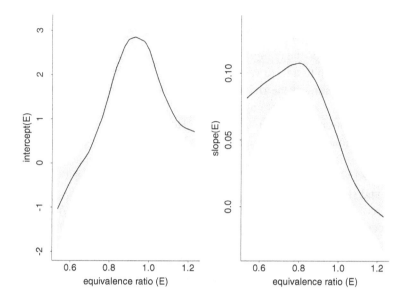

varying coefficient models that were just discussed are used for interactions between two continuous covariates when the effect of one covariate can be modeled
parametrically. There are two approaches to the important case not yet discussed,
completely nonparametric continuous-by-continuous two-way interactions:

(1) tensor products of spline bases;
(2) bivariate radial basis functions.

In the tensor product approach, the interaction basis functions are products of
spline basis functions in one variable multiplied by spline basis functions in the
other variable; see Section 13.2. We have seen tensor products before: they were
used for discrete-by-continuous interactions. In that case, the basis functions of
the discrete variable (the indicator functions of its levels) were multiplied by the
spline basis functions of the continuous variable. Also, varying coefficient models are tensor product models. The basis function of the linear regression in x,
namely x itself, was multiplied by spline basis functions in s to obtain the interaction basis functions.

We have had good success with tensor product models of continuous-by-
continuous interactions in some examples – for instance, the biomonitoring example in Section 1.4. In fact, the fit in Figure 1.8 uses tensor product splines.
The reason that this example worked well with tensor product interactions was
that the predictor variables (latitude and longitude) were distributed reasonably
evenly on a *rectangular* region with sides parallel to the coordinate axes. In other
examples, where the covariates are on more irregular regions (such as the scallops
data in the next chapter), tensor product interaction models have led to frustrating
experiences. For this reason, we focus on alternatives to tensor product models
of continuous-by-continuous interactions.

Radial basis models for two-way interactions will be discussed in Chapter 13.

12.6 Bibliographical Notes

Sophisticated models for handling interactions have been developed for smoothing splines. Examples include Wahba (1986, 1988), Chen (1993), and Gu (2002).

Varying coefficient models were first proposed by Hastie and Tibshirani (1993). Several contributions have since been made to the topic. Examples include Carroll, Ruppert, and Welsh (1998), Kauermann and Tutz (1999), and Cai, Fan, and Li (2000).

13

Bivariate Smoothing

13.1 Introduction

The previous chapter dealt with interactions between a continuous variable and a categorical variable. *Bivariate smoothing* is essentially concerned with interactions between two continuous variables. In its purest form, bivariate smoothing is free of any structural assumptions on the way in which the two continuous variables affect the mean response. Just as Chapters 3 and 5 dealt with flexible smoothing of scatterplots, bivariate smoothing deals with flexible smoothing of "point clouds". Figure 13.1 shows such a point cloud. The response is a monotone transformation, $y' = \log(y + 1)$, of the sizes (in number of scallops) of scallop catches recorded in a 1990 survey cruise in the Atlantic continental shelf off Long Island, New York. These are plotted against longitude and latitude. Interest centers on the mean response as a general bivariate function of longitude and latitude.

The scallop data were obtained from Lange et al. (1994), corresponding to the contribution "Geostatistical Estimates of Scallop Abundance" by Ecker and Heltshe (1994).

Figure 13.2 shows a bivariate smooth of these data as an image plot. It reveals certain regions suggestive of higher scallop abundance, which is of obvious interest to the fishing industry. This surface estimate was obtained through an extension of the penalized spline approach to scatterplot smoothing. The details of this extension are the focus of this chapter.

Bivariate smoothing is of central interest in a number of application areas such as mining, hydrology, and public health. Indeed, the name *geostatistics* describes

Figure 13.1
Transformed scallop catches plotted against geographical location.

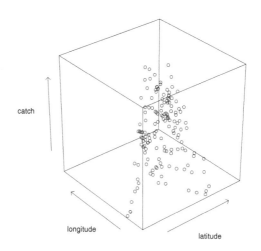

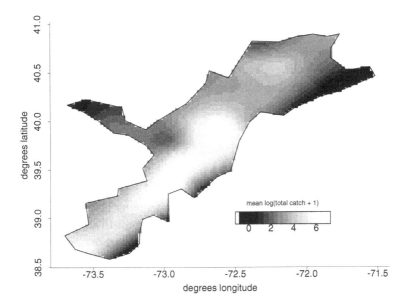

Figure 13.2 Bivariate smooth corresponding to scallop data depicted in Figure 13.1. Lighter areas correspond to higher catches.

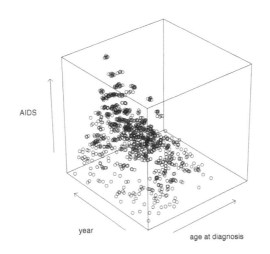

Figure 13.3 Incidence of AIDS against yearly age (age at diagnosis for cases) and calendar year.

the process of converting geographically referenced responses, such as those depicted in Figure 13.1, to maps such as Figure 13.2. The main tool of geostatistics, *kriging,* has close connections with penalized spline smoothing, as we will see shortly.

A nongeographical bivariate smoothing example is depicted in Figure 13.3. The response is incidence of acquired immune deficiency syndrome (AIDS), specifically the number of new diagnoses in a year among Italian men who have sex with men. The predictors are yearly age (age at diagnosis for cases) and calendar year. There is reason to believe that these two predictors interact (Marschner and Bosch 1998). In pure technical terms, the problems of the scallop and Italian AIDS examples are identical: fit a surface through a point cloud. There is, however, a subtle difference. In the scallop example, longitude and latitude have

The Italian AIDS data were kindly provided by Dr. Rino Bellocco of Karolinska Institutet, Sweden.

no particular interpretation and are just the conventional axes for geographically referenced data. Any perpendicular axes will do. But this is not the case for the AIDS data, where the two predictors are meaningful. As we will see in subsequent sections, this difference between geographical and nongeographical bivariate smoothing has led to a divergence of approaches to the problem. When viewed in terms of basis functions, however, the various approaches will be seen to be quite similar to each other.

13.2 Choice of Bivariate Basis Functions

As seen in earlier chapters, penalized spline smoothing relies on a set of basis functions that permit the handling of nonlinear structure. Bivariate smoothing requires bivariate basis functions, but the extension from one dimension to two can be done in at least two distinct ways. One is based on products while the other is based on rotation.

Suppose that s and t are two continuous predictors of the response variable y that possibly interact. The general bivariate smoothing model is

$$y_i = f(s_i, t_i) + \varepsilon_i,$$

where f is a real-valued *bivariate* function. In linear regression it is common to model interactions by adding the product term $\gamma_s s_i t_i$ to the linear additive model:

$$y_i = \beta_0 + \beta_s s_i + \beta_t t_i + \gamma_s s_i t_i + \varepsilon_i.$$

The natural extension for truncated lines is

$$y_i = \beta_0 + \beta_s s_i + \sum_{k=1}^{K^s} u_k^s (s_i - \kappa_k^s)_+ + \beta_t t_i + \sum_{k=1}^{K^t} u_k^t (t_i - \kappa_k^t)_+$$

$$+ \gamma s_i t_i + \sum_{k=1}^{K^s} v_k^s s_i (t_i - \kappa_k^t)_+ + \sum_{k=1}^{K^t} v_k^t t_i (s_i - \kappa_k^s)_+$$

$$+ \sum_{k=1}^{K^s} \sum_{k'=1}^{K^t} v_{kk'}^{st} (s_i - \kappa_k^s)_+ (t_i - \kappa_k^t)_+ + \varepsilon_i. \tag{13.1}$$

Model (13.1) is obtained from the basis functions

$$1, s, (s - \kappa_1^s)_+, \ldots, (s - \kappa_{K^s}^s)_+,$$

$$1, t, (t - \kappa_1^t)_+, \ldots, (t - \kappa_{K^t}^t)_+$$

by forming all pairwise products. The resulting basis is often referred to as a *tensor product* basis.

Figure 13.4 shows the basis functions corresponding to (13.1) with

$$\kappa_1^s = \kappa_1^t = 0.3 \quad \text{and} \quad \kappa_2^s = \kappa_2^t = 0.6.$$

A possible drawback of tensor product splines is their dependence on the orientation of the coordinate axes. For example, if (13.1) were used to fit a surface

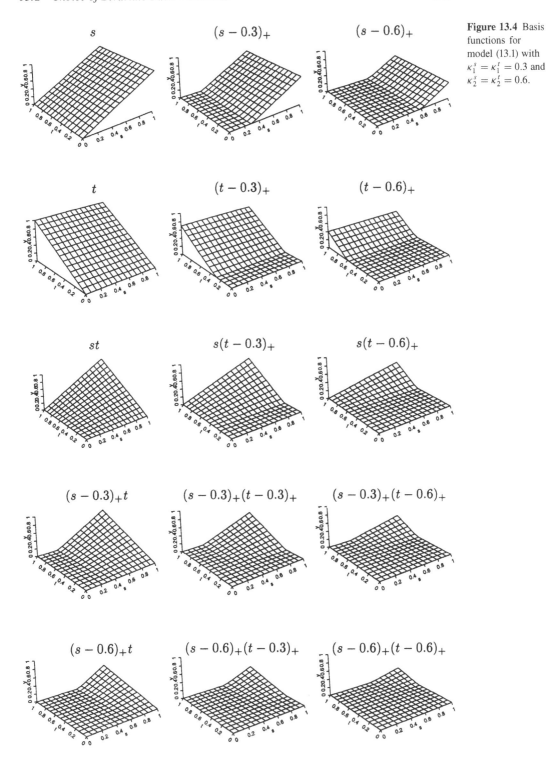

Figure 13.4 Basis functions for model (13.1) with $\kappa_1^s = \kappa_1^t = 0.3$ and $\kappa_2^s = \kappa_2^t = 0.6$.

Figure 13.5 Radial
basis functions
corresponding to
the knots used in
Figure 13.4.

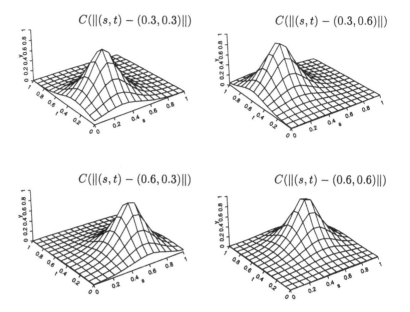

to the scallop data of Figure 13.1 then the result would change if the geographical locations were measured on axes with different orientations. Rotational invariance can be achieved through the use of *radial basis functions*. These are basis functions of the form

$$C(\|(s,t) - (\kappa^s, \kappa^t)\|)$$

for some univariate function C. Since the value of the function at (s,t) depends only on the distance from the knot (κ^s, κ^t), the function is radially symmetric about this point. Figure 13.5 shows the radial basis functions corresponding to the knots used in Figure 13.4. We use the same knots only as an illustration, not because the knots appropriate for one basis are appropriate for the other. In this case $C(x) = \exp(-x/0.1)(1 + x/0.1)$, which is a special case of an important family of radial basis functions defined soon in equation (13.6).

In a nongeographical application like the Italian AIDS example, rotational invariance is not a big problem, but for geographical smoothing it is at least reassuring to know that the answers are independent of axis orientation. Indeed, bivariate smoothing based on radial basis functions arises from the traditional method used in geographical circles, where bivariate smoothing is known as *kriging*. The following section reviews this approach.

13.3 Kriging

The term *kriging* is derived from the name of a South African mining engineer, D. G. Krige, who conducted seminal research in spatial interpolation in the 1960s (see e.g. Krige 1966) driven by the need to map ore grade from drill samples taken at different geographical locations. Parallel research was conducted in the Paris

School of Mines (e.g. Matheron 1965), and early work along these lines is attributed by Webster (1998) to Kolmogorov (1941). The gist of their approach is assuming that the spatial measurements are realizations of a stochastic process with parsimonious covariance structure and then using best prediction theory (Section 4.3) to construct a map over the region of interest. We will summarize the basics of kriging here and then relate it to spline regression in the next section.

Let

$$(\mathbf{x}_1, y_1), \ldots, (\mathbf{x}_n, y_n)$$

be a set of data where the responses $y_i \in \mathbb{R}$ and where $\mathbf{x}_i \in \mathbb{R}^d$ are predictors. The nonparametric regression model for these data is

$$y_i = f(\mathbf{x}_i) + \varepsilon_i,$$

where the ε_i are uncorrelated and with common variance σ_ε^2. As we have discussed in earlier chapters, splines provide a flexible means of modeling the smooth function f and can be used to interpolate over a given subregion of $\mathbb{R}^d$.

The simple kriging model for the same data is

$$y_i = \mu + S(\mathbf{x}_i) + \varepsilon_i,$$

where $\{S(\mathbf{x}) : \mathbf{x} \in \mathbb{R}^d\}$ is a zero mean stationary stochastic process in $\mathbb{R}^d$ that is independent of the ε_i. Interpolation at a general point $\mathbf{x}_0 \in \mathbb{R}^d$ is obtained as

$$\hat{y}_0 = \bar{y} + \hat{S}(\mathbf{x}_0), \tag{13.2}$$

> Stationarity of S means that, for all $\mathbf{h} \in \mathbb{R}^d$, the joint probability distribution of $S(\mathbf{x})$ and $S(\mathbf{x} + \mathbf{h})$ is the same for all $\mathbf{x} \in \mathbb{R}^d$.

where $\hat{S}(\mathbf{x}_0)$ is the best linear predictor of $S(\mathbf{x}_0)$ based on the data in $\mathbf{y}$. We can use (13.2) to interpolate over any given subregion of $\mathbb{R}^d$. Therefore, kriging provides a way of fitting a surface to the point cloud $(\mathbf{x}_i, y_i)$, $1 \le i \le n$. Recall from Section 4.3 that the best linear predictor (BLP) of $S(x_0)$ is the one of the form

$$\hat{S}(\mathbf{x}_0) = \mathbf{a}^\mathsf{T}\mathbf{y} + b$$

that minimizes

$$\mathsf{E}[\{\hat{S}(\mathbf{x}_0) - S(\mathbf{x}_0)\}^2].$$

From (4.5) the solution is

$$\hat{S}(\mathbf{x}_0) = \mathbf{c}_0^\mathsf{T}(\mathbf{C} + \sigma_\varepsilon^2\mathbf{I})^{-1}(\mathbf{y} - \mu\mathbf{1}),$$

where

$$\mathbf{C} \equiv \mathrm{Cov}\{[S(\mathbf{x}_1), \ldots, S(\mathbf{x}_n)]^\mathsf{T}\}$$

and

$$\mathbf{c}_0 = [\mathrm{Cov}\{S(\mathbf{x}_0), S(\mathbf{x}_1)\}, \ldots, \mathrm{Cov}\{S(\mathbf{x}_0), S(\mathbf{x}_n)\}]^\mathsf{T}.$$

It is therefore apparent that the covariance structure of S is all that is needed for obtaining the BLP. The usual approach is to postulate a parsimonious model for $\mathrm{Cov}\{S(\mathbf{x}), S(\mathbf{x} + \mathbf{h})\}$ (which by stationarity depends only on $\mathbf{h}$), estimate the parameters to obtain estimates $\hat{\mathbf{C}}$ and $\hat{\mathbf{c}}_0$, and then apply the kriging formula

$$\hat{y}_0 = \bar{y} + \hat{\mathbf{c}}_0^\mathsf{T}(\hat{\mathbf{C}} + \hat{\sigma}_\varepsilon^2\mathbf{I})^{-1}(\mathbf{y} - \bar{y}\mathbf{1}). \tag{13.3}$$

A common assumption for simplification of $\mathrm{Cov}\{S(\mathbf{x}), S(\mathbf{x} + \mathbf{h})\}$ is the assumption of *isotropy*:

Table 13.1 Some correlation functions and corresponding spectral densities.

	Correlation function	Spectral density
Exponential	$e^{-|r|}$	$\frac{1}{\pi(1+x^2)}$
Gaussian	e^{-r^2}	$\frac{1}{\sqrt{2\pi}}e^{-x^2/2}$
Triangular	$(1-|r|)_+$	$\frac{1-\cos(x)}{\pi x^2}$
Spherical	$\left(1-\frac{3}{2}|r|+\frac{1}{2}|r|^3\right)I_{[-1,1]}(r)$	not simple

$$\text{Cov}\{S(\mathbf{x}), S(\mathbf{x}+\mathbf{h})\} \text{ depends only on } \|\mathbf{h}\|. \tag{13.4}$$

This says that covariance between sites that are $\|\mathbf{h}\|$ units apart is the same, regardless of direction and the sites' location. This is a stronger assumption than stationarity, which says that this covariance is independent of location but may depend on direction. For example, if spatial data are stationary then all pairs of points with one a mile due east of the other have the same covariance, but pairs such that one is a mile due north of the other could have a different covariance. If the data are isotropic then pairs a mile part east-to-west would have the same covariance as pairs one mile apart north-to-south.

Condition (13.4) implies that

$$\mathbf{C} = \left[C(\|\mathbf{x}_i - \mathbf{x}_j\|)\right]_{1 \le i,j \le n},$$

where

$$C(r) \equiv \sigma_S^2 C_0(r), \qquad \sigma_S^2 \equiv \text{Var}\{S(\mathbf{x})\}.$$

Here $C_0(r)$ satisfies $C_0(0) = 1$, and if one selects a class of models for $C_0(r)$ then it should be chosen to ensure that $\mathbf{C}$ is positive definite. The functions C and C_0 are respectively called the *covariance function* and *correlation function* of the isotropic process S.

Characterization of the class of functions that result in valid covariance matrices is of interest when it comes to choosing C_0 in practice. There is a particularly simple solution to this problem known as *Bochner's theorem*. Essentially it states that

C_0 is a valid correlation function if and only if it is the characteristic function of a symmetric random variable.

This means that C_0 must be expressible as

$$C_0(r) = \int_{-\infty}^{\infty} e^{irx} f_X(x)\,dx,$$

where f_X is the density of a continuous random variable X and $f_X(-x) = f_X(x)$. This density is sometimes called the *spectral density* of the process S (e.g. Stein 1999). Some examples are listed in Table 13.1 and graphed in Figure 13.6.

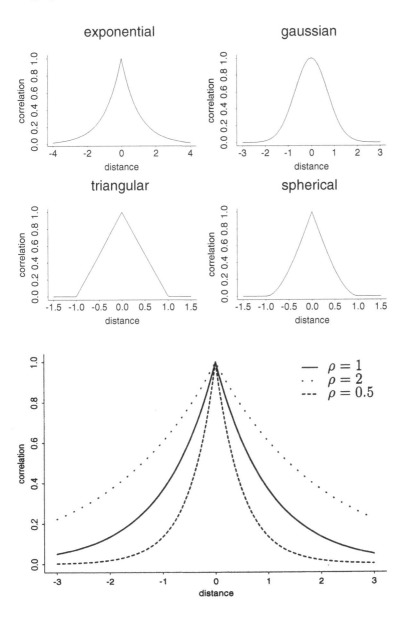

Figure 13.6
Correlation functions from Table 13.1.

Figure 13.7
Exponential correlation function for three different values of the range parameter ρ.

We will now focus on the exponential correlation function

$$C_0(r) = e^{-|r|}$$

because of its particularly simple form. It may be extended in a number of ways. The first is to add a scale parameter to allow for appropriate adjustment when the scale of the x_i change. This results in

$$C_0(r) = e^{-|r/\rho|} \qquad (13.5)$$

for some $\rho > 0$, which is known as the *range* parameter. Figure 13.7 illustrates this covariance function for various values of ρ. Larger values of ρ correspond

Table 13.2 Matérn covariance functions for $\nu = m + 1/2$, $m = 0, 1, 2, 3$.

ν	$C_0(r)$
1/2	$e^{-|r|}$
3/2	$e^{-|r|}(1 + |r|)$
5/2	$e^{-|r|}\left(1 + |r| + \frac{1}{3}|r|^2\right)$
7/2	$e^{-|r|}\left(1 + |r| + \frac{2}{5}|r|^2 + \frac{1}{15}|r|^3\right)$

Figure 13.8 Matérn correlation function for four different values of the smoothness parameter ν.

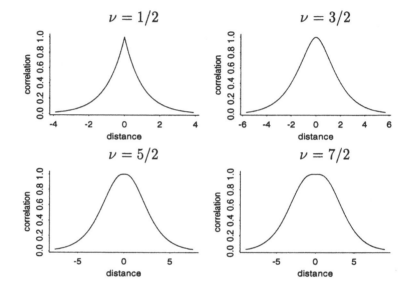

to longer range correlations being present, while the opposite is the case for smaller ρ.

However, the fact that $C_0(r)$ has a cusp at $r = 0$ means that the krige predictor (as a linear combination of $C(\|\mathbf{x}_i - \mathbf{x}_0\|)$, $1 \leq i \leq n$) will have discontinuous partial derivatives. Therefore, it is reasonable to ask for the option that $C(r)$ be a smoother function. Recent literature (e.g., Kent, 1998; Stein 1999) advocate the use of the *Matérn family,* for which the covariance functions have the general form

An alternative parameterization to (13.6) (e.g. Handcock and Wallis 1994) uses $2\sqrt{\nu}/\rho$ in place of $1/\rho$.

$$C_M(r) \equiv \frac{\sigma_S^2}{2^{\nu-1}\Gamma(\nu)} \left(\frac{r}{\rho}\right)^\nu \mathcal{K}_\nu\left(\frac{r}{\rho}\right), \quad \sigma_S^2, \rho, \nu > 0. \quad (13.6)$$

Here $\mathcal{K}_\nu$ is the modified Bessel function of order ν (such functions do not have a closed form for general ν). Bessel functions and modified Bessel functions are so-called special functions in mathematics (Abramowitz and Stegun 1974). They do not have closed-form expressions for general ν, but algorithms and software exist (e.g., the Fortran 77 library SPECFUN and several functions in MATLAB). However, if $\nu = m + \frac{1}{2}$ for $m = 0, 1, 2, \ldots$ then it does have a simple form; see Table 13.2. The functions in this table are plotted in Figure 13.8. The Matérn correlation function with parameter ν corresponds to the characteristic function of a t-density with 2ν degrees of freedom. Stein (1999) strongly advocates use of this family of covariance functions for kriging. Reasons include well-behaved likelihood surfaces and a wide range of smoothness, controlled by $0 < v < \infty$.

The classical approach to selecting C_0 and its parameters, as well as σ_S^2 and σ_ε^2, is through a *variogram* analysis. The *semivariogram* (half the so-called variogram) of the spatial process $\{y(\mathbf{x}) : \mathbf{x} \in \mathbb{R}^d\}$ is defined as

$$\gamma(\mathbf{h}) = \tfrac{1}{2} \text{Var}\{y(\mathbf{x} + \mathbf{h}) - y(\mathbf{x})\}. \tag{13.7}$$

For the isotropic process

$$y(\mathbf{x}) = \mu + S(\mathbf{x}) + \varepsilon(\mathbf{x}),$$

where $\text{Cov}\{S(\mathbf{x}), S(\mathbf{x} + \mathbf{h})\} = C(\|\mathbf{h}\|)$, the semivariogram depends only on the interpoint distance $r \equiv \|\mathbf{h}\|$. In this case we can work with $\gamma_I(r) \equiv \gamma(\|\mathbf{h}\|)$, which can be shown to be

$$\gamma_I(r) = \sigma_\varepsilon^2 + C(0) - C(r), \quad r > 0. \tag{13.8}$$

Note that from (13.7) we have $\gamma_I(0) = 0$, but from (13.8) and the continuity of C (which we assume),

$$\lim_{r \downarrow 0} \gamma_I(r) = \sigma_\varepsilon^2,$$

which in geostatistics is called the *nugget*. Estimators of γ (e.g. Cressie 1993) are then used to choose σ_ε^2 as well as the covariance function C and its parameters.

An alternative to variogram analysis that has received an increasing amount of attention in recent years is likelihood-based estimation of the covariance parameters (Mardia and Marshall 1984; O'Connell and Wolfinger 1997; Stein 1999). Stein (1999), in particular, argues against the use of empirical semivariograms on the grounds that they are too untrustworthy. He recommends Matérn family kriging with σ_S^2, σ_ε^2, ρ, and v all chosen via (restricted) maximum likelihood. This is a tall order, not only because of the number of parameters that need to be estimated simultaneously but also because ρ and v enter in a nonlinear fashion. The benefits of likelihood-based estimation of these two parameters is questionable (see e.g. Nychka 2000, sec. 13.3.4), and there are significant computational advantages to having them chosen in a simpler fashion.

13.3.1 *The Kriging Algorithm*

When all of the pieces of this section are assembled, the following *kriging algorithm* results:

(1) Specify the covariance function C and σ_ε^2 via either variogram analysis or likelihood-based estimation.

(2) Construct the estimated covariance matrix

$$\hat{\mathbf{C}} = \left[C(\|\mathbf{x}_i - \mathbf{x}_j\|) \right]_{1 \le i, j \le n}.$$

(3) Set up a mesh of $\mathbf{x}_0$ values in the subregion of $\mathbb{R}^d$ of interest.

Figure 13.9 An
example scatterplot
for $n = 9$.

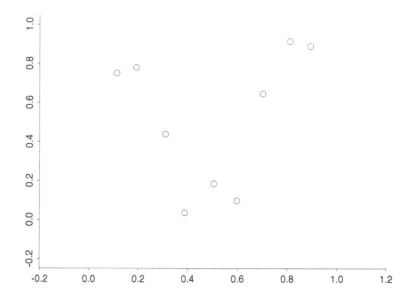

(4) For each $\mathbf{x}_0$ in the mesh:
   (a) compute $\mathbf{c}_0 = [C(\|\mathbf{x}_0 - \mathbf{x}_i\|)]_{1 \le i \le n}$;
   (b) compute

$$\hat{y}(\mathbf{x}_0) = \bar{y} + \mathbf{c}_0^{\mathsf{T}}(\hat{\mathbf{C}} + \hat{\sigma}_\varepsilon^2 \mathbf{I})^{-1}(\mathbf{y} - \bar{y}\mathbf{1}).$$

(5) Plot the $\hat{y}(\mathbf{x}_0)$ values against $\mathbf{x}_0$ to obtain the map.

Finally, we note that this algorithm assumes isotropy of S. There has been a great deal of work on *anisotropic* kriging (e.g. Cressie 1993), where the assumption of isotropy is relaxed.

13.4 General Radial Smoothing

Kriging provides one means of radial smoothing, but not the only one. In this section we show that it is a subset of a family of radial smoothers that also includes smoothing splines. We call the family *general radial smoothers*. We will begin by showing how they differ from the penalized spline smoothers used throughout the earlier chapters. First we will work in one dimension. Because of the radial nature of the smoothing, the higher-dimensional extension is trivial.

Consider the problem of smoothing the scatterplot shown in Figure 13.9. Figure 13.10 shows the truncated line basis for this problem. Note, however, that the knots are at the data $x_1, \ldots, x_n$ with $n = 9$. We will stick with full knots for now so that the connections between kriging and splines will be more apparent. One sidelight concerning the full knot sequence depicted in Figure 13.10 is that the leftmost and rightmost knots are redundant, and the latter leads to a zero column in the design matrix. However, this is not a problem for penalized splines because the ridge regression for the fit is still defined when the smoothing parameter is positive.

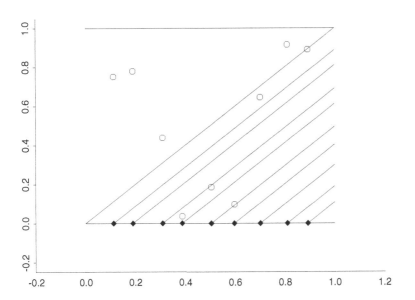

Define

$$\mathbf{X} = [1 \ x_i]_{1 \le i \le n} \quad \text{and} \quad \mathbf{Z} = \left[(x_i - x_j)_+ \right]_{1 \le i, j \le n}. \tag{13.9}$$

We showed in Chapter 3 that smoothing of the scatterplot can be achieved by setting

$$\hat{\mathbf{y}} = \mathbf{X}\hat{\boldsymbol{\beta}} + \mathbf{Z}\hat{\mathbf{u}},$$

where

$$\begin{bmatrix} \hat{\boldsymbol{\beta}} \\ \hat{\mathbf{u}} \end{bmatrix} = \operatorname*{argmin}_{\boldsymbol{\beta}, \mathbf{u}} \left(\| \mathbf{y} - \mathbf{X}\boldsymbol{\beta} - \mathbf{Z}\mathbf{u} \|^2 + \lambda^2 \begin{bmatrix} \boldsymbol{\beta} \\ \mathbf{u} \end{bmatrix}^{\mathsf{T}} \mathbf{D} \begin{bmatrix} \boldsymbol{\beta} \\ \mathbf{u} \end{bmatrix} \right)$$

and where $\mathbf{D} = \operatorname{diag}(0, 0, 1, \dots, 1)$. Recall from Section 4.9 that $\hat{\boldsymbol{\beta}}$ and $\hat{\mathbf{u}}$ correspond to the EBLUP in the mixed model

$$\mathbf{y} = \mathbf{X}\boldsymbol{\beta} + \mathbf{Z}\mathbf{u} + \boldsymbol{\varepsilon}, \quad \operatorname{Cov} \begin{bmatrix} \mathbf{u} \\ \boldsymbol{\varepsilon} \end{bmatrix} = \begin{bmatrix} \sigma_u^2 \mathbf{I} & \mathbf{0} \\ \mathbf{0} & \sigma_\varepsilon^2 \mathbf{I} \end{bmatrix}, \quad \lambda^2 = \frac{\hat{\sigma}_\varepsilon^2}{\hat{\sigma}_u^2}.$$

Now consider the transformation of the truncated line basis shown in Figure 13.11. These are obtained by taking linear combinations of the columns of $\mathbf{X}$ and $\mathbf{Z}$ in such a way that $\mathbf{X}$ remains unchanged and $\mathbf{Z}$ becomes the radially symmetric matrix

$$\mathbf{Z}_R = \left[|x_i - x_j| \right]_{1 \le i, j \le n}.$$

The transformation can be expressed in terms of an $(n + 2) \times (n + 2) = 11 \times 11$ matrix $\mathbf{L}$ for which

$$[\mathbf{X} \ \mathbf{Z}_R] = [\mathbf{X} \ \mathbf{Z}]\mathbf{L}.$$

The vector of fitted values for this new basis is

$$\hat{\mathbf{y}} = \mathbf{X}\hat{\boldsymbol{\beta}}_R + \mathbf{Z}_R\hat{\mathbf{u}}_R,$$

where $\hat{\boldsymbol{\beta}}_R$ and $\hat{\mathbf{u}}_R$ are given by

Figure 13.11 A linear
transformation of the
basis functions of
Figure 13.10.

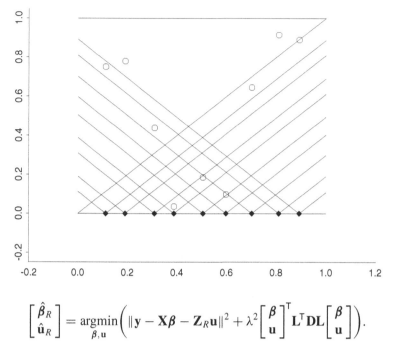

$$\begin{bmatrix} \hat{\boldsymbol{\beta}}_R \\ \hat{\mathbf{u}}_R \end{bmatrix} = \underset{\boldsymbol{\beta},\mathbf{u}}{\operatorname{argmin}} \left(\|\mathbf{y} - \mathbf{X}\boldsymbol{\beta} - \mathbf{Z}_R \mathbf{u}\|^2 + \lambda^2 \begin{bmatrix} \boldsymbol{\beta} \\ \mathbf{u} \end{bmatrix}^{\mathsf{T}} \mathbf{L}^{\mathsf{T}} \mathbf{DL} \begin{bmatrix} \boldsymbol{\beta} \\ \mathbf{u} \end{bmatrix} \right).$$

While $\mathbf{Z}_R$ exhibits radial symmetry, the same is not true for the penalty

$$\lambda^2 \begin{bmatrix} \boldsymbol{\beta} \\ \mathbf{u} \end{bmatrix}^{\mathsf{T}} \mathbf{L}^{\mathsf{T}} \mathbf{DL} \begin{bmatrix} \boldsymbol{\beta} \\ \mathbf{u} \end{bmatrix}. \tag{13.10}$$

Also, penalty (13.10) does not have a simple multivariate extension. An alternative penalty that does is simply $\lambda \mathbf{u}^{\mathsf{T}} \mathbf{Z}_R \mathbf{u}$. This leads to the criterion

$$\begin{bmatrix} \hat{\boldsymbol{\beta}} \\ \hat{\mathbf{u}} \end{bmatrix} = \underset{\boldsymbol{\beta},\mathbf{u}}{\operatorname{argmin}} (\|\mathbf{y} - \mathbf{X}\boldsymbol{\beta} - \mathbf{Z}_R \mathbf{u}\|^2 + \lambda \mathbf{u}^{\mathsf{T}} \mathbf{Z}_R \mathbf{u}). \tag{13.11}$$

On face value the penalty $\lambda \mathbf{u}^{\mathsf{T}} \mathbf{Z}_R \mathbf{u}$ might seem somewhat arbitrary and with no justification other than its radial symmetry. However, it can be shown that such a choice corresponds to the *thin plate spline* family of smoothers (see e.g. Green and Silverman 1994) and, in the nonparametric regression model

Cubic smoothing
splines, mentioned
in Chapter 3, are
members of the family
of thin plate splines.

$$y_i = f(x_i) + \varepsilon_i,$$

corresponds to minimization of

$$\sum_{i=1}^n \{y_i - f(x_i)\}^2 + \lambda \int_{-\infty}^{\infty} \{f'(x)\}^2 \, dx. \tag{13.12}$$

The penalty in (13.12) is the integral of a squared derivative – in this case the first derivative, which is appropriate for a linear spline. Cubic smoothing splines correspond to penalization of the second derivative. However, it is possible to penalize the mth derivative for any m such that $2m > d$, where d is the dimension of x; that is, $d = 1$ here and $d = 2$ for bivariate smoothing, which we consider next (Green and Silverman 1994; Nychka 2000). Note that $m = 1$ is possible only for 1-dimensional smoothing.

Can estimator (13.11) be derived from a mixed model? It would appear that the solution to (13.11) corresponds to the EBLUP in the "mixed model"

$$\mathbf{y} = \mathbf{X}\boldsymbol{\beta} + \mathbf{Z}_R\mathbf{u} + \boldsymbol{\varepsilon}, \quad \text{Cov}\begin{bmatrix} \mathbf{u} \\ \boldsymbol{\varepsilon} \end{bmatrix} = \begin{bmatrix} \sigma_u^2\mathbf{Z}_R^{-1} & \mathbf{0} \\ \mathbf{0} & \sigma_\varepsilon^2\mathbf{I} \end{bmatrix}.$$

However, this is not a valid mixed model since it implies that

$$\text{Cov}(\mathbf{Z}_R\mathbf{u}) = \sigma_u^2\mathbf{Z}_R$$

even though $\mathbf{Z}_R$ is *not* a proper covariance matrix – it is not necessarily positive definite. There are at least three possible ways out, as shown in each of the following three subsections.

13.4.1 *Generalized Covariance Functions*

The first way to rectify the lack of positive definiteness of $\mathbf{Z}_R$ is to replace it by

$$\mathbf{Z}_P = \left[-|x_i - x_j| + |x_i| + |x_j| \right]_{1 \leq i, j \leq n}.$$

This matrix is positive semidefinite and usually positive definite, as will be assumed, so

$$\mathbf{y} = \mathbf{X}\boldsymbol{\beta}_R + \mathbf{Z}_P\mathbf{u}_R + \boldsymbol{\varepsilon}, \quad \text{Cov}\begin{bmatrix} \mathbf{u} \\ \boldsymbol{\varepsilon} \end{bmatrix} = \begin{bmatrix} \sigma_u^2\mathbf{Z}_P^{-1} & \mathbf{0} \\ \mathbf{0} & \sigma_\varepsilon^2\mathbf{I} \end{bmatrix}$$

is a valid mixed model. Moreover, as shown in Section 13.9, $\hat{\boldsymbol{\beta}}$ and $\hat{\mathbf{u}}$ in (13.11) are unaffected by replacement of $\mathbf{Z}_R$ by $\mathbf{Z}_P$ (French, Kammann, and Wand 2001). The same is true for the fitted values. The function

$$C(r) = -|r|$$

is sometimes referred to as a *generalized covariance function,* since it is possible to add increments which make it a valid covariance function but which cancel in the BLUP computations and have no effect on the final answer (Kitanidis 1997).

13.4.2 *Positive Definitization*

For a general square matrix $\mathbf{M}$, there exists a square root $\mathbf{M}^{1/2}$. The matrices

$$\mathbf{M}^{1/2}(\mathbf{M}^{1/2})^\mathsf{T} \quad \text{and} \quad (\mathbf{M}^{1/2})^\mathsf{T}\mathbf{M}^{1/2}$$

are both positive semidefinite, and positive definite if $\mathbf{M}$ is nonsingular.

We can also obtain a valid mixed model by using the *positive definitization* of $\mathbf{Z}_R$,

$$\left(\left[|x_i - x_j| \right]_{1 \leq i, j \leq n}^{1/2} \right)^\mathsf{T} \left[|x_i - x_j| \right]_{1 \leq i, j \leq n}^{1/2}.$$

The radial properties are maintained, and the hope is that it makes little practical difference to the smooth.

The square root of a matrix can be defined by its *singular value decomposition*; Section A.2.11 of the Appendix gives the details. Note that this definition differs from that of the symmetric matrix square root of the MATLAB language.

Figure 13.12
Basis functions
corresponding
to $\mathbf{Z}_E \equiv$
$[e^{-|x_i-x_j|/\rho}]_{1\leq i,j\leq n}$.

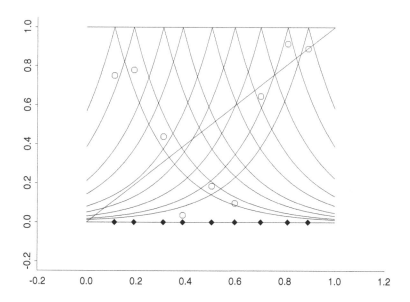

13.4.3 *Proper Covariance Matrices*

The final way out is to simply use proper covariance functions such as

$$e^{-|x_i-x_j|/\rho}$$

since, as we discussed in Section 13.3,

$$\mathbf{Z}_E \equiv \left[e^{-|x_i-x_j|/\rho} \right]_{1\leq i,j\leq n}$$

is a proper covariance matrix (see Figure 13.12). But this, essentially, just gets us back to kriging. When viewed this way kriging and smoothing splines are seen to be quite closely related. For fitting at the data, they are both just penalized regressions with radial basis functions. Some subtle differences exist for fitting at other values of $x_0 \in \mathbb{R}$. An advantage of kriging is that the mixed model representation is immediate because the $\mathbf{Z}$ matrix is positive definite. As we pointed out at the end of Section 13.3, a disadvantage is that the range parameter ρ, and perhaps a smoothness parameter, needs to be chosen.

13.4.4 *Low-Rank Radial Smoothers*

Each of the radial smoothers described in the preceding subsections are BLUPs for mixed models of the form

$$\mathbf{y} = \mathbf{X}\boldsymbol{\beta} + \mathbf{Z}_C\mathbf{u} + \boldsymbol{\varepsilon},$$

where $\mathbf{X}$ is defined by (13.9),

$$\text{Cov}(\mathbf{u}) = \sigma_u^2 (\mathbf{Z}_C^{-1/2})(\mathbf{Z}_C^{-1/2})^{\mathsf{T}}, \tag{13.13}$$

and

$$\mathbf{Z}_C \equiv \left[C(|x_i - x_j|) \right]_{1 \le i, j \le n}$$

for some real-valued function C possibly containing parameters. The fitted functions involve linear combinations of

$$C(|x - x_j|), \quad 1 \le j \le n,$$

for $x \in \mathbb{R}$.

Such a smoother is *full-rank* as defined in Section 3.12. As we have argued a number of times in the preceding chapters, there are considerable computational payoffs from using low-rank approximations. We now describe a low-rank extension for radial smoothers.

Let $\kappa_1, \ldots, \kappa_K$ be a set of knot locations. Then an approximation based on the smaller set of basis functions

$$C(|x - \kappa_k|), \quad 1 \le k \le K,$$

arises from fitting the mixed model

$$\mathbf{y} = \mathbf{X}\boldsymbol{\beta} + \mathbf{Z}_K \mathbf{u} + \boldsymbol{\varepsilon}, \quad \text{Cov}(\mathbf{u}) = \sigma_u^2 (\boldsymbol{\Omega}_K^{-1/2})(\boldsymbol{\Omega}_K^{-1/2})^{\mathsf{T}},$$

where $\mathbf{X}$ is given by (13.9),

$$\mathbf{Z}_K \equiv \left[C(|x_i - \kappa_k|) \right]_{\substack{1 \le k \le K \\ 1 \le i \le n}}, \quad \text{and} \quad \boldsymbol{\Omega}_K \equiv \left[C(|\kappa_k - \kappa_{k'}|) \right]_{1 \le k, k' \le K};$$

we observe that $\mathbf{u}$ is now a $K \times 1$ random vector. Using the transformation $\mathbf{Z} = \mathbf{Z}_K \boldsymbol{\Omega}_K^{-1/2}$, the final model can be written as

$$\mathbf{y} = \mathbf{X}\boldsymbol{\beta} + \mathbf{Z}\mathbf{u} + \boldsymbol{\varepsilon}, \quad \text{Cov} \begin{bmatrix} \mathbf{u} \\ \boldsymbol{\varepsilon} \end{bmatrix} = \begin{bmatrix} \sigma_u^2 \mathbf{I} & \mathbf{0} \\ \mathbf{0} & \sigma_\varepsilon^2 \mathbf{I} \end{bmatrix}. \tag{13.14}$$

This form allows fitting through standard mixed model software.

How do each of these alternatives compare to the ordinary penalized spline smoother? In lieu of a thorough simulation study, Figure 13.13 shows the results of applying some of the kriging-type smoothers to the LIDAR data. In each case 12 degrees of freedom are used and the range parameter is chosen so that the basis functions cover approximately the range of the data. The fits are virtually identical and, when overlaid, are indistinguishable from one another (there are some slight differences near the boundaries).

Radial smoothers in one dimension have little to add compared with ordinary penalized splines. Their strength is their simple extendibility to multivariate predictors, as we will now show.

13.4.5   *Higher-Dimensional Radial Smoothers*

So far, we have only discussed radial smoothers in one dimension. Our real interest in them is for higher-dimensional smoothing, a topic we can now take up.

Because of its radial form, dependence on the data is entirely through the point-to-point distances

Figure 13.13
Comparison of four
radial smooths of the
LIDAR data with a
linear and quadratic
penalized spline
smooth; 12 degrees of
freedom are used in
each case.

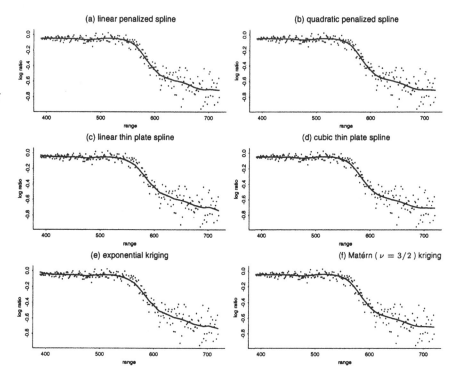

$$|x_i - \kappa_k|, \quad 1 \le i \le n, \ 1 \le k \le K. \tag{13.15}$$

Therefore, extension to $\mathbf{x}_i \in \mathbb{R}^d$ essentially involves replacing (13.15) with

$$\|\mathbf{x}_i - \kappa_k\|, \quad 1 \le i \le n, \ 1 \le k \le K.$$

For $\mathbf{x}_i \in \mathbb{R}^d$ $(1 \le i \le n)$ and $\kappa_k \in \mathbb{R}^d$ $(1 \le k \le K)$, approximate thin plate splines of higher dimension can be obtained by taking $\mathbf{X}$ to have columns spanning the space of all d-dimensional polynomials in the components of the $\mathbf{x}_i$ with degree less than m:

$$\mathbf{Z} = \left[C(\|\mathbf{x}_i - \kappa_k\|) \right]_{\substack{1 \le i \le n \\ 1 \le k \le K}} \left[C(\|\kappa_k - \kappa_{k'}\|) \right]_{1 \le k, k' \le K}^{-1/2},$$

where

$$C(\mathbf{r}) = \begin{cases} \|\mathbf{r}\|^{2m-d} & \text{for } d \text{ odd,} \\ \|\mathbf{r}\|^{2m-d} \log \|\mathbf{r}\| & \text{for } d \text{ even,} \end{cases}$$

and m is an integer satisfying $2m - d > 0$ that controls the smoothness of $C(\cdot)$ (see discussion following (13.12)). Note the addition of the $\log \|\mathbf{r}\|$ factor for even dimensions. In the full knot case, this arises from the multivariate extension of squared derivative penalties such as that given in (13.12).

Alternatively, one could use radial basis functions corresponding to the proper covariance functions described in Section 13.3. For example, the two simplest members of the Matérn class (for any $d \ge 1$) are

$$C(\mathbf{r}) = \begin{cases} \exp(-\|\mathbf{r}\|/\rho), & \nu = \tfrac{1}{2}, \\ \exp(-\|\mathbf{r}\|/\rho)(1 + \|\mathbf{r}\|/\rho), & \nu = \tfrac{3}{2}. \end{cases}$$

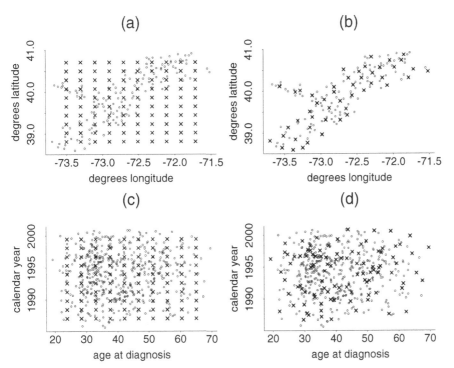

Figure 13.14 Knot selection based on rectangular lattices (panels (a) and (c)) and space filling algorithms ((b) and (d)) for the scallop and Italian AIDS data; o = data and x = knot.

13.4.6   *Choice of Knots*

In the full-rank case the knots correspond to the predictors, but in the low-rank case a set of $K < n$ knots in $\mathbb{R}^d$ needs to be chosen. One approach is to put down a rectangular lattice of knots as in panels (a) and (c) of Figure 13.14. However, this has the tendency to waste a lot of knots. For example, for the scallop data only about 40% of the knots are needed. A reasonable alternative strategy is to have the knots "mimic" the distribution of the predictor space. In one dimension, a simple solution is

$$\kappa_k = (k/K)\text{th sample quantile of the unique } x_i.$$

However, the notion of quantile does not have a straightforward extension beyond $d = 1$ dimension. A pointer on how to reasonably handle $d > 1$ arises from the fact that, for $d = 1$, sample quantiles correspond to *maximal separation* of K points among the unique x_i. In higher dimensions, *space filling* designs (Johnson, Moore, and Ylvisaker 1990; Nychka and Saltzman 1998) are based on a maximal separation principle. Panels (b) and (d) of Figure 13.14 show the results of applying such an algorithm to the scallop and Italian AIDS data. We see that there is no wastage of knots, which will likely lead to better approximation in sparse regions of the prediction space. On the other hand, space filling knot selection is harder to implement. The FUNFITS module (Nychka et al. 1998) provides some software for this, although it can be slow for large n and K. A remedy for this problem is to apply the space filling algorithm to a random sample of the x_i.

At the time of writing, the FUNFITS module is available at the Internet site (www.cgd.ucar. edu/stats/ Funfits/index. shtml).

Figure 13.15
Bivariate smooths
of the scallop data
with (a) 5 degrees
of freedom,
(b) 10 degrees
of freedom,
(c) 20 degrees
of freedom, and
(d) 40 degrees of
freedom. Lighter
areas correspond to
higher catches.

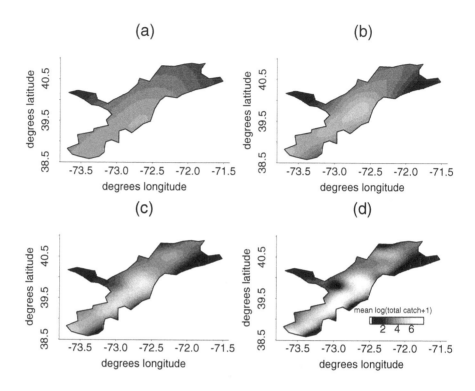

13.4.7 *Degrees of Freedom*

Although the notion of degrees of freedom is usually confined to univariate smoothers, the principle extends naturally to most higher-dimensional smoothers, including those described in the current section. Since the fitted values can be written in the form

$$\hat{\mathbf{y}} = \mathbf{X}\hat{\boldsymbol{\beta}} + \mathbf{Z}\hat{\mathbf{u}},$$

there is an $n \times n$ matrix $\mathbf{S}$ for which

$$\hat{\mathbf{y}} = \mathbf{S}\mathbf{y}$$

and we can define

$$df_{\text{fit}} = \text{tr}(\mathbf{S}).$$

A planar fit uses 3 degrees of freedom, whereas an unpenalized fit based on K radial basis functions uses $K + 3$ degrees of freedom. All df_{fit} values in between are possible via appropriate tweaking of the smoothing parameter λ or the variance ratio $\sigma_\varepsilon^2/\sigma_u^2$. Figure 13.15 shows bivariate smooths of the scallop data with four different degrees-of-freedom values.

13.5 Default Automatic Bivariate Smoother

Based on considerations summarized in the previous three sections, we have arrived at a "default" automatic bivariate smoother with which we are reasonably happy. It is a low-rank thin plate spline with REML smoothing parameter choice and space filling knot selection. The attractions are

- quite fast for even large sample sizes,
- stable even when the design has lots of sparse regions, and
- implementable using mixed model software once the knots have been selected.

The most difficult aspect is space filling selection of the knots, but publicly available software is available for this (Nychka et al. 1998).

The full prescription of this automatic bivariate smoother for inputs $\mathbf{x}_1, \ldots, \mathbf{x}^n \in \mathbb{R}^2$ and $y_1, \ldots, y_n \in \mathbb{R}$ is as follows.

(1) Choose the number of knots to be

$$K = \max\{20, \min(n/4, 150)\}.$$

(2) If $n > 1500$ then take a random sample of the $\mathbf{x}_i$ of size 1500. Do not do this if $n \leq 1500$. Apply the space filling algorithm to the sample to obtain knots $\kappa_1, \ldots, \kappa_K \in \mathbb{R}^2$.

(3) Form the matrices $\mathbf{X} = [1 \quad \mathbf{x}_i^{\mathsf{T}}]_{1 \leq i \leq n}$,

$$\mathbf{Z}_K = \left[\|\mathbf{x}_i - \kappa_k\|^2 \log\|\mathbf{x}_i - \kappa_k\| \right]_{\substack{1 \leq k \leq K \\ 1 \leq i \leq n}},$$

and

$$\mathbf{\Omega} = \left[\|\kappa_k - \kappa_{k'}\|^2 \log\|\kappa_k - \kappa_{k'}\| \right]_{1 \leq k, k' \leq K}.$$

(4) Find the singular value decomposition of $\mathbf{\Omega}$,

$$\mathbf{\Omega} = \mathbf{U} \operatorname{diag}(\mathbf{d}) \mathbf{V}^{\mathsf{T}},$$

and use this to obtain the matrix square root of $\mathbf{\Omega}$:

$$\mathbf{\Omega}^{1/2} = \mathbf{U} \operatorname{diag}(\sqrt{\mathbf{d}}) \mathbf{V}^{\mathsf{T}}.$$

(5) Compute

$$\mathbf{Z} = \mathbf{Z}_K \mathbf{\Omega}^{-1/2}.$$

(6) Use mixed model software to fit the mixed model

$$\mathbf{y} = \mathbf{X}\boldsymbol{\beta} + \mathbf{Z}\mathbf{u} + \boldsymbol{\varepsilon}, \quad \begin{bmatrix} \mathbf{u} \\ \boldsymbol{\varepsilon} \end{bmatrix} \sim N\left(\begin{bmatrix} \mathbf{0} \\ \mathbf{0} \end{bmatrix}, \begin{bmatrix} \sigma_u^2 \mathbf{I} & \mathbf{0} \\ \mathbf{0} & \sigma_\varepsilon^2 \mathbf{I} \end{bmatrix} \right),$$

with σ_u^2 and σ_ε^2 chosen via REML. Let $\boldsymbol{\beta}$ and $\mathbf{u}$ be the resulting EBLUPs of $\boldsymbol{\beta}$ and $\mathbf{u}$.

(7) Set up a mesh of $\mathbf{x}_0$ values in the subregion of $\mathbb{R}^2$ of interest.

(8) For each $\mathbf{x}_0$ in the mesh,

    (a) form $\mathbf{X}_0 = [1 \quad \mathbf{x}_0^{\mathsf{T}}]$ and

$$\mathbf{Z}_0 = [\|\mathbf{x}_0 - \kappa_k\|^2 \log\|\mathbf{x}_0 - \kappa_k\|]_{1 \leq k \leq K} (\mathbf{\Omega}^{1/2})^{-1};$$

(b) compute

$$\hat{y}(\mathbf{x}_0) = \mathbf{X}_0\hat{\boldsymbol{\beta}} + \mathbf{Z}_0\hat{\mathbf{u}}.$$

(9) Plot the $\hat{y}(\mathbf{x}_0)$ values against $\mathbf{x}_0$.

Note that using REML implies that normality of the response can be reasonably assumed. If this is not reasonable (owing to e.g. outliers or heavy skewness) then the algorithm described here may need modification, or perhaps the response should be transformed.

13.6 Geoadditive Models

A basic geostatistical study concerns mapping the mean of a response of interest, y, based on data

$$(y_i, \texttt{longitude}_i, \texttt{latitude}_i), \quad 1 \leq i \leq n.$$

However, in many such studies data on other variables are likely to impact y_i and could have a confounding effect. For example, if y_i measures occurrence of Alzheimer's disease in a particular city then suburbs with more retirees are likely to have higher prevalence, regardless of environmental exposures. Age will thus have a confounding effect, so a proper assessment of environmental effects on Alzheimer's disease will require controlling for age. In the Upper Cape Cod reproductive study described in Kammann and Wand (2003), possible confounders are smoking level and maternal age. Yet some of these have nonlinear relationships with birthweight. Under the additivity assumption we can handle such covariate effects by combining the ideas of additive models and kriging. We call the result a *geoadditive* model.

For clarity, consider the situation where (s_i, t_i), $1 \leq i \leq n$, represent continuous confounders and $\mathbf{x}_i \in \mathbb{R}^2$. A geoadditive model for such data is

$$y_i = f(s_i) + g(t_i) + h(\mathbf{x}_i) + \varepsilon_i, \quad s_i, t_i \in \mathbb{R}, \ \mathbf{x}_i \in \mathbb{R}^2. \tag{13.16}$$

This can be written as a mixed model

$$\mathbf{y} = \mathbf{X}\boldsymbol{\beta} + \mathbf{Z}\mathbf{u} + \boldsymbol{\varepsilon},$$

where

$$\mathbf{X} = [1 \ \ s_i \ \ t_i \ \ \mathbf{x}_i^\mathsf{T}]_{1 \leq i \leq n}$$

and $\mathbf{Z}$ is obtained by concatenating matrices containing spline basis functions to handle f, g, and h, respectively. For f and g, either linear spline or radial basis functions could be used. For h we prefer radial basis functions with default choices as given in Section 13.5. Through appropriate linear transformation of the basis functions (see e.g. steps (4) and (5) of Section 13.5) it is possible to achieve the canonical covariance structure for $\mathbf{u}$,

$$\text{Cov}(\mathbf{u}) = \begin{bmatrix} \sigma_f^2\mathbf{I} & \mathbf{0} & \mathbf{0} \\ \mathbf{0} & \sigma_g^2\mathbf{I} & \mathbf{0} \\ \mathbf{0} & \mathbf{0} & \sigma_h^2\mathbf{I} \end{bmatrix},$$

and fit the model simultaneously with mixed model software. Kammann and Wand (2003) give a fuller description of geoadditive models and illustrate their use on the Upper Cape Cod reproductive data.

13.7 Additive Plus Interaction Models

Several types of interactions were discussed in Chapter 12, but continuous-by-continuous interactions were postponed until this chapter. We now have machinery sufficient to handle this type of interaction. In fact, model (13.16) is a special case of a model with additive terms and continuous-by-continuous interactions. More complex additive models with interaction terms are easy to construct, thanks again to the modularity of spline modeling.

Suppose that we have the four predictors s, t, u, v, that s enters additively, that t and u possibly interact, and that t and v also possibly interact. Then our model is

$$y_i = f(s_i) + g(t_i, u_i) + h(t_i, v_i) + \varepsilon_i, \quad s_i, t_i, u_i, v_i \in \mathbb{R}. \tag{13.17}$$

To implement this model, we would take a univariate spline basis in s and bivariate radial bases in both (t, u) and (t, v). To view this model as a mixed linear model, the polynomial terms in each basis – plus a column of ones for the intercept – would be put together to form the matrix $\mathbf{X}$. The polynomial terms would be the linear terms t, u, and v and, if we model f as a pth-degree spline, $s, \ldots, s^p$. The other terms, truncated power function in s and radial basis function in (t, u) and (t, v), would form $\mathbf{Z}$.

13.8 Generalized Bivariate Smoothing

Given the earlier sections and Chapter 10, extension of bivariate smoothing and geoadditive models to the generalized response situation is relatively straightforward. For example, if the response y is binary then the analogue of (13.16) is

$$\text{logit}\{P(y_i = 1)\} = f(s_i) + g(t_i) + h(\mathbf{x}_i); \tag{13.18}$$

this can be fit through a mixed model of the form

$$\text{logit}\{P(y_i = 1|\mathbf{u})\} = (\mathbf{X}\boldsymbol{\beta} + \mathbf{Zu})_i,$$

where $\mathbf{u}$ is a random effects vector with covariance structure exemplified by that given in (13.13). In the case where all covariate effects are linear, (13.18) essentially corresponds to the model proposed by Diggle et al. (1998).

Such models are a special case of generalized linear mixed models described in Section 10.8 and can be fit using algorithms there. The formulation and implementation of Bayesian approaches to such models is described in Chapter 16.

13.9 Appendix: Equivalence of BLUP using $\mathbf{Z}_R$ and $\mathbf{Z}_P$

The equivalence of the BLUP based on $\mathbf{Z}_P$ or $\mathbf{Z}_R$ can be established through examination of the criterion for BLUP given, for example, in McCulloch and Searle

(2001). For each $1 \leq i \leq n$, we seek ℓ for which $\ell^\mathsf{T}\mathbf{y}$ is unbiased for $(\mathbf{X}\boldsymbol{\beta}+\mathbf{Z}_P\mathbf{u})_i$. This is equivalent to

$$\mathbf{X}^\mathsf{T}\ell = \mathbf{X}^\mathsf{T}\mathbf{e}_i,$$

where $\mathbf{e}_i$ is the $n \times 1$ vector with 1 in the ith position and 0s elsewhere. For $\mathbf{X}$ as given in (13.9), the unbiasedness condition becomes

$$\ell^\mathsf{T}\mathbf{1} = 1 \quad \text{and} \quad \ell^\mathsf{T}\mathbf{x} = x_i,$$

where $\mathbf{x} = [x_1, \ldots, x_n]^\mathsf{T}$.

Let m_0 and m_1 be Lagrange multipliers that impose the unbiasedness conditions. For ℓ to minimize the variance of the prediction error $\ell^\mathsf{T}\mathbf{y} - (\mathbf{X}\boldsymbol{\beta}+\mathbf{Z}_P\mathbf{u})_i$, we need to minimize

$$
\begin{aligned}
\text{Var}&\{\ell^\mathsf{T}\mathbf{y} - (\mathbf{X}\boldsymbol{\beta} + \mathbf{Z}_P\mathbf{u})_i\} + m_0(\ell^\mathsf{T}\mathbf{1} - 1) + m_1(\ell^\mathsf{T}\mathbf{x} - x_i) \\
&= \ell^\mathsf{T}(\sigma_u^2\mathbf{Z}_P + \sigma_\varepsilon^2\mathbf{I})\ell - 2\sigma_u^2\ell^\mathsf{T}\mathbf{Z}_P\mathbf{e}_i + \sigma_u^2(\mathbf{Z}_P)_{ii} \\
&\quad + m_0(\ell^\mathsf{T}\mathbf{1} - 1) + m_1(\ell^\mathsf{T}\mathbf{x} - x_i) \\
&= \ell^\mathsf{T}(\sigma_u^2\mathbf{Z}_R + \sigma_\varepsilon^2\mathbf{I})\ell - 2\sigma_u^2\ell^\mathsf{T}\mathbf{Z}_R\mathbf{e}_i + \sigma_u^2(\mathbf{Z}_R)_{ii} \\
&\quad + m_0(\ell^\mathsf{T}\mathbf{1} - 1) + m_1(\ell^\mathsf{T}\mathbf{x} - x_i) + 2\sigma_u^2(\ell^\mathsf{T}\mathbf{1} - 1)(\ell^\mathsf{T}|\mathbf{x}| - |x_i|).
\end{aligned}
$$

It is clear from this that the first unbiasedness constraint will cause the last term to vanish and that minimization over ℓ is unaffected by replacement of $\mathbf{Z}_P$ by $\mathbf{Z}_R$. Therefore, for fixed $\lambda = \sigma_\varepsilon^2/\sigma_u^2$, only $\mathbf{Z}_R$ matters.

Note that similar arguments have appeared in the literature to demonstrate equivalences between smoothing splines and kriging (e.g., Kimeldorf and Wahba 1971; Duchon 1976; Cressie 1990).

13.10 Bibliographical Notes

Reviews of multivariate thin plate splines may be found in Green and Silverman (1994) and Gu (2000). Kriging is summarized in several books, including Cressie (1993), Kitanidis (1997), and Stein (1999). Nychka (2000) describes connections between kriging and splines.

14

Variance Function Estimation

14.1   Introduction

In Chapters 1–6 we concentrated on the estimation of regression functions for the the mean of the response y as a function of a predictor x. In the simplest case, the regression function was modeled semiparametrically using linear regression splines, so that

$$f(x) = \mathsf{E}(y|x) = \beta_0 + \beta_1 x + \sum_{k=1}^{K} u_k (x - \kappa_k)_+.$$

The function was fit using the linear mixed model (see Section 4.9) under the assumption that

$$g(x) = \mathrm{Var}(y|x) = \sigma_\varepsilon^2. \tag{14.1}$$

In many examples, the assumption (14.1) of constant conditional variance is unrealistic. For example, consider the LIDAR data in Figure 14.1. It is clear from this figure that the variability in the `logratio` is much smaller when `range` = 400 than it is when `range` = 700. This situation of the variability changing as the predictor changes is known as *heteroscedasticity*; constant conditional variance

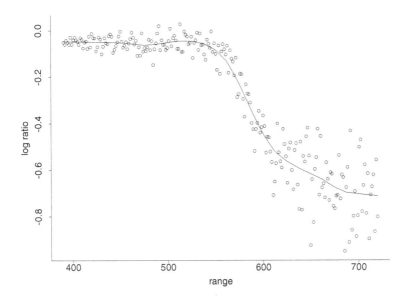

Figure 14.1 LIDAR data with scatterplot smooth. Note how the variability in the response (`logratio`) increases as the predictor (`range`) increases. This is known as heteroscedasticity.

is known as *homoscedasticity*. A comprehensive review of heteroscedasticity is given in Carroll and Ruppert (1988). Details of fitting models for variances in the general parametric case, including nonlinear least squares, were discussed in Section 10.7.

The purpose of this chapter is to give a brief introduction to the problem of heteroscedasticity in smoothing problems and to indicate how one can both understand and adjust for such nonconstant variance. We make no pretense that this is the last word on the subject, and it indeed remains a field of active research; see for example Opsomer et al. (1999).

In the linear mixed model, not accounting for heteroscedasticity does not necessarily invalidate the estimate of the regression function. The estimator that assumes homoscedasticity is unbiased and often is reasonably efficient under heteroscedasticity. This is a well-known robustness property of generalized least squares (Section 4.5). There are, however, three main reasons for trying to understand how the variability changes with the predictor.

It is possible to understand and adjust for nonconstant variances in smoothing and semiparametric problems.

1. *The regression function is estimated most efficiently when nonconstant variance is accounted for.* This is generally the least important reason, because the robustness property of generalized least squares suggests that one will not do too badly in terms of estimation. However, in examples where the response standard deviation (conditional on the predictors) varies over several orders of magnitude, not accounting for heteroscedasticity can be disastrous.

A major purpose of understanding and adjusting nonconstant variation is validity of inference. In addition, the efficiency of inference is enhanced – for example, smaller standard error bars.

2. *Ignoring heteroscedasticity may lead to incorrect inferences.* This is probably the most important reason for thinking about heteroscedasticity.

(a) For example, the structure of the data in Figure 14.1 makes it clear that, if we were to try to draw a confidence interval for the true regression function, we should be much less certain about what will happen for large values of range ($= x$) than for small values.

(b) Carroll and Ruppert (1988) point out that calibration inference (predicting an x from a y) is adversely affected by ignoring heteroscedasticity.

(c) In semiparametric models with a parametric component (as in Chapter 7), inference about the regression parameter is not valid unless heteroscedasticity is accounted for. In our experience, this can often mean the difference between a statistically significant and a statistically nonsignificant outcome.

3. In some cases, *understanding how variability changes with the predictor may be of intrinsic interest*. We have encountered instances in biology where the effect of a treatment is to cause an increase in variance rather than an increase in mean. In financial engineering, the variances of returns on assets are of fundamental importance. For example, the famous Black–Scholes formula gives the price of a call option, which gives its owner the right (but not the obligation) to purchase a stock at a specified price and future date. The price of the option depends on the variance of the stock returns but not on the expected value.

There are three general approaches to handling the problem of heteroscedasticity. First, one can *change the model*. For example, with positive data such as measured amounts or counts, one would expect nonconstant variance. The Poisson model for count data states that the variance equals the mean, while the Gamma model states that the variance is proportional to the square of the mean. In such circumstances, the data may be modeled most naturally as a generalized linear mixed model (Chapter 10). However, there are many problems where natural generalized linear mixed models do not capture the variability adequately. For example, in radioimmunoassays and ELISA assays, the variability is generally somewhere in between that suggested by Poisson variation and Gamma variation.

Generalized linear models (GLMs) can impose a particular form to the variances. For example, in the Gamma GLM, the variance is proportional to the square of the mean.

A second approach is to *transform the response data* to make the variation more constant. As described in Section 2.9, we tried various transformations of the LIDAR response `logratio` and were unable to find one that stabilized the variance.

Response transformation is often a simple way of removing nonconstant variation.

The third and final approach is to *model the variance* as a function of the predictor. We take up this approach in the following section. As discussed in Carroll and Ruppert (1988), transformations are appropriate only when the conditional variance of the response is a function of the response's conditional expectation. The same is true of GLMs. If the conditional variance is a function of a predictor variable rather than a function of the conditional expectation, then modeling the conditional variance is necessary. An example is the LIDAR data, for which the conditional variance of `logratio` is a function of `range` but not a function of $E(\texttt{logratio}|\texttt{range})$.

14.2   Formulation

The idea of *variance function estimation* is to allow the variance to be a function of the predictors. In other words, this means treating the variance as if it were a regression function.

There are many ways to implement variance function estimation. In this section, we illustrate one approach that is appropriate for the LIDAR data, where there is no natural generalized linear mixed model and there is no natural data transformation for `logratio`. The idea is to model the logarithm of the variance function as a linear mixed model. Thus, we extend (14.1) to

$$g(x) = \exp\left(\gamma_0 + \gamma_1 x + \sum_{k=1}^{K} v_k (x - \kappa_k)_+\right), \qquad (14.2)$$

where we use the exponential function to ensure that the variance function $g(x)$ is positive. Models such as (14.2) are often called logspline models, since the logarithm of the right-hand side of (14.2) is a spline.

This section discusses problems for which the variance is not modeled as a known function of the mean; instead, it is modeled separately.

As in the rest of the book, the terms involving the nonlinear part of (14.2) must be penalized to ensure stable estimation. The simplest approach is to do exactly what we have done before – namely, penalize by assuming that these nonlinear coefficients are random effects, so that

$$v_k \text{ i.i.d. } N(0, \sigma_v^2).$$

We can thus write the entire model as a double mixed model,

$$\mathbf{y}|\mathbf{u}, \mathbf{v} \sim N[\mathbf{X}\boldsymbol{\beta} + \mathbf{Z}\mathbf{u}, \text{diag}\{\exp(\mathbf{X}\boldsymbol{\gamma} + \mathbf{Z}\mathbf{v})\}], \tag{14.3}$$

with the random effects being doubled as well:

$$\begin{bmatrix} \mathbf{u} \\ \mathbf{v} \end{bmatrix} \sim N\left(\begin{bmatrix} \mathbf{0} \\ \mathbf{0} \end{bmatrix}, \begin{bmatrix} \sigma_u^2 \mathbf{I} & \mathbf{0} \\ \mathbf{0} & \sigma_v^2 \mathbf{I} \end{bmatrix} \right).$$

Model (14.3) can in principle be fit by maximum likelihood, but computational implementation appears to be challenging and no less difficult than fitting a generalized linear mixed model (Chapter 10). The reason for this is as follows. If the vector $\mathbf{f}$ of mean function values were known then the squared errors $(\mathbf{y} - \mathbf{f})^2$ would follow a generalized linear mixed model of Gamma type, so that

$$(\mathbf{y} - \mathbf{f})^2|\mathbf{v} \sim \text{Gamma}\{\tfrac{1}{2}, 2\exp(\mathbf{X}\boldsymbol{\gamma} + \mathbf{Z}\mathbf{v})\}, \tag{14.4}$$

where the notation

$$\mathbf{z} \sim \text{Gamma}(\mathbf{r}, \mathbf{s})$$

means that each z_i has density

$$f(z) = \frac{1}{s_i^{r_i} \Gamma(r_i)} z^{r_i - 1} e^{-z/s_i}.$$

As an alternative, the following iterative algorithm is easier to implement.

(1) Fit a standard linear mixed model to $\mathbf{y}$ and $\mathbf{x}$, and call the fitted function $\hat{\mathbf{f}}$.
(2) Form the squared residuals $\hat{\mathbf{r}}^2 = (\mathbf{y} - \hat{\mathbf{f}})^2$.
(3) Fit to the squared residuals the generalized linear mixed model Gamma$\{\tfrac{1}{2}, 2\exp(\mathbf{X}\boldsymbol{\gamma} + \mathbf{Z}\mathbf{v})\}$. Call the fitted function $\exp(\hat{\mathbf{g}})$.
(4) Fit a heteroscedastic mixed model: pretending that the vector of estimated variance function values, $\hat{\mathbf{g}}$, is the actual variance function, fit the model

$$\mathbf{y}|\mathbf{u} \sim N[\mathbf{X}\boldsymbol{\beta} + \mathbf{Z}\mathbf{u}, \text{diag}\{\exp(\hat{\mathbf{g}})\}]. \tag{14.5}$$

This is achieved by using the algorithms given in Section 4.4 and following sections, where the error matrix $\mathbf{R} = \text{diag}\{\exp(\hat{\mathbf{g}})\}$. Special software is *not* required for this step. Let $\hat{\mathbf{f}}$ be the resulting regression function.
   (a) To implement this step with standard software, let $\mathbf{R}^{-1/2}$ be the diagonal matrix whose elements are diag$\{\exp(-\hat{\mathbf{g}}/2)\}$. Then run standard mixed model software but with "response" $\mathbf{y}_* = \mathbf{R}^{-1/2}\mathbf{y}$ and with design matrices $\mathbf{X}_* = \mathbf{R}^{-1/2}\mathbf{X}$ and $\mathbf{Z}_* = \mathbf{R}^{-1/2}\mathbf{Z}$.
   (b) Call the resulting fitted "function" $\mathbf{f}_*$. The estimate of $\mathbf{f}$ is $\hat{\mathbf{f}} = \mathbf{R}^{1/2}\mathbf{f}_*$.
(5) Return to step (2) and iterate.

14.3   Application to the LIDAR Data

Figure 14.2 gives the fit and variability bar for the LIDAR data when constant variance is assumed; Figure 14.3 gives the same plot but after modeling the variances. Note how the former overestimates the variability when range is near 400

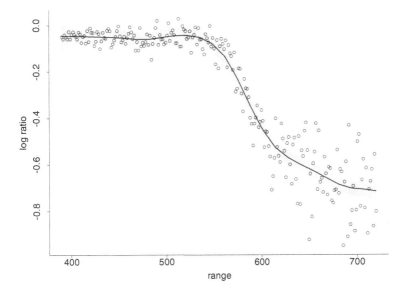

Figure 14.2 Fit and variability bar for the LIDAR data when the variances are assumed to be constant.

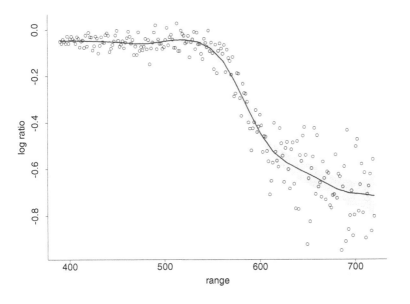

Figure 14.3 Fit and variability bar for the LIDAR data when the variances are assumed to be heteroscedastic (variability bar based on variance function estimate).

while it underestimates the variability when range is near 700. This difficulty with inference is characteristic of regression problems when heteroscedasticity is ignored.

The most common way to assess whether a fitted variance function is adequate is to plot the normalized absolute values of the residuals, $|\mathbf{y} - \hat{\mathbf{f}}|/\sqrt{\hat{\mathbf{g}}}$, against the predictor x and then smooth the resulting fit to see if there is any unexplained variability (see Section 2.9). This plot is given in Figures 14.4 and 14.5. Figure 14.4 shows an obvious trend, thus indicating that the assumption of a constant conditional variance is not supported by the data. On the other hand, Figure 14.5 shows

Figure 14.4 Smooth of the standardized absolute residuals in the LIDAR data, where standardization assumes homoscedasticity. The obvious trend indicates that the constant fitted variance function does not explain the variability in the problem.

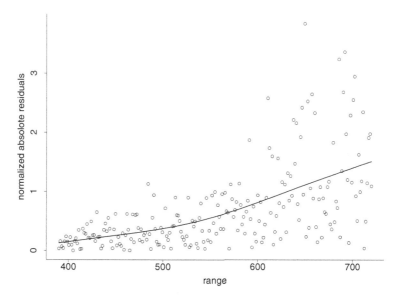

Figure 14.5 Smooth of the standardized absolute residuals in the LIDAR data, where standardization assumes heteroscedasticity that is modeled nonparametrically. The lack of a major trend indicates that the fitted variance function adequately explains the variability in the problem.

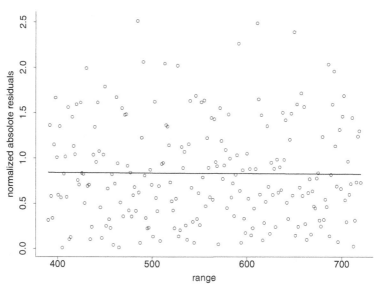

no obvious trend, thus indicating that the double mixed model has explained both the mean and the variance functions adequately.

14.4   Quasilikelihood and Variance Functions

It is relatively easy to do smoothing when variation is not constant but is related to the mean via a known functional form.

In Section 14.2 we allowed the variances to take an unspecified form. In many problems, such as for overdispersion models and generalized linear models, the variance is linked to the mean via a known function and an unknown parameter. That is, when the mean is $f(x)$, the variance is

$$g(x) \equiv \mathrm{Var}(y|x) = V\{f(x), \boldsymbol{\theta}\}, \tag{14.6}$$

where $V(\cdot)$ is a known function and $\boldsymbol{\theta}$ is an unknown parameter. For example, a common model for overdispersion in count data is $V\{f(x), \boldsymbol{\theta}\} = \theta_0 f^{\theta_1}(x)$, where $\theta_1 > 1$ or $\theta_1 = 1$ and $\theta_0 > 1$, so there is more dispersion than for the Poisson model, where $\theta_0 = \theta_1 = 1$. In assays, the same model applies except $1 \leq \theta_1 \leq 2$. More detailed discussion is given in Carroll and Ruppert (1988, chaps. 1–3); see Section 10.7 for the parametric case. We call these models *quasilikelihood and variance function models*.

The algorithm for estimating $f(x)$ in quasilikelihood and variance function models is the same as that described in Section 14.2, except that the vector of variances $\hat{g}$ is estimated in a different way. If $\boldsymbol{\theta}$ is known and if $\hat{\mathbf{f}}$ is the current estimate of $\mathbf{f}$, then

$$\hat{g} = V\{f(x), \boldsymbol{\theta}\}.$$

It thus remains to estimate $\boldsymbol{\theta}$. This is done by a method we call pseudolikelihood, and it involves estimating $\boldsymbol{\theta}$ by pretending that $\hat{\mathbf{f}}$ is the actual mean vector (not an estimate) and that $\mathbf{y}$ is normally distributed. Thus, in a sample of size n, we choose $\boldsymbol{\theta}$ so that it maximizes

$$-\frac{1}{2}\sum_{i=1}^{n}\log[V\{\hat{f}(x_i), \boldsymbol{\theta}\}] - \frac{1}{2}\sum_{i=1}^{n}\{y_i - \hat{f}(x_i)\}^2 \Big/ V\{\hat{f}(x_i), \boldsymbol{\theta}\}.$$

There is a REML version of the pseudolikelihood estimator; see Carroll and Ruppert (1988) for a detailed discussion.

14.5 Bibliographical Notes

A comprehensive introduction to variance function estimation is Carroll and Ruppert (1988). The ideas behind variance function estimation as a general technique appear in Carroll and Ruppert (1982). Davidian and Carroll (1987) present a unified treatment of variance function estimation that sheds light on earlier work that treats special cases.

15

Measurement Error

15.1 Introduction

Error in predictors means that the observed data regression function often looks nothing like the actual regression function.

Measurement error in predictors causes loss of information and biases and even misleading conclusions for inference.

Figure 15.1 depicts a case where the true regression function is $f(x) = \sin(2x)$ and the responses y *exactly* fit the true function so that $y = f(x)$, but instead of observing x we observe $w = x + v$, where the measurement error v is normally distributed with mean 0 and variance 1. Moreover, the sample size is $n = 200$ and the true but unobserved values of x are normally distributed with mean 0 and variance 1. Plotted are the observed (y, w) data and the true function. Note how the observed data look nothing at all like the true function. The essential point here is that measurement error causes the observed data to lose its features.

Figure 15.2 is the same plot, except that now – instead of adding measurement error to the covariate with the same variance as that of the covariate – we add measurement error to the response with the same variance as that of the response. Note how the $\sin(2x)$ is readily identified. This emphasizes that the error

Figure 15.1
Simulated data with measurement error *in the covariate* x. The measurement error variance is the same as the variance of x. The true regression function $\sin(2x)$ is plotted. Here the observed error-prone data look nothing like the true regression function.

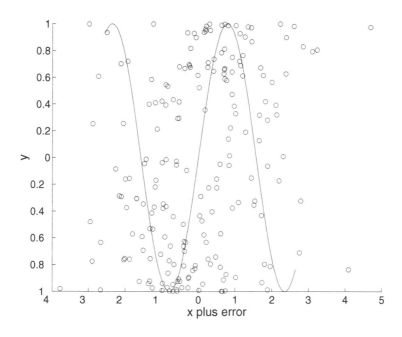

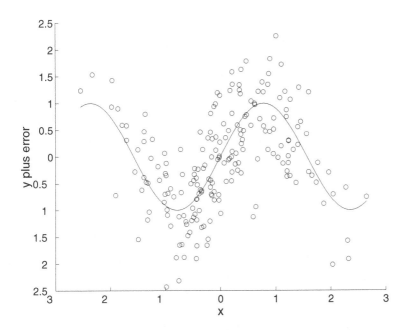

Figure 15.2
Simulated data with
measurement error *in
the response* y. The
measurement error
variance is the same
as the variance of y.
Here the observed
error-prone data look
much like the true
regression function
$\sin(2x)$.

in the covariate has a much greater potential to distort the features of the regression function.

The purpose of this section is to discuss briefly some methods that can be used to obtain information about the true regression function. Remarkably, even though the observed data plot looks nothing at all like the actual regression function, we can obtain a reasonable estimate of the regression function – at least for much of the range of the x.

15.2 Formulation

In this section we illustrate how one might fit a nonparametric regression function with an error-prone covariate using mixed model methods. Ganguli, Staudenmayer, and Wand (2001) show how to extend these methods to the additive model of Chapter 8. Their work builds upon the Bayesian work of Berry, Carroll, and Ruppert (2002).

In the example of Section 15.5 we will fit an additive model with measurement error, but to keep the exposition simple we here present the methods only for univariate regression.

Let $\mathbf{y}$ be an $n \times 1$ vector of responses, let $\mathbf{x}$ be the $n \times 1$ vector of predictors, and let $f(x)$ be the regression function. If $\mathbf{x}$ were observable, we showed in Section 4.9 how to estimate the regression function using mixed model methods; repeating equation (4.30), the mixed model representation is

Measurement error
models take the form
of an intractable and
complex generalized
linear mixed model
(GLMM).

$$\mathbf{y} = \mathbf{X}\boldsymbol{\beta} + \mathbf{Z}\mathbf{u} + \boldsymbol{\varepsilon}, \qquad \mathrm{Cov}\begin{bmatrix} \mathbf{u} \\ \boldsymbol{\varepsilon} \end{bmatrix} = \begin{bmatrix} \sigma_u^2 \mathbf{I} & \mathbf{0} \\ \mathbf{0} & \sigma_\varepsilon^2 \mathbf{I} \end{bmatrix}. \qquad (15.1)$$

In a measurement error model, instead of observing $\mathbf{x}$ we observe $\mathbf{w}$, which we now take to be unbiased measures of $\mathbf{x}$, so that

$$\mathbf{w} = \mathbf{x} + \mathbf{v}, \quad \mathbf{v} \stackrel{\text{ind.}}{\sim} N(\mathbf{0}, \sigma_v^2 \mathbf{I}), \tag{15.2}$$

where the measurement error variance σ_v^2 is assumed known for purposes of exposition. Of course, $\sigma_v^2 = 0$ corresponds to $\mathbf{x}$ being observed. In practice, σ^2 is unknown and is replaced by an estimate from, for example, replicate measurements.

Measurement error models are formally the same as latent variable models, but the model for the measurement error leads to techniques that differ from latent variable modeling.

The difficulty with measurement error models is that $\mathbf{x}$ is unobservable and hence *latent*. To understand the main hurdle, it is useful to think of the unobserved $\mathbf{x}$ as random effects. Just as for the GLMMs described in Section 10.8, computation of maximum likelihood and REML estimates requires that these latent variables be integrated out. The difficulty with doing this is strictly computational: the regression spline formulation means that the latent variables enter (15.1) nonlinearly and so are impossible to integrate analytically.

Thus, special computational tools are required to compute regression spline estimates in measurement error models. This has been done by Berry et al. (2002) using Bayesian techniques and by Ganguli et al. (2001) using the EM (expectation maximization) algorithm. Since we have not yet described Bayesian methods, we will indicate how the EM algorithm can be used in this context. We will assume that the x are normally distributed with mean μ_x and variance σ_x^2. Since we are assuming that the measurement error variance is known, good starting estimates for (μ_x, σ_x^2) are the mean of the w and the variance of the w minus σ_v^2.

The assumption that x is normally distributed makes the computations somewhat easier, but it is not required. Some additional model robustness can be achieved by positing more flexible distributions, for example, mixtures of normals (Carroll, Roeder, and Wasserman 1999) or an altered Gaussian family (Davidian and Gallant 1993).

15.3 The Expectation Maximization (EM) Algorithm

We discussed the Monte Carlo EM algorithm briefly in the context of GLMMs; see Section 10.8.5.

The vector of unknown parameters is $\boldsymbol{\psi} = (\boldsymbol{\beta}, \sigma_u^2, \sigma_\varepsilon^2, \mu_x, \sigma_x^2)$. Denote the observed data as $\mathbf{Y}_{\text{obs}} = [\mathbf{y}, \mathbf{w}]$ and the "complete" data as $\mathbf{Y}_{\text{comp}} = [\mathbf{y}, \mathbf{w}, \mathbf{x}, \mathbf{u}]$. The log-likelihood for the complete data is denoted by $\ell_{\text{comp}}(\boldsymbol{\psi})$, while the log-likelihood for the observed data is denoted by $\ell_{\text{obs}}(\boldsymbol{\psi})$. In our case, the former is easily computed (see below) whereas the latter involves intractable numerical integration to integrate out the unobserved $\mathbf{X}$ and $\mathbf{u}$.

Indeed, except for a constant, the complete data log-likelihood for a sample of size n and K knots is given by

$$\begin{aligned}
\ell_{\text{comp}}(\boldsymbol{\psi}) = {}& -(2\sigma_\varepsilon^2)^{-1} \|\mathbf{y} - \mathbf{X}\boldsymbol{\beta} - \mathbf{Z}\mathbf{u}\|^2 - \|\mathbf{x} - \mu_x \mathbf{1}\|^2/(2\sigma_x^2) \\
& - \|\mathbf{w} - \mathbf{x}\|^2/(2\sigma_v^2) - \|\mathbf{u}\|^2/(2\sigma_u^2) \\
& - (n/2)\log(\sigma_\varepsilon^2) - (K/2)\log(\sigma_u^2) - (n/2)\log(\sigma_v^2). \quad (15.3)
\end{aligned}$$

The observed data log-likelihood $\ell_{\text{obs}}(\boldsymbol{\psi})$ is the integral of the exponential of (15.3) over both $\mathbf{x}$ and $\mathbf{u}$. One might think that this integral can be computed analytically, but in fact it cannot, as we now show.

Define $\mathbf{V}(\mathbf{x}, \sigma_u^2, \sigma_\varepsilon^2) = \sigma_u^2 \mathbf{Z}\mathbf{Z}^\mathsf{T} + \sigma_\varepsilon^2 \mathbf{I}$. Define $\phi(\mathbf{y}, \mathbf{x}, \sigma_u^2, \sigma_\varepsilon^2, \boldsymbol{\beta})$ to be the multivariate normal density function for $\mathbf{y}$ with mean $\mathbf{X}\boldsymbol{\beta}$ and covariance matrix $\mathbf{V}(\mathbf{x}, \sigma_u^2, \sigma_\varepsilon^2)$. Then the complete data likelihood is proportional to

$$(\sigma_x^2 \sigma_v^2)^{-n/2} \int \phi(\mathbf{y}, \mathbf{x}, \sigma_u^2, \sigma_\varepsilon^2, \boldsymbol{\beta}) \exp\left(\frac{-\|\mathbf{x} - \mu_x \mathbf{1}\|^2}{2\sigma_x^2}\right) \exp\left(\frac{-\|\mathbf{w} - \mathbf{x}\|^2}{2\sigma_v^2}\right) d\mathbf{x}.$$

(15.4)

Note that the integral in (15.4) is an n-dimensional integral. The presence of $\mathbf{x}$ in the marginal covariance matrix $\mathbf{V}(\mathbf{x}, \sigma_u^2, \sigma_\varepsilon^2)$ means that the integral in (15.4) cannot be computed analytically, and since it is of high dimension, it cannot be computed by standard numerical recipes. This results in the need for the EM algorithm or Bayesian methods using Markov chain Monte Carlo (Chapter 16).

The EM algorithm is an iterative method. Let the current estimates of the parameters be $\boldsymbol{\psi}^{(t)}$. In the EM algorithm, the "expectation" or E-step consists of evaluating the expectation of the complete data log-likelihood as a function of $\boldsymbol{\psi}$ conditional on the observed data, where the conditional expectation is taken assuming that the current value $\boldsymbol{\psi}^{(t)}$ is the true value. In symbols, the E-step consists of calculating

$$Q(\boldsymbol{\psi}; \boldsymbol{\psi}^{(t)}) = \mathsf{E}[\ell_{\text{comp}}(\boldsymbol{\psi}) | \mathbf{Y}_{\text{obs}}; \boldsymbol{\psi}^{(t)}].$$

Next, the "maximization" or M-step consists of calculating

$$\boldsymbol{\psi}^{(t+1)} = \underset{\boldsymbol{\psi}}{\text{argmax}} \, Q(\boldsymbol{\psi}; \boldsymbol{\psi}^{(t)}).$$

As seen for example in McCulloch (1997) or Booth and Hobert (1999), the hard step is the E-step; the M-step is typically simple.

In principle, in order to implement the E-step we need to compute the joint distribution of $(\mathbf{x}, \mathbf{u})$ given the observed data. In order to speed up the computations, we use a method suggested by van Dyk (2000), which in our context reduces to sampling only from the distribution of $\mathbf{x}$ given $(\mathbf{y}, \mathbf{w}, \mathbf{u})$. This is facilitated by the following fact, which arises from (15.3).

Fact 1: The density of $[\mathbf{x}|\mathbf{u}, \mathbf{y}, \mathbf{w}]$ is proportional to

$$\exp\left(-\frac{1}{2\sigma_\varepsilon^2}\|\mathbf{y} - \mathbf{X}\boldsymbol{\beta} - \mathbf{Z}\mathbf{u}\|^2 - \frac{1}{2\sigma_x^2}\|\mathbf{x} - \mu_x \mathbf{1}\|^2 - \frac{1}{2\sigma_v^2}\|\mathbf{w} - \mathbf{x}\|^2\right). \quad (15.5)$$

As a result, conditional quasirandom variates from $\mathbf{x}$'s conditional distribution can be generated using the Metropolis–Hastings algorithm. (A convenient reference is Robert and Casella 1999.) This fact is also used by Berry et al. (2002).

We now specify the complete algorithm.

The Metropolis–Hastings algorithm can be used for computing numerically the expectations required for Monte Carlo EM.

(1) Start with an initial $\boldsymbol{\psi}^{(0)}$. Set $t = 0$.
(2) Use the Metropolis–Hastings algorithm and Fact 1 to draw m samples from the distribution of $\mathbf{x}$ given $(\mathbf{u}, \mathbf{y}, \mathbf{w})$ evaluated at $\boldsymbol{\psi}^{(t)}$. Call these samples $(\mathbf{x}_{1t}, \ldots, \mathbf{x}_{mt})$.
  (a) This is the most time-consuming step in the EM algorithm, because generating the samples requires evaluation of (15.5).

(b) The choice of the number of samples m is a problem that remains difficult to solve. McCulloch (1997) and Booth and Hobert (1999) discuss this issue in the context of GLMMs. In our calculations, we have started with $m = 50$ and increased it by 10 for each interaction of the EM algorithm, to a maximum of 500.

(3) Let

$$\mathbf{P} = \begin{bmatrix} \mathbf{X}^\mathsf{T}\mathbf{X} & \mathbf{X}^\mathsf{T}\mathbf{Z} \\ \mathbf{Z}^\mathsf{T}\mathbf{X} & \mathbf{Z}^\mathsf{T}\mathbf{Z} + (\hat{\sigma}_\varepsilon^2/\sigma_u^2)\mathbf{I}_K \end{bmatrix}.$$

(4) As before, let $\mathbf{C} = [\mathbf{X}, \mathbf{Z}]$. With the results from step (1), compute Monte Carlo estimates of the conditional expectations of

$$\mathbf{P}, \quad \mathbf{C}^\mathsf{T}\mathbf{y}, \quad \text{and} \quad \mathbf{C}^\mathsf{T}\mathbf{C}$$

given $(\mathbf{y}, \mathbf{w}, \mathbf{u})$. We denote estimates of these quantities by

$$\hat{\mathbf{P}}, \quad \widehat{\mathbf{C}^\mathsf{T}\mathbf{y}}, \quad \text{and} \quad \widehat{\mathbf{C}^\mathsf{T}\mathbf{C}}.$$

For example, $\hat{\mathbf{P}}$ is computed by constructing the $\mathbf{X}$ design matrix for each of $(\mathbf{x}_{1t}, \ldots, \mathbf{x}_{mt})$, say $(\mathbf{X}_{1t}, \ldots, \mathbf{X}_{mt})$, and then $\hat{\mathbf{P}} = m^{-1} \sum_{j=1}^{m} \mathbf{X}_{jt}^\mathsf{T}\mathbf{X}_{jt}$.

(5) Holding the estimates from the previous step fixed, run several iterations of an update scheme. For instance, using the standard EM algorithm to compute REML estimates (see e.g. Dempster, Rubin, and Tsutakawa 1981), the $(k+1)$th nested updates are as follows.

(a) Set

$$\begin{bmatrix} \beta^{(k+1)} \\ \mathbf{b} \end{bmatrix} = \hat{\mathbf{P}}^{-1}\widehat{\mathbf{C}^\mathsf{T}\mathbf{y}}.$$

(b) Set

$$\sigma_u^{2(k+1)} = \frac{\mathbf{u}^\mathsf{T}\mathbf{u} + \text{tr}(\sigma_\varepsilon^{2(k)}\hat{\mathbf{P}}^{-1})}{K}.$$

(c) Set

$$\sigma_\varepsilon^{2(k+1)} = \frac{\mathbf{y}^\mathsf{T}\mathbf{y} - 2\widehat{\mathbf{y}^\mathsf{T}\mathbf{C}}\begin{bmatrix} \beta^{(k+1)} \\ \mathbf{u} \end{bmatrix}}{n}$$

$$+ \frac{\begin{bmatrix} \beta^{(k+1)} \\ \mathbf{u} \end{bmatrix}^\mathsf{T} \widehat{\mathbf{C}^\mathsf{T}\mathbf{C}} \begin{bmatrix} \beta^{(k+1)} \\ \mathbf{u} \end{bmatrix}}{n} + \frac{\sigma_\varepsilon^{2(k)}\,\text{tr}(\widehat{\mathbf{C}^\mathsf{T}\mathbf{C}}\hat{\mathbf{P}}^{-1})}{n}.$$

(6) Calculate $\mu_x^{(t+1)}$ and $\sigma_x^{2(t+1)}$ using standard point estimates based on step (4)'s Monte Carlo data. Specifically, $\mu_x^{(t+1)}$ is the mean across all components of $(\mathbf{x}_{1t}, \ldots, \mathbf{x}_{mt})$. Also, $\sigma_x^{2(t+1)}$ is the mean across all components of $(\mathbf{x}_{jt} - \mu_x^{(t)})^2$.

We terminate the algorithm after plots of the current estimates of the regression function appear to stabilize (Wei and Tanner 1990).

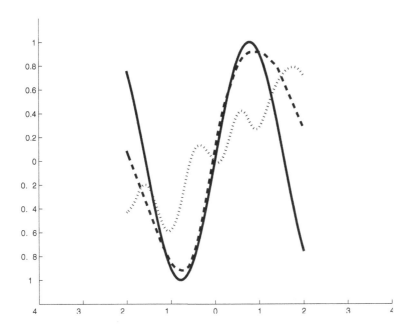

Figure 15.3
Simulated data
with true regression
function sin(2x)
(large solid line),
the fitted regression
function that ignores
measurement error
(light dotted line),
and the fitted function
that accounts for
measurement error
(large dotted line).

15.4 Simulated Example Revisited

Figure 15.3 displays the results of the EM algorithm accounting for measurement error, along with the true function and the naive function that ignores the measurement error.

In plotting this function, one should be aware of a limitation. Specifically, since the unobserved x are distributed as Normal$(0, 1)$, approximately 95% of them lie between -2.0 and 2.0. In the actual simulated data, all but twelve x-observations lie in this interval. The net effect is that we cannot expect to obtain a particularly good estimate of the regression function outside this range, even for fairly large data sets.

Remarkably, as seen in Figure 15.3, correcting for measurement error comes very close to recouping the actual regression function, even though the observed data look nothing like the regression function. We are not so bold as to suggest that this will happen with all functions; indeed, the results of Fan and Truong (1993) strongly suggest that there must be some functions that are impossible to estimate efficiently. However, simulations in Berry et al. (2002) suggest that it is often possible to do much better in regression function estimation by accounting for rather than ignoring measurement error.

In measurement error models, the observed predictor is more variable than the unobserved true predictor. This means that function fitting is accurate only on a subinterval of the observed data.

15.5 Sensitivity Analysis Example

Here we consider briefly a somewhat more complex problem. We fit a simple additive model to an air pollution–mortality data set recorded in Milan, Italy, from 1980 to 1989 (Zanobetti et al. 2000) and analyze its sensitivity to measurement

Figure 15.4 Air
pollution data
sensitivity analysis.

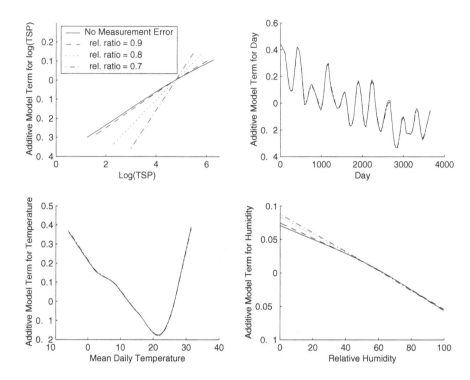

error in total suspended particles. In the example, we have no information about
the measurement error variance, so here our purpose is to show that models such
as this one can be sensitive to covariate measurement error. Also see Dominici,
Zeger, and Samet (2000) for a discussion of measurement error in air pollution–
mortality studies.

Let i index day and let $\mathtt{TSP}_i$, $\mathtt{day}_i$, $\mathtt{temp}_i$, $\mathtt{humid}_i$, and $\mathtt{mort}_i$ be (respectively)
the measured total suspended particles, sequential day number, average tempera-
ture, average relative humidity, and mortality (from natural causes – International
Classification of Diseases, 9th rev., pp. 1–799) count on day i. Our additive model
regression equation is:

$$\mathrm{E}\{\log(\mathtt{mort}_i)\} = \beta_0 + f_1\{\log(\mathtt{TSP}_i)\} + f_2(\mathtt{day}_i) + f_3(\mathtt{temp}_i) + f_4(\mathtt{humid}_i).$$

We are interested in assessing the sensitivity of estimates to measurement error
in $\log\{\mathtt{TSP}_i\}$. Let v be the measurement error with variance σ_v^2. We have no information about the size of measurement error variance, so here we varied what
is called the reliability ratio:

$$\text{reliability ratio} = \frac{\mathrm{var}\{\log(\mathtt{TSP}_i)\}}{\mathrm{var}\{\log(\mathtt{TSP}_i)\} + \sigma_v^2}.$$

We took values 1.0 (no measurement error), 0.9, 0.8, and 0.7. Ganguli et al. (2001)
developed EM algorithm methods for additive models that are subject to measurement error.

Figure 15.4 shows that any estimate of the function that relates $\log(\mathrm{TSP})$ to
mortality appears to be sensitive to measurement error. This suggests that an

analysis of these data would benefit from measurement error modeling and the collection of validation data that allow σ_v^2 to be estimated.

15.6   Bibliographical Notes

Comprehensive surveys of the measurement error literature are given by Fuller (1987) for the linear model and by Carroll, Ruppert, and Stefanski (1995) for general regression models. Some work has been done on GLMMs; see Wang et al. (1998). Recent discussions of semiparametric modeling in the presence of covariate measurement error are given by Berry et al. (2002) and Ganguli et al. (2001).

16

Bayesian Semiparametric Regression

16.1 Introduction

Classical statistics treats parameters as fixed unknown quantities. Bayesian statistics is based on a different philosophy; parameters are treated as random variables. The probability distribution of a parameter characterizes knowledge about the parameter's value, and this distribution changes as new data are acquired. The mixed models of classical statistics have a Bayesian flavor because some parameters are treated as random. However, in a mixed model both the fixed effects and the variance components are treated as nonrandom unknowns. Bayesians go one step beyond mixed models in that they treat *all* parameters as random. In this chapter we take the mixed model formulation of Section 4.9 and extend it to a fully Bayesian model.

Bayesian statistics differs from classical statistics in two important respects:

(1) the use of the prior distribution to characterize knowledge of the parameter values prior to data collection; and
(2) the use of the posterior distribution – that is, the conditional distribution of the parameters given the data – as the basis of inference.

Some statisticians are uneasy about the use of priors, but when done with care, the use of priors is quite sensible. In some situations, we might have strong prior beliefs that will influence our analysis. For example, suppose we needed to estimate the probability that a toss of a coin comes up heads. If we inspect the coin and see that it is not bent or otherwise unusual, we might be rather certain that the probability is close to $\frac{1}{2}$. Even with some data – say, 7 heads out of 10 tosses – we might rather estimate the probability by some value much closer to $\frac{1}{2}$ than to the observed frequency. In other situations, where we feel that we know little or nothing about the parameters, we can chose a "vague" or "noninformative" prior – for example, a density that is uniform over a large but bounded set containing all reasonable values of the parameter. As an noninformative prior, one could also use an "improper prior" (i.e., one with infinite mass) such as the uniform distribution on the real line. By stating a prior, we make clear how much, *or how little,* we believe we know a priori about the parameter.

In contrast to the controversy about the use of priors, there is little argument that the posterior provides a powerful inferential machinery. For example, in the context of semiparametric spline modeling, we will see that it allows one to as-sess the effects of uncertainty in the smoothing parameters upon a smooth fit.

In a Bayesian analysis, the likelihood of classical statistics become the condi-tional density of the data given the parameters. Bayesians add to this the prior density of the parameters. The product of the prior and the likelihood is the joint density of the data and the parameters. The marginal density of the data is ob-tained by integrating out the parameters from the joint density. Bayes theorem states that the conditional distribution of the parameters given the data, which is called the posterior distribution, is the ratio of the joint density of the data and the parameters to the marginal density of the data.

The calculations required by Bayesian inference – in particular, the need to integrate the parameters out of the joint density – have been a serious obstacle to the application of Bayesian methods. However, new Monte Carlo techniques such as Markov chain Monte Carlo and importance sampling have provided pow-erful methods for attacking this problem. In this chapter we use the Gibbs sam-pler (an MCMC technique) to provide a fully Bayesian approach to smoothing by penalized splines. Although we discuss only univariate splines, the MCMC methodology is readily adapted to additive and interaction models.

With the advent of Gibbs sampling and other Markov chain Monte Carlo (MCMC) methods, Bayesian methods are becoming more widespread. In particular, the BUGS software for MCMC allows some otherwise complicated analyses to be carried out routinely.

16.2 General Framework

Throughout this chapter we will use the "[ ]" notation for probability densities that has become standard in the MCMC literature. For example, $[W]$ is the den-sity of a random variable W, while $[W|U, V]$ is the conditional density of W given U and V.

Let $\mathcal{D}$ be the observed data and let θ be the unknown parameter vector. Then $[\theta]$ is the prior density of θ and $[\mathcal{D}|\theta]$ is the likelihood. The joint density of $(\mathcal{D}, \theta)$ is $[\theta][\mathcal{D}|\theta]$, and the marginal density of $\mathcal{D}$ is $[\mathcal{D}] = \int [\mathcal{D}|\theta][\theta] \, d\theta$. The posterior density of θ is

$$[\theta|\mathcal{D}] = \frac{[\mathcal{D}|\theta][\theta]}{\int [\mathcal{D}|\theta][\theta] \, d\theta}. \tag{16.1}$$

When the dimension of θ is high, deterministic methods of numerical inte-gration are unfeasible for calculating the integral in the denominator of (16.1). Notice, however, that this denominator is just a normalizing factor; that is, it does not depend on θ. There are methods of sampling from a density that is known only up to a constant of proportionality. These methods allow us to sample from the posterior without calculating the denominator. Thus, for example, if $\theta_1, \ldots, \theta_N$ is such a sample then $N^{-1} \sum_{i=1}^{N} \theta_i$ provides an estimate of the posterior mean.

The denominator of (16.1) is required to compute such quantities as the posterior mean and Bayesian confidence intervals.

MCMC is the most common method of implementing Bayesian techniques.

16.2.1 *Markov Chain Monte Carlo*

Among the most popular methods of sampling from a posterior are the MCMC methods. The idea is to sample from a Markov chain whose stationary distribu-tion is equal to the posterior. Suppose that we can partition θ into subvectors:

The componentwise conditional distributions are often called the *complete conditionals.*

$$\theta = (\theta_1, \ldots, \theta_M),$$

where for each $m = 1, \ldots, M$ it is easy to sample from

$$[\theta_m | \mathcal{D}, \theta_j, \ j \neq m].$$

Then we can form a Markov chain by starting at an arbitrary value

$$\theta^{(0)} = (\theta_1^{(0)}, \ldots, \theta_M^{(0)})$$

and then sampling from

$$[\theta_1 | \mathcal{D}, \theta_2^{(0)}, \ldots, \theta_M^{(0)}]$$

to obtain $\theta_1^{(1)}$, then from $[\theta_2 | \mathcal{D}, \theta_1^{(1)}, \theta_3^{(0)}, \ldots, \theta_M^{(0)}]$ to obtain $\theta_2^{(1)}$, and so forth. After one cycle we have $\theta^{(1)} = (\theta_1^{(1)}, \ldots, \theta_M^{(1)})$. Continuing in this manner we obtain a Markov chain $\theta^{(0)}, \theta^{(1)}, \ldots$. Under weak assumptions, this chain will converge to a stationary distribution that is the posterior (Tierney 1994). MCMC works well when one can sample easily from these complete conditional distributions. Up to a multiplicative constant, the conditional density $[\theta_m | \mathcal{D}, \theta_j, \ j \neq m]$ is simply the posterior density with only θ_m varying. Often the numerator of the posterior in (16.1), when viewed as a function of θ_m, is seen to be proportional to some familiar density (e.g., a normal or gamma). Then, since the denominator does not depend on θ, the conditional posterior density of θ_m must be equal – and not just proportional – to that density.

There are a number of important implementational issues that are discussed in the books listed in Section 16.7 and in the journal literature. These include:

(1) the length of the "burn in period", which is the initial part of the chain that is discarded to eliminate the effects of the starting value;
(2) whether one should use one long chain or multiple starting values;
(3) how starting values should be chosen (this is an especially important question if one uses multiple short chains);
(4) diagnosis of convergence of the chain;
(5) partitioning of the parameter vectors;
(6) the use of centering of covariates and auxiliary variables to improve convergence.

16.2.2   *Credible Sets*

Let θ_1 be a subvector of θ and suppose we would like to know the likely values of θ_1. A frequentist's confidence set for θ_1 is a random set with the following property: if sampling is repeated indefinitely then a known fraction, $1 - \alpha$, of the sets generated will contain θ_1. The quantity $1 - \alpha$ is called the "confidence coefficient" or the "coverage probability". Coverage probability is a probability only under repeated sampling, not for a given data set. Therefore, a frequentist will *not* say that the probability that θ_1 is in the confidence set constructed for a particular data set is $1 - \alpha$. Rather, the frequentist will state that this probability is either 0 or 1 but it is not known which. A confidence interval is the special case of a confidence set where θ_1 is univariate and the confidence set is an interval.

A Bayesian credible set is an analog of a confidence set. A credible set for $\boldsymbol{\theta}_1$ Credible sets are easy
is a set of values of that subparameter with probability $1 - \alpha$ under the posterior to compute from
distribution (Berger 1985). Unlike a frequentist with a confidence set, a Bayesian MCMC output.
will say that $1 - \alpha$ is the probability that $\boldsymbol{\theta}_1$ is in the credible set for a particular
data set. To appreciate the distinction between confidence and credible sets, let
$\boldsymbol{\theta}$ be a parameter and let $\mathcal{S}(\mathcal{D})$ be a subset of the parameter space that depends
on data $\mathcal{D}$. Then $\mathcal{S}(\mathcal{D})$ is a confidence set if

$$P\{\boldsymbol{\theta} \in \mathcal{S}(\mathcal{D})|\boldsymbol{\theta}\} = 1 - \alpha \quad \text{for all } \boldsymbol{\theta}. \tag{16.2}$$

Since the probability in (16.2) is conditional on $\boldsymbol{\theta}$, it depends only on the likelihood, not on the posterior, and can be computed either within or outside the
Bayesian framework. Here $\mathcal{S}(\mathcal{D})$ is a credible set if

$$P\{\boldsymbol{\theta} \in \mathcal{S}(\mathcal{D})|\mathcal{D}\} = 1 - \alpha \quad \text{for all } \mathcal{D}. \tag{16.3}$$

The probability in (16.3) is conditional on the data, so it is a posterior probability
and is meaningful only within the Bayesian framework. Within that framework,
we can take the expectations of the conditional probabilities in either (16.2) or
(16.3), and in either case we obtain the same result:

$$P\{\boldsymbol{\theta} \in \mathcal{S}(\mathcal{D})\} = 1 - \alpha.$$

Although credible sets and confidence sets are developed from entirely different philosophies, often a Bayesian credible set can be used as a confidence set
because we can show that the former's coverage probability (in the frequentist's
sense) is close to $1 - \alpha$, that is, (16.2) holds approximately.

A $1 - \alpha$ credible interval for a univariate subparameter can be easily constructed from the output of a MCMC analysis. The left endpoint of the interval is
the $\alpha/2$ sample quantile of the realizations of that subparameter from the chain.
Similarly, the right endpoint is the $1 - \alpha/2$ sample quantile. Often the marginal posterior distribution of a subparameter is nearly normal. In this case the
"normal theory" credible interval is the posterior mean of that subparameter plus
and minus $z(1 - \alpha/2)$ times its posterior standard deviation.

16.3   Scatterplot Smoothing

Recall the mixed model representation of a penalized spline:

$$\mathbf{y} = \mathbf{X}\boldsymbol{\beta} + \mathbf{Z}\mathbf{u} + \boldsymbol{\varepsilon}, \qquad \text{Cov}\begin{bmatrix} \mathbf{u} \\ \boldsymbol{\varepsilon} \end{bmatrix} = \begin{bmatrix} \sigma_u^2 \mathbf{I} & \mathbf{0} \\ \mathbf{0} & \sigma_\varepsilon^2 \mathbf{I} \end{bmatrix}, \tag{16.4}$$

where $\mathbf{X}\boldsymbol{\beta}$ is the pure polynomial component of the spline and $\mathbf{Z}\mathbf{u}$ is the component with spline basis functions. Letting $(\boldsymbol{\beta}, \mathbf{u}, \sigma_u^2, \sigma_\varepsilon^2)$ be the parameter vector, the mixed model specifies a $N(0, \sigma_u^2 \mathbf{I})$ prior on $\mathbf{u}$ as well as the likelihood,
$[\mathbf{y}|\boldsymbol{\beta}, \mathbf{u}, \sigma_u^2, \sigma_\varepsilon^2]$. To specify a complete Bayesian model, we also need a prior distribution on $(\boldsymbol{\beta}, \sigma_u^2, \sigma_\varepsilon^2)$. Assuming that little is known about $\boldsymbol{\beta}$, it makes sense
to put an improper uniform prior on $\boldsymbol{\beta}$. Or, if a proper prior is desired, one could

use a $N(0, \sigma_\beta^2 \mathbf{I})$ prior with σ_β^2 so large that, for all intents and purposes, the normal distribution is uniform on the range of $\boldsymbol{\beta}$. Therefore, we will use $[\boldsymbol{\beta}] \equiv 1$. More informative priors could be used, of course, in contexts where they seem desirable.

A random variable *W* has an inverse gamma distribution with parameters *A* and *B* if W^{-1} has a Gamma(A, B) distribution; common notation is $W \sim \mathrm{IG}(A, B)$. One generates an inverse gamma random variable by reciprocating a gamma random variable. The latter has mean A/B and variance A/B^2. The coefficient of variation, defined to be the ratio of the standard deviation to the mean, is $1/\sqrt{A}$.

We will assume that the prior on σ_ε^2 is *inverse gamma* with parameters A_ε and B_ε – denoted $\mathrm{IG}(A_\varepsilon, B_\varepsilon)$ – so that its density is

$$[\sigma_\varepsilon^2] = \frac{B_\varepsilon^{A_\varepsilon}}{\Gamma(A_\varepsilon)} (\sigma_\varepsilon^2)^{-(A_\varepsilon+1)} \exp\left(-\frac{B_\varepsilon}{\sigma_\varepsilon^2}\right). \tag{16.5}$$

If $A_\varepsilon > 1$ then the mean of this random variable is finite and equals $B_\varepsilon/(A_\varepsilon - 1)$; if $A_\varepsilon > 2$ then its variance is finite and equals $B_\varepsilon^2/\{(A_\varepsilon - 1)^2(A_\varepsilon - 2)\}$. We may also write this as

$$\sigma_\varepsilon^2 \sim \mathrm{IG}(A_\varepsilon, B_\varepsilon).$$

Further, we assume that

$$\sigma_u^2 \sim \mathrm{IG}(A_u, B_u).$$

Here A_ε, B_ε, A_u, and B_u are "hyperparameters" that determine the priors and must be chosen by the statistician. These hyperparameters must be strictly positive in order for the priors to be proper. If A_ε and B_ε were zero, then $[\sigma_\varepsilon^2]$ would be proportional to the improper prior $1/\sigma_\varepsilon^2$, which is equivalent to $\log(\sigma_\varepsilon)$ having an improper uniform prior. Therefore, choosing A_ε and B_ε both close to zero (say, both equal to 0.1) gives an essentially noninformative, but proper, prior. The same reasoning applies to A_u and B_u.

The model we have constructed is a *hierarchical Bayes* model, where the random variables are arranged in a hierarchy such that distributions at each level are determined by the random variables in the previous levels. At the bottom of the hierarchy are the known hyperparameters. At the next level are the fixed effects parameters and variance components whose distributions are determined by the hyperparameters. At the level above this are the random effects, $\mathbf{u}$ and $\boldsymbol{\varepsilon}$, whose distributions are determined by the variance components. The top level contains the data, $\mathbf{y}$.

Except for a constant of proportionality, denoted by "$\propto$", the posterior distribution is equal to

$$[\boldsymbol{\beta}, \mathbf{u}, \sigma_\varepsilon^2, \sigma_u^2 | \mathbf{y}] \propto [\mathbf{y}|\boldsymbol{\beta}, \mathbf{u}, \sigma_\varepsilon^2][\mathbf{u}|\sigma_u^2][\sigma_u^2][\boldsymbol{\beta}][\sigma_\varepsilon^2]. \tag{16.6}$$

This is the numerator of (16.1); the denominator of (16.1) is the constant of proportionality, which cannot be computed easily but is still required for inference.

If we isolate the part of (16.6) that depends on $(\boldsymbol{\beta}, \mathbf{u})$ then we see that the conditional posterior of $(\boldsymbol{\beta}, \mathbf{u})$ given $(\sigma_\varepsilon^2, \sigma_u^2)$ – that is, the complete conditional – is proportional to

$$\exp\left\{-\frac{1}{2\sigma_\varepsilon^2}\left(\|\mathbf{y} - \mathbf{X}\boldsymbol{\beta} - \mathbf{Z}\mathbf{u}\|^2 + \frac{\sigma_\varepsilon^2}{\sigma_u^2}\|\mathbf{u}\|^2\right)\right\}. \tag{16.7}$$

The term in parentheses in (16.7) is a nonnegative quadratic function of $(\boldsymbol{\beta}, \mathbf{u})$ and so (16.7) is proportional to a multivariate normal density. By the usual technique of "completing the square", it may be shown that

$$[\boldsymbol{\beta}, \mathbf{u} | \sigma_\varepsilon^2, \sigma_u^2, \mathbf{y}] \sim \mathrm{N}\left\{ \left(\mathbf{C}^\mathsf{T}\mathbf{C} + \frac{\sigma_\varepsilon^2}{\sigma_u^2}\mathbf{D} \right)^{-1} \mathbf{C}^\mathsf{T}\mathbf{y}, \sigma_\varepsilon^2 \left(\mathbf{C}^\mathsf{T}\mathbf{C} + \frac{\sigma_\varepsilon^2}{\sigma_u^2}\mathbf{D} \right)^{-1} \right\}. \quad (16.8)$$

Here, as before, $\mathbf{C} = [\mathbf{X}, \mathbf{Z}]$ and $\mathbf{D}$ is a diagonal matrix with $p + 1$ zeros followed by K ones on the diagonal; this corresponds to the $p + 1$ polynomial coefficients and the K knots. Note that the posterior mean in (16.8) is the penalized spline with smoothing parameter equal to $\sigma_\varepsilon^2/\sigma_u^2$; see (4.31). Also, the covariance in (16.8) is the same covariance matrix as (6.13), though in (6.13) the variance components are treated as fixed unknown constants. Thus, as part of the MCMC chain, one generates $(\boldsymbol{\beta}, \mathbf{u})$ from the current values of $(\sigma_\varepsilon^2, \sigma_u^2)$ according to the multivariate normal distribution, with mean and covariance matrix given by (16.8).

It is important to generate random variables from (16.8) and not just substitute in the mean. The latter quick and dirty method will underestimate the variance of the fit, sometimes badly.

The complete conditional for σ_ε^2 is proportional to

$$(\sigma_\varepsilon^2)^{-(n/2+A_\varepsilon+1)} \exp\left\{ -\frac{1}{\sigma_\varepsilon^2} \left(\frac{1}{2}\|\mathbf{y} - \mathbf{X}\boldsymbol{\beta} - \mathbf{Z}\mathbf{u}\|^2 + B_\varepsilon \right) \right\}. \quad (16.9)$$

Therefore, comparing (16.5) to (16.9) shows that

$$[\sigma_\varepsilon^2 | \mathbf{y}, \boldsymbol{\beta}, \mathbf{u}, \sigma_u^2] \sim \mathrm{IG}\left(A_\varepsilon + \tfrac{1}{2}n, \ B_\varepsilon + \tfrac{1}{2}\|\mathbf{y} - \mathbf{X}\boldsymbol{\beta} - \mathbf{Z}\mathbf{u}\|^2 \right).$$

By the same reasoning,

$$[\sigma_u^2 | \mathbf{y}, \boldsymbol{\beta}, \mathbf{u}, \sigma_\varepsilon^2] \sim \mathrm{IG}\left(A_u + \tfrac{1}{2}K, \ B_u + \tfrac{1}{2}\|\mathbf{u}\|^2 \right). \quad (16.10)$$

To sample from the posterior, we iterate N times (for some fixed N) through the following three steps.

(1) Sample $(\boldsymbol{\beta}, \mathbf{u})$ from the multivariate normal distribution:

$$\mathrm{N}\left\{ \left(\mathbf{C}^\mathsf{T}\mathbf{C} + \frac{\sigma_\varepsilon^2}{\sigma_u^2}\mathbf{D} \right)^{-1} \mathbf{C}^\mathsf{T}\mathbf{y}, \sigma_\varepsilon^2 \left(\mathbf{C}^\mathsf{T}\mathbf{C} + \frac{\sigma_\varepsilon^2}{\sigma_u^2}\mathbf{D} \right)^{-1} \right\}.$$

(2) Sample σ_u^2 from $\mathrm{IG}\left(A_u + \tfrac{1}{2}K, \ B_u + \tfrac{1}{2}\|\mathbf{u}\|^2 \right)$.
(3) Sample σ_ε^2 from $\mathrm{IG}\left(A_\varepsilon + \tfrac{1}{2}n, \ B_\varepsilon + \tfrac{1}{2}\|\mathbf{y} - \mathbf{X}\boldsymbol{\beta} - \mathbf{Z}\mathbf{u}\|^2 \right)$.
(4) Return to step (1) and iterate.

This Markov chain can be started at $(\boldsymbol{\beta}^{(0)}, \mathbf{u}^{(0)})$ equal to a penalized spline estimate, σ_ε^2 estimated as in Chapter 3, and $\sigma_u^2 = \sigma_\varepsilon^2/\lambda$ (where λ is chosen by GCV; see Section 4.9, where it is shown that $\lambda = \sigma_\varepsilon^2/\sigma_u^2$). Alternatively, $(\boldsymbol{\beta}^{(0)}, \mathbf{u}^{(0)}, (\sigma_\varepsilon^{(0)})^2, (\sigma_u^{(0)})^2)$ can be estimated from the mixed model formulation of penalized splines discussed in Section 4.9. The exact choice of starting values in not important, since the chain converges quickly to the stationary distribution and the beginning of the chain is discarded as a burn-in period.

Notice that, given $(\boldsymbol{\beta}, \mathbf{u}, \mathbf{y})$, it follows that σ_ε^2 and σ_u^2 are independent. Therefore, the net effect of steps (2) and (3) is to sample $(\sigma_\varepsilon^2, \sigma_u^2)$ from its conditional distribution given $(\boldsymbol{\beta}, \mathbf{u}, \mathbf{y})$. Also, because of this independence, interchanging the order of steps (2) and (3) has no effect on the algorithm. The MCMC iterates between sampling the regression coefficients $(\boldsymbol{\beta}, \mathbf{u})$ given the variance components $(\sigma_\varepsilon^2, \sigma_u^2)$ and vice versa – all conditional on the data, $\mathbf{y}$.

The Bayes estimate under squared-error loss of any subvector of the parameter vector is the mean of its posterior distribution. Uncertainty about that set of parameters can be measured by its posterior covariance matrix. A simple (but somewhat inefficient) method for estimating the posterior mean and covariance of a set of parameters is the sample mean and covariance of these parameters from the N iterations of the MCMC. Often, as is the case here, more efficient estimates can be found via Rao–Blackwell techniques; see Section 16.6.

The Rao–Blackwell theorem provides a method for improving the efficiency of an estimator by computing its conditional expectation given a sufficient statistic. See Section 16.6.

A simple method of approximating the posterior of $(\boldsymbol{\beta}, \mathbf{u})$ is to use (16.8) with σ_ε^2 and σ_u^2 replaced by estimators from, for example, a mixed model analysis. Methods such as this one are often called "empirical Bayes". In an empirical Bayes analysis, one starts with a prior containing unknown parameters, replaces the parameters with estimates, and then performs a Bayesian analysis with the previously unknown parameters now regarded as fixed. The phrase "empirical Bayes" is disliked by some Bayesians, since empirical Bayes methods are neither truly Bayesian (Deely and Lindley 1981) nor any more empirical than fully Bayesian methods. Although we prefer the term "approximate Bayes", we will use the term "empirical Bayes" because it is so well known.

The literature on MCMC discusses a number of important topics such as appropriate burn-in periods, monitoring convergence to the stationary distribution, and the question of how many independent chains to use. In our particular application, we have found that burn-in periods and monitoring for convergence are not needed. Also, a single chain of 1000–3000 iterations seems sufficient. The reason why it is relatively easy to implement an MCMC here is that we are iterating between sampling from $(\boldsymbol{\beta}, \mathbf{u})$ given $(\sigma_\varepsilon^2, \sigma_u^2)$ and vice versa. Moreover, the distribution of the regression coefficients is relatively insensitive to the values of the variance components, and vice versa. Also, with the starting values we suggest – that is, using estimates from a mixed model analysis – the chain essentially starts in the stationary distribution or at least quite close to it.

16.3.1 *Application to LIDAR Data*

We implemented a Bayesian spline analysis of the LIDAR data using 30-knots penalized quadratic splines. Assume the usual nonparametric regression notation,

$$y_i = f(x_i) + \varepsilon_i,$$

throughout this section.

Results from 3000 iterations of an MCMC sampler are shown in Figure 16.1. Figure 16.1(a) shows five realizations from the posterior of $(\boldsymbol{\beta}, \mathbf{u})$. One can see that there is little variation in the curves except at the boundaries, where the high boundary variance is evident. All five curves follow roughly the same pattern. They are nearly horizontal until range is about 550, then decrease rapidly until range is about 620, and then decrease a bit more slowly. The behavior of these realizations corroborates the finding of Section 6.8.3 that there is a strong plume of mercury in the region where range is between 550 and 620 and a "shoulder" on this plume between range $= 620$ and 700 (see Figure 6.13).

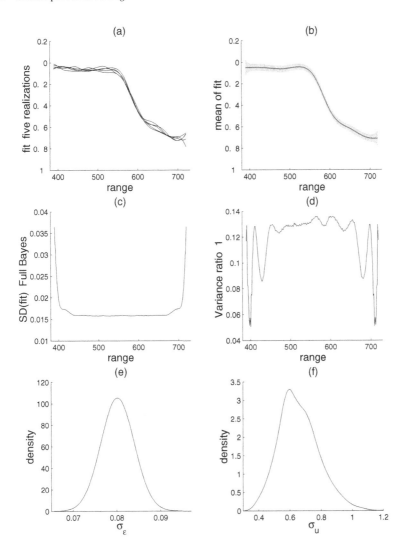

Figure 16.1 Bayesian analysis of the LIDAR data using a 3000-iterate Gibbs sampler and 30-knot quadratic splines. (a) Five realizations from the posterior of $f(x)$. (b) Posterior mean of $f(x)$ and 95% credible intervals. (c) Standard deviation of the posterior distribution of $f(x)$. (d) Relative error in approximating the posterior standard deviation of $f(x)$ by an approximate (or empirical) Bayes method. (e) Density estimate from smoothing a 200-bin histogram of 3000 realizations of the posterior of σ_ε. (f) Estimate of the posterior density of σ_u.

In panel (b) of Figure 16.1, the average of the 3000 values of (16.8) is used as a point estimate of the true curve. Similarly, in panel (c), the posterior standard deviation of $f(x)$ is shown. The higher variance at the boundaries is clear.

In panel (d) we compare the fully Bayesian estimate of the posterior variance of $f(x)$ to an approximate Bayesian analysis discussed in Section 16.2.2, where $(\sigma_\varepsilon^2, \sigma_u^2)$ are treated as known. In this panel, the relative error – or, more precisely, the ratio of the two posterior variance estimates minus 1 – is plotted. The plot shows that the approximate Bayesian method underestimates the posterior variance by about 10% in the interior and at the extreme boundary region, and by somewhat less in a region near the boundary. Thus, ignoring the variability of the smooth fit (due to variability in the smoothing parameters) yields a reasonable approximation. Confidence intervals, such as in Chapter 6, or approximate Bayesian credible intervals that ignore this variability in the smoothing parameters should

Figure 16.2
Iterations of the
Gibbs sampler for the
LIDAR data. (a) Plot
of estimated mean at
`range` = 620 at every
30th iteration. (b) Plot
of estimated mean
at `range` = 620 at
iterations 1501–1600.
In both panels, the
solid horizontal line is
at the sample mean of
all 3000 iterations.

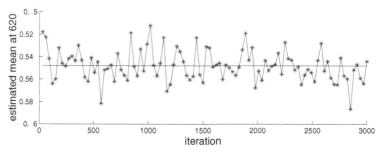

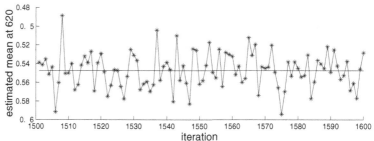

have reasonably accurate coverage probability in this example. However, a fully Bayesian credible interval will achieve better coverage probability. Moreover, the accuracy of the approximate Bayesian method will be less in other examples, such as the Canadian age and income data analysis of the next subsection.

Panels (e) and (f) of Figure 16.1 show density estimates based on 3000 realizations of σ_ε and σ_u. The estimates were obtained by smoothing a 200-bin histogram as described in Section 11.8. Estimates of their posterior densities could also be obtained by the more efficient methods based on *Rao–Blackwellization*; see Section 16.6. Since these variance components are of less interest, such Rao–Blackwellized density estimates are not normally needed. An advantage of direct density estimation is that it illustrates the frequency distribution of the actual MCMC output.

When applying MCMC to a Bayesian analysis it is important to check whether the chain has converged to the stationary distribution, that is, to the posterior. It is also essential to check whether the number of iterations was sufficient. The number of iterates needed for an accurate estimation of the posterior depends strongly on the correlation between iterates. Strong correlation causes the chain to move slowly through the posterior. When the correlation is very strong, hundreds of thousands of iterations might be needed. Fortunately, this is not the case here; the iterates behave as if they are uncorrelated – the primary reason for this felicitous situation is that, in step (1), we sample ($\boldsymbol{\beta}, \mathbf{u}$) as a block.

As an example of the behavior of the chain, Figure 16.2 contains plots of $\hat{f}(620)$ as a function of the iteration number. Panel (a) plots every 30th iteration (i.e., iterations 30, 60, ..., 2970, 3000), and panel (b) shows 100 consecutive iterations (iterations 1501 to 1600). Notice that no trend appears in panel (a); the lack of a trend is an indication that the chain starts in the stationary distribution. Also, there is no indication of autocorrelation in either panel. Autocorrelation, if it existed, should be evident in the consecutive observations in panel (b) being somewhat similar to each other. In fact, the sample autocorrelation function of the 3000 iterations is -0.013, -0.008, 0.024, and 0.018 at lags 1 to 4. Autocorrelations greater than $2/\sqrt{3000} = 0.0365$ would be considered significantly different from zero, but the absolute value of all four autocorrelations are below that cut-off. In summary, the realizations from the chain of the posterior of $\hat{f}(620)$ behave like independent and identically distributed observations from that posterior. We have found similar independence, or near independence, when plotting $\hat{f}$ at other values of range and when plotting σ_ε^2 and σ_u^2. This is an ideal situation, and is somewhat unusual for MCMC methods. Often, MCMC suffers from high auto-correlations between iterations, and either very long chains or many independent chains are needed to obtain a good representation of the posterior. Because of the near independence between iterations here, a single chain of 3000 observations is more than sufficient to obtain an accurate estimate of the posterior.

As discussed in Section 2.3.2, the k-lag sample autocorrelation of a time series $\{Z_1, \ldots, Z_n\}$ is the correlation between the pairs $(Z_t, Z_{t+k})_{t=1}^{n-k}$. For an sequence of independent observations, the autocorrelations will be zero at all lags except for sampling error. If one examines many autocorrelation coefficients, some will be large in absolute value by chance. Large absolute autocorrelations at small lags are a sign of dependence.

16.3.2   *Application to Age and Income Data*

Realizations from the posterior such as shown in Figure 16.1(a) can help us judge whether a feature seen in a smooth is "real" or perhaps due merely to chance.

Consider the Canadian age and income data introduced in Section 5.3.2.1. There is evidence that the expected income begins to decrease after age 50; see Figure 5.4 and the discussion in Section 6.8.

Figure 16.3 shows the same output from an MCMC analysis as Figure 16.1 except now for the age and income data. Notice that the variance ratio minus 1 in panel (d) is typically about 0.25. This value is larger than for the LIDAR data and is an indication that one might wish to account for uncertainty in the smoothing parameter when making inferences about the age and income data.

Figure 16.4 is a histogram of the 3000 realizations of $f(50) - f(60)$, the difference in log-income at ages 50 and 60. The difference was positive in 99.5% of the 3000 realizations and in 99.1% of the realizations if one compares ages 50 and 65 instead of 50 and 60. The reason for slightly less positive differences when comparing age 65, rather than 60, to age 50 is the serious variance inflation at the boundary. These percentages are the posterior probabilities that income declines from age 50 to 60 (or 65), and since they are close to 1 they provide strong evidence of a decline in expected income after age 50. To see where the decline begins, 24 realizations are plotted in Figure 16.5, six realizations per panel. The realizations peak almost anywhere between 50 and 57, indicating that the location of the true peak cannot be established with any certainty using this amount of data.

16.4   **Linear Mixed Models**

The Gibbs sampling scheme given on page 281 for scatterplot smoothing is really just fitting a mixed model with a single variance component. In this section we briefly describe the extension to linear mixed models with more than one variance

Figure 16.3 Bayesian analysis of the Canadian age and income data using a 3000-iterate Gibbs sampler and 15-knot quadratic spline. (a) Five realizations from the posterior of $f(x)$. (b) Posterior mean of $f(x)$ and 95% credible intervals. (c) Standard deviation of the posterior distribution of $f(x)$. (d) Relative error in approximating the posterior standard deviation of $f(x)$ by an approximate (or empirical) Bayes method. (e) Density estimate from smoothing a 200-bin histogram of 3000 realizations of the posterior of σ_ε. (f) Estimate of the posterior density of σ_u.

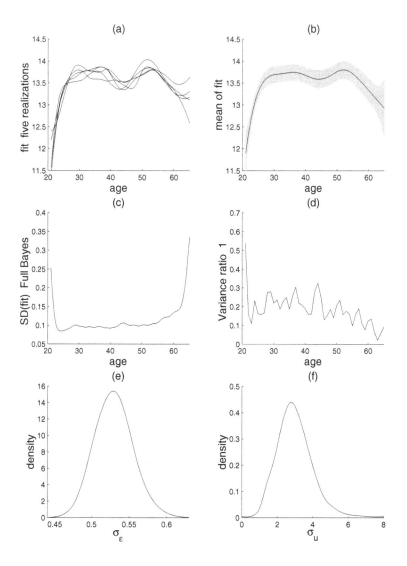

component. This extension allows Bayesian analyses of the Gaussian response models described in Chapters 7, 8, 9, 12, and 13.

The Gaussian linear mixed model with several variance components is

$$\mathbf{y} = \mathbf{X}\boldsymbol{\beta} + \mathbf{Z}\mathbf{u} + \boldsymbol{\varepsilon}, \qquad \mathrm{Cov}\begin{bmatrix} \mathbf{u} \\ \boldsymbol{\varepsilon} \end{bmatrix} = \begin{bmatrix} \mathbf{G} & \mathbf{0} \\ \mathbf{0} & \sigma_\varepsilon^2 \mathbf{I} \end{bmatrix}, \qquad (16.11)$$

where

$$\mathbf{G} \equiv \underset{1 \le \ell \le L}{\mathrm{blockdiag}} \, \sigma_{u\ell}^2 \mathbf{I} \quad \text{and} \quad \mathbf{u} = [\mathbf{u}_1^{\mathsf{T}}, \dots, \mathbf{u}_L^{\mathsf{T}}]^{\mathsf{T}} \qquad (16.12)$$

is the partition of $\mathbf{u}$ such that $\mathrm{Cov}(\mathbf{u}_\ell) = \sigma_{u\ell}^2 \mathbf{I}$. Let q_ℓ, $1 \le \ell \le L$, denote the number of entries in $\mathbf{u}_\ell$ and suppose that the prior distribution for $\sigma_{u\ell}^2$ is $\mathrm{IG}(A_{u\ell}, B_{u\ell})$. As in Section 16.3, we will assume a uniform improper prior on $\boldsymbol{\beta}$. Then an appropriate Gibbs sampling scheme for fitting (16.11) is as follows.

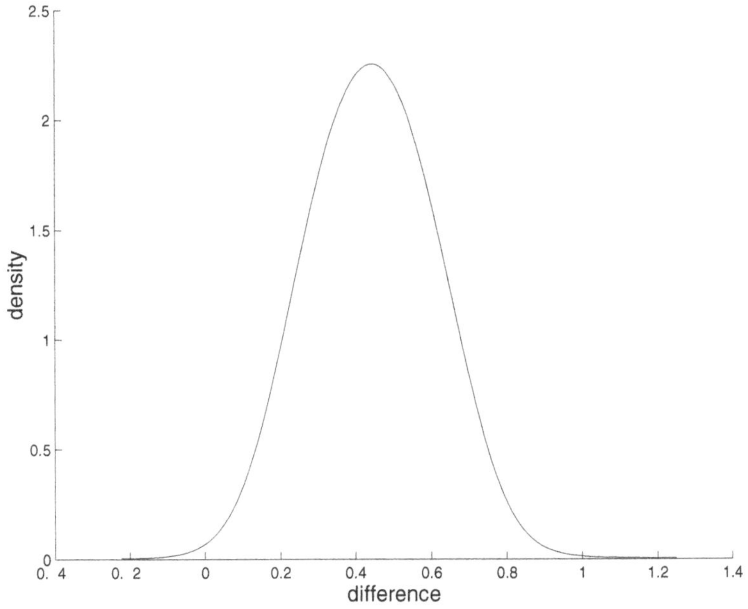

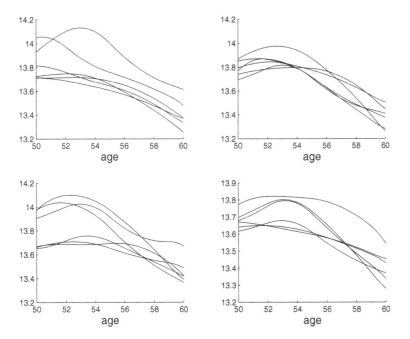

Figure 16.4
Histogram of 3000
realizations of
$f(50) - f(60)$.

Figure 16.5 Twenty-four realizations of $f(\texttt{age})$, $50 <$ age < 65. The realizations are shown in groups of six so that the individual curves are more easily seen. Since the realizations are i.i.d., group membership is arbitrary.

(1) Sample $(\boldsymbol{\beta}, \mathbf{u})$ from the multivariate normal distribution

$$\text{N}\{(\mathbf{C}^{\mathsf{T}}\mathbf{C} + \sigma_{\varepsilon}^2 \mathbf{B})^{-1}\mathbf{C}^{\mathsf{T}}\mathbf{y}, \sigma_{\varepsilon}^2 (\mathbf{C}^{\mathsf{T}}\mathbf{C} + \sigma_{\varepsilon}^2 \mathbf{B})^{-1}\},$$

where $\mathbf{C} \equiv [\mathbf{X} \ \mathbf{Z}]$ and $\mathbf{B} \equiv \text{blockdiag}(\mathbf{0}, \mathbf{G}^{-1})$.

(2) For $1 \leq \ell \leq L$, sample $\sigma_{u\ell}^2$ from $\text{IG}\big(A_{u\ell} + \tfrac{1}{2}q_{\ell}, \ B_{u\ell} + \tfrac{1}{2}\|\mathbf{u}_{\ell}\|^2\big)$.

(3) Sample σ_{ε}^2 from $\text{IG}\big(A_{\varepsilon} + \tfrac{1}{2}n, \ B_{\varepsilon} + \tfrac{1}{2}\|\mathbf{y} - \mathbf{X}\boldsymbol{\beta} - \mathbf{Z}\mathbf{u}\|^2\big)$.

(4) Return to step (1) and iterate.

16.5   Generalized Linear Mixed Models

Now suppose that the distribution of the response given the covariates is not Gaussian but rather has a distribution in another exponential family – for example, is binomial. Generalized linear models (GLMs) that allow such response distributions were discussed in Chapter 10. In particular, the generalized linear mixed models (GLMMs) of Section 10.8 are useful here since the random effects can include the spline coefficients that should be penalized as well as, for example, random subject effects. However, as discussed in Section 10.8, GLMMs present computational challenges well beyond those of fixed effects generalized linear models. MCMC algorithms for GLMMs require the same order of computational effort as the non-Bayesian exact methods presented in Section 10.8 such as Monte Carlo EM (though approximate methods based on the Laplace method are much less intensive than MCMC algorithms).

As in Chapter 10, to keep the exposition simple we will consider the theory only for canonical exponential families with scale parameter $\phi = 1$ (e.g., logistic regression). Using the notation of Section 10.8, let the density of $\mathbf{y}$ given $(\boldsymbol{\beta}, \mathbf{u})$ be

$$[\mathbf{y}|\boldsymbol{\beta}, \mathbf{u}] = \exp\{\mathbf{y}^{\mathsf{T}}(\mathbf{X}\boldsymbol{\beta} + \mathbf{Z}\mathbf{u}) - \mathbf{1}^{\mathsf{T}}b(\mathbf{X}\boldsymbol{\beta} + \mathbf{Z}\mathbf{u}) + \mathbf{1}^{\mathsf{T}}c(\mathbf{y})\} \qquad (16.13)$$

and assume that

$$\mathbf{u} \sim \mathrm{N}(\mathbf{0}, \mathbf{G})$$

as in (16.12).

The sampling scheme for $\sigma_{u\ell}^2$, $1 \le \ell \le L$, remains the same as for the Gaussian case:

$$\left[\sigma_{u\ell}^2 | \boldsymbol{\beta}, \mathbf{u}, \mathbf{y}, \underset{1 \le \ell' \le L, \ell' \ne \ell}{\sigma_{u\ell'}^2}\right] \sim \mathrm{IG}\left(A_{u\ell} + \tfrac{1}{2}q_\ell, \ B_{u\ell} + \tfrac{1}{2}\|\mathbf{u}_\ell\|^2\right).$$

However,

$$[\boldsymbol{\beta}, \mathbf{u}|\mathbf{y}, \sigma_{u1}^2, \ldots, \sigma_{uL}^2]$$
$$\propto \exp\{\mathbf{y}^{\mathsf{T}}(\mathbf{X}\boldsymbol{\beta} + \mathbf{Z}\mathbf{u}) - \mathbf{1}^{\mathsf{T}}b(\mathbf{X}\boldsymbol{\beta} + \mathbf{Z}\mathbf{u}) - \tfrac{1}{2}\mathbf{u}^{\mathsf{T}}\mathbf{G}^{-1}\mathbf{u}\}, \quad (16.14)$$

and sampling is somewhat difficult because this conditional density is not in any standard family. In such instances, more involved algorithms such as *adaptive rejection sampling* (Gilks and Wild 1992) or the *Metropolis–Hastings algorithm* (Metropolis et al. 1953; Hastings 1970) are required to complete the sampling scheme. We will focus on the latter.

The Metropolis–Hastings algorithm can be used to generate a Markov chain with a given stationary distribution when the density of that target distribution is known only up to a constant of proportionality – for example, $[\boldsymbol{\beta}, \mathbf{u}|\mathbf{y}, \sigma_{u1}^2, \ldots, \sigma_{u\ell}^2]$ here. The Metropolis–Hastings algorithm allows one to sample from a convenient distribution (e.g., a normal distribution) and then to "accept" this new observation depending upon its probability at the target distribution. We will not discuss the theory behind the Metropolis–Hastings algorithm, nor will we introduce the most general form of the algorithm. A good introduction to the general algorithm is given by Chib and Greenberg (1995). Often a pure Metropolis–Hastings algorithm is not used and instead the steps of Metropolis–Hastings are put into a more complex MCMC. That is what we propose here.

Let $(\boldsymbol{\beta}_c, \mathbf{u}_c)$ be the current value of the $(\boldsymbol{\beta}, \mathbf{u})$ in the chain. We generate a trial value for the new $(\boldsymbol{\beta}, \mathbf{u})$, call it $(\boldsymbol{\beta}_t, \mathbf{u}_t)$, that is normally distributed with mean $(\boldsymbol{\beta}_c, \mathbf{u}_c)$. The choice of covariance matrix is discussed in the next paragraph. Let R be the ratio of (16.14) evaluated at $(\boldsymbol{\beta}_t, \mathbf{u}_t)$ to the same density evaluated at $(\boldsymbol{\beta}_c, \mathbf{u}_c)$. Then with probability $\min(1, R)$ we "accept" $(\boldsymbol{\beta}_t, \mathbf{u}_t)$ – that is, we replace $(\boldsymbol{\beta}_c, \mathbf{u}_c)$ with $(\boldsymbol{\beta}_t, \mathbf{u}_t)$. Otherwise, the current value of $(\boldsymbol{\beta}, \mathbf{u})$ is retained.

> In GLMMs or even GLMs, generating observations of $(\boldsymbol{\beta}, \mathbf{u})$ from their complete conditionals generally requires special tools such as adaptive rejection sampling or the Metropolis–Hastings algorithm.

For the covariance of the trial values $(\boldsymbol{\beta}_t, \mathbf{u}_t)$ we recommend a multiple of the approximate conditional covariance matrix of the penalized quasilikelihood estimate of $(\boldsymbol{\beta}, \mathbf{u})$, namely (10.23). The scalar multiplier, call it τ, should be sufficiently small so that we accept the trial values with reasonably high probability; as $\tau \to 0$, $(\boldsymbol{\beta}_t, \mathbf{u}_t)$ will approach $(\boldsymbol{\beta}_c, \mathbf{u}_c)$ and the acceptance probability will converge to unity. However, τ should not be too small, either. If τ is small then, since the trial $(\boldsymbol{\beta}_t, \mathbf{u}_t)$ is close to $(\boldsymbol{\beta}_c, \mathbf{u}_c)$, the chain will move slowly even if nearly all trial values are accepted. We have found $0.2 \le \tau \le 0.4$ to work reasonably well in our examples.

In summary, here is our algorithm for fitting Bayesian generalized linear mixed models.

(1) Set the tuning parameter τ. We recommend $0.2 \le \tau \le 0.4$.
(2) Obtain initial estimates of $\boldsymbol{\beta}_c, \mathbf{u}_c, \sigma_{u1}^2, \ldots, \sigma_{uL}^2$ via penalized quasi-likelihood and set

$$\boldsymbol{\Sigma} = (\mathbf{C}^{\mathsf{T}} \mathbf{W} \mathbf{C} + \mathbf{B})^{-1} \mathbf{C}^{\mathsf{T}} \mathbf{W} \mathbf{C} (\mathbf{C}^{\mathsf{T}} \mathbf{W} \mathbf{C} + \mathbf{B})^{-1},$$

where $\mathbf{C} = [\mathbf{X}, \mathbf{Z}]$ and $\mathbf{W} = \mathrm{diag}\{b''(\mathbf{X}\boldsymbol{\beta}_c + \mathbf{Z}\mathbf{u}_c)\}$.

(3) Generate

$$\begin{bmatrix} \boldsymbol{\beta}_t \\ \mathbf{u}_t \end{bmatrix} \sim \mathrm{N}\left(\begin{bmatrix} \boldsymbol{\beta}_c \\ \mathbf{u}_c \end{bmatrix}, \tau \boldsymbol{\Sigma} \right)$$

and compute

$$R = \frac{\exp\{\mathbf{y}^{\mathsf{T}}(\mathbf{X}\boldsymbol{\beta}_t + \mathbf{Z}\mathbf{u}_t) - \mathbf{1}^{\mathsf{T}} b(\mathbf{X}\boldsymbol{\beta}_t + \mathbf{Z}\mathbf{u}_t) - \frac{1}{2} \mathbf{u}_t^{\mathsf{T}} \mathbf{G}^{-1} \mathbf{u}_t\}}{\exp\{\mathbf{y}^{\mathsf{T}}(\mathbf{X}\boldsymbol{\beta}_c + \mathbf{Z}\mathbf{u}_c) - \mathbf{1}^{\mathsf{T}} b(\mathbf{X}\boldsymbol{\beta}_c + \mathbf{Z}\mathbf{u}_c) - \frac{1}{2} \mathbf{u}_c^{\mathsf{T}} \mathbf{G}^{-1} \mathbf{u}_c\}}.$$

Now generate U, a random uniform variate between 0 and 1.
  (a) If $U \le R$ then replace $(\boldsymbol{\beta}_c, \mathbf{u}_c)$ by $(\boldsymbol{\beta}_t, \mathbf{u}_t)$.
  (b) If $U > R$ then leave $(\boldsymbol{\beta}_c, \mathbf{u}_c)$ unchanged.
(4) For $1 \le \ell \le L$, sample $\sigma_{u\ell}^2$ from $\mathrm{IG}\left(A_{u\ell} + \frac{1}{2} q_\ell, B_{u\ell} + \frac{1}{2} \|\mathbf{u}_\ell\|^2\right)$.
(5) Return to step (3) and iterate.

16.5.1 *Probit Mixed Models*

If the entries of $\mathbf{y}$ are binary, then the Metropolis–Hastings algorithm just described can be used to fit a logistic mixed model by taking $b(x) = \log(1 + e^x)$. However, in this special case an even easier implementation is possible using the probit model (see e.g. Chib and Greenberg 1995). The probit model was discussed in Section 10.3, and Figure 10.2 shows that the probit and logistic models are very nearly the same.

The probit model begins with an auxiliary variable $a = [a_1, \ldots, a_n]^\mathsf{T}$, which has a multivariate normal distribution with mean $\mathbf{X}\boldsymbol{\beta} + \mathbf{Z}\mathbf{u}$ and identity covariance matrix. If a were observed then we would have exactly the problem described in Section 16.3 – with the simplification that $\sigma_\varepsilon^2 = 1$. The link to the observed binary response data $\mathbf{y}$ is that y_i is the indicator that $a_i > 0$. With this convention, the observed binary data satisfy

$$P(y_i = 1 | \mathbf{u}) = \Phi\{(\mathbf{X}\boldsymbol{\beta} + \mathbf{Z}\mathbf{u})_i\},$$

where Φ is the standard normal cumulative distribution function. Note that

$$[a_i | \boldsymbol{\beta}, \mathbf{u}, \sigma_{u1}^2, \ldots, \sigma_{uL}^2]$$
$$\propto \{I(a_i > 0)\}^{y_i=1}\{I(a_i \le 0)\}^{y_i=0} \exp\left[-\tfrac{1}{2}\{a_i - (\mathbf{X}\boldsymbol{\beta} + \mathbf{Z}\mathbf{u})_i\}^2\right].$$

This means that, for those components of $\mathbf{y}$ that equal 1, the corresponding components of a have the distribution of a normal random variable with mean $\mathbf{X}\boldsymbol{\beta} + \mathbf{Z}\mathbf{u}$ but *truncated from the left at* 0. Also, for those components of $\mathbf{y}$ that equal 0, the corresponding components of a have the distribution of a normal random variable with mean $\mathbf{X}\boldsymbol{\beta} + \mathbf{Z}\mathbf{u}$ but *truncated from the right at* 0. Robert (1995) describes algorithms for generating truncated normal random variables.

The full algorithm, given starting values, is as follows.

(1) Sample $(\boldsymbol{\beta}, \mathbf{u})$ from the multivariate normal distribution

$$\mathrm{N}\{(\mathbf{C}^\mathsf{T}\mathbf{C} + \mathbf{B})^{-1}\mathbf{C}^\mathsf{T}a, \ (\mathbf{C}^\mathsf{T}\mathbf{C} + \mathbf{B})^{-1}\},$$

where $\mathbf{C} \equiv [\mathbf{X} \ \mathbf{Z}]$ and $\mathbf{B} \equiv \mathrm{blockdiag}(\mathbf{0}, \mathbf{G}^{-1})$.

(2) For $1 \le \ell \le L$, sample $\sigma_{u\ell}^2$ from $\mathrm{IG}\left(A_{u\ell} + \tfrac{1}{2}q_\ell, \ B_{u\ell} + \tfrac{1}{2}\|\mathbf{u}_\ell\|^2\right)$.

(3) For $1 \le i \le n$:

    (a) if $y_i = 0$, sample a_i from a $\mathrm{N}\{(\mathbf{X}\boldsymbol{\beta} + \mathbf{Z}\mathbf{u})_i, 1\}$ density truncated over $(-\infty, 0)$;

    (b) if $y_i = 1$, sample a_i from a $\mathrm{N}\{(\mathbf{X}\boldsymbol{\beta} + \mathbf{Z}\mathbf{u})_i, 1\}$ density truncated over $(0, \infty)$.

    Set $a = [a_1, \ldots, a_n]^\mathsf{T}$.

(4) Return to step (1) and iterate.

16.5.1.1 *Union and Wages Data Revisited*

Here we illustrate Bayesian methods in the union and wages example discussed in Chapter 11 (see especially Section 11.2 and Figure 11.2). In Chapter 11 we fit a logistic nonparametric regression model to these data. Here we fit a linear spline with 20 knots. The Bayesian fit was based on a chain of 10,000 samples, of which the first 5000 were removed as a burn-in period.

Figure 16.6 gives the posterior mean fit and compares it to the frequentist fit: note how the two differ hardly at all. Figure 16.7 gives the posterior mean fit along with the 95% credible interval. Note that this graph is almost identical to the corresponding frequentist plot, Figure 11.2.

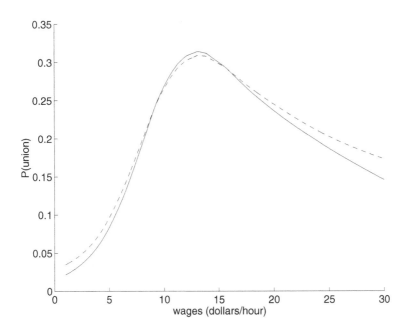

Figure 16.6
Comparison of the frequentist (solid line) and posterior mean Bayesian (dashed line) probit fits to the union and wages data, using a linear regression spline with 20 knots.

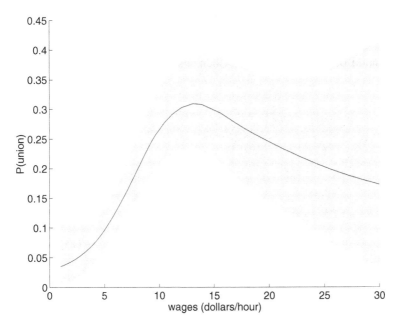

Figure 16.7 Bayesian probit fit to the union and wages data with 95% credible interval, using a linear regression spline with 20 knots.

16.6 Rao–Blackwellization

This section deals with Rao–Blackwellization of the MCMC. Readers not interested in the details of the implementation can safely skip this section.

Consider the scatterplot smoothing setting of Section 16.3. The posterior expectation $(\boldsymbol{\beta}, \mathbf{u})$ can be estimated from the MCMC simulation by averaging the conditional mean in (16.8) over the realizations of $(\sigma_u^2, \sigma_\varepsilon^2)$. The MCMC realizations of $(\boldsymbol{\beta}, \mathbf{u})$ have these conditional means, so averaging the conditional means rather than the realizations themselves reduces Monte Carlo variance. Let N denote the number of realizations from the MCMC algorithm. An efficient estimate of the posterior covariance matrix of $(\boldsymbol{\beta}, \mathbf{u})$ is obtained by the following steps:

(1) the sample covariance of the N conditional means of $(\boldsymbol{\beta}, \mathbf{u})$ in (16.8) estimates $\mathrm{Cov}[\mathsf{E}\{(\boldsymbol{\beta}, \mathbf{u})|\sigma_u^2, \sigma_\varepsilon^2, \mathbf{y}\}]$;

(2) averaging the N covariance matrices in (16.8) estimates $\mathsf{E}[\mathrm{Cov}\{(\boldsymbol{\beta}, \mathbf{u})|\sigma_u^2, \sigma_\varepsilon^2, \mathbf{y}\}]$;

(3) by (A.8), summing the two provides an efficient estimate of $\mathrm{Cov}\{(\boldsymbol{\beta}, \mathbf{u})|\mathbf{y}\}$.

Since the mean of the $\mathrm{IG}(a, b)$ distribution is $b/(a - 1)$ for $a > 1$, the conditional mean of σ_u^2 in step (2) is $(B_u + \frac{1}{2}\|\mathbf{u}\|^2)/(A_u + \frac{1}{2}K - 1)$. Averaging this over the N simulated values of $\mathbf{u}$ provides an estimate of σ_u^2. If an estimate of the posterior density of σ_u^2 is desired, say for plotting, then the $\mathrm{IG}(A_u + \frac{1}{2}K, B_u + \frac{1}{2}\|\mathbf{u}\|^2)$ density can be averaged over the N values of $\mathbf{u}$. Estimates of σ_ε^2 and its posterior density can be found in the same way.

16.7   Bibliographical Notes

Recent books on Bayesian statistics that contain advice on Markov chain Monte Carlo include Gelman et al. (1995), Tanner (1996), Robert and Casella (1999), Carlin and Louis (2000), Ibrahim, Chen, and Lipsitz (2001), and Liu (2001).

17

Spatially Adaptive Smoothing

17.1 Introduction

The penalty parameter λ of a penalized spline controls the trade-off between bias and variance. In this chapter we introduce a method of fitting penalized splines wherein λ varies spatially in order to accommodate possible spatial nonhomogeneity of the regression function. In other words, λ is allowed to be a function of the independent variable x. Allowing λ to be a function of spatial location can improve mean squared error (MSE; see Section 3.11) and the accuracy of inference.

Suppose we are using a quadratic spline that has constant curvature between knots. If the regression function has rapid changes in curvature, then a small value of λ is needed so that the second derivative of the fitted spline can take jumps large enough to accommodate these changes in curvature. Conversely, if the curvature changes slowly, then a large value of λ will not cause large bias and will reduce df_{fit} and the variance of the fitted values. The problem is that a regression function may be spatially nonhomogeneous with some regions of rapidly changing curvature and other regions of little change in curvature. A single value of λ is not suitable for such functions. The inferiority – in terms of MSE – of splines having a single smoothing parameter is shown in a simulation study by Wand (2000). In that study, for regression functions with significant spatial inhomogeneity, penalized splines with a single smoothing parameter were not competitive with knot selection methods. In an empirical study by Ruppert (2002), the spatially adaptive penalized splines described in this chapter were found to be at least as efficient as knot selection methods.

Another problem with having only a single smoothing parameter concerns inference. As seen in Chapters 4 and 16, penalized splines are BLUPs if one assumes a certain mixed model or Bayes estimates for particular priors. A single smoothing parameter corresponds to a spatially homogeneous mixed model distribution or Bayesian prior. For a penalized spline, the mixed model assumption (or prior) is that the knot coefficients $u_1, \ldots, u_K$ are independent $N(0, \sigma_u^2)$ for a single parameter σ_u^2. Such priors on $u_1, \ldots, u_K$ are not appropriate for spatially heterogeneous f. Consider the confidence intervals based on the posterior variance of $f(\cdot)$ discussed in Section 16.2.2. As Nychka (1988) shows, the resulting confidence bands have good *average* (over x) coverage probabilities but do not have accurate pointwise coverage probabilities in areas of high oscillations or other "features" in f.

17.2   A Local Penalty Method

Here is a simple approach to spatially varying λ. Choose another set of the knots, $\{\kappa_k^*\}_{k=1}^M$, where M is smaller than K and such that $\{\kappa_1^* = \kappa_1 < \cdots < \kappa_M^* = \kappa_K\}$. The penalty at one of these "subknots", say κ_k^*, is set equal to a parameter λ_k^*. The penalties at the original knots, $\{\kappa_k\}_{k=1}^K$, are determined by interpolation of the penalties at the $\{\kappa_k^*\}_{k=1}^M$. The interpolation is on the log-penalty scale to ensure positivity of the penalties. Thus we have a penalty $\lambda(\kappa_k)$ at each κ_k, but these penalties depend only upon $\lambda = (\lambda_1^*, \ldots, \lambda_M^*)^\mathsf{T}$. Therefore, $\{\lambda(\kappa_1), \ldots, \lambda(\kappa_K)\}$ is a function of λ. This function need not be derived explicitly but rather can be computed by using an interpolation algorithm; we used MATLAB's built-in linear interpolator. One could, of course, use other interpolation methods (e.g., cubic interpolation). If linear interpolation is used, then $\log\{\lambda(\cdot)\}$ is a linear spline with knots at $\{\kappa_k^*\}_{k=1}^M$.

As in Chapter 3, let $\mathbf{y} = [y_1, \ldots, y_n]^\mathsf{T}$ and let $\mathbf{C}$ be a matrix with ith row equal to

$$\mathbf{C}_i = [1 \;\; x_i \;\; \cdots \;\; x_i^p \;\; (x_i - \kappa_1)_+^p \;\; \cdots \;\; (x_i - \kappa_K)_+^p]. \tag{17.1}$$

Let $\mathbf{D}(\lambda)$ be a diagonal matrix whose first $p + 1$ diagonal elements are 0 and whose remaining diagonal elements are $\lambda^2(\kappa_1), \ldots, \lambda^2(\kappa_K)$, which depend only on λ. Then, as in Section 3.5, penalized spline estimates of

$$\nu \equiv [\boldsymbol{\beta}^\mathsf{T} \;\; \mathbf{u}^\mathsf{T}]^\mathsf{T}$$

are given by

$$\hat{\nu}(\lambda) = \{\mathbf{C}^\mathsf{T}\mathbf{C} + \mathbf{D}(\lambda)\}^{-1}\mathbf{C}^\mathsf{T}\mathbf{y}. \tag{17.2}$$

The smoothing parameter $\lambda = (\lambda_1^*, \ldots, \lambda_M^*)$ can be determined by minimizing the generalized cross-validation statistic

$$\mathrm{GCV}(\lambda) = \frac{\|\mathbf{y} - \mathbf{C}\hat{\nu}(\lambda)\|^2}{\{1 - df_\mathrm{fit}(\lambda)/n\}^2}.$$

Here

$$df_\mathrm{fit}(\lambda) = \mathrm{tr}[\{\mathbf{C}^\mathsf{T}\mathbf{C} + \mathbf{D}(\lambda)\}^{-1}\mathbf{C}^\mathsf{T}\mathbf{C}] \tag{17.3}$$

is the degrees of freedom of the fit.

A search over an M-dimensional grid is not recommended because of computational cost. Rather, we recommend that one start with $\lambda_1^*, \ldots, \lambda_M^*$, each equal to the best global value of λ chosen by minimizing GCV as in Chapter 5. Then each λ_k^* is varied, with the others fixed, over a 1-dimensional grid centered at the current value of λ_k^*. On each such step, λ_k^* is replaced by the λ-value minimizing GCV on this grid. This minimizing of GCV over each λ_k^* is repeated a total of N_iter times. Although minimizing GCV over the λ_k^* one at a time in this manner does not guarantee finding the global minimum of GCV over $\lambda_1^*, \ldots, \lambda_M^*$, our simulations show that this procedure is effective in selecting a satisfactory amount of local smoothing. The minimum GCV global λ is a reasonably good starting value for the smoothing parameters, and each step of our algorithm improves on this start in the sense of lowering GCV. Since each λ_k^* controls the penalty over only a small range of x, the optimal value of one λ_k^* should depend only slightly upon the other λ_k^*. We believe this is why the one-at-a-time search strategy works effectively.

17.3 Completely Automatic Algorithm

The local penalty method has three tuning parameters: the number of knots, K; the number of subknots, M; and the number of iterations, N_{iter}. The exact values of the tuning parameters are not crucial provided they are within certain acceptable ranges – the crucial parameter is λ, which is selected by GCV, not the user. However, users may want a completely automatic algorithm that requires no user-specified parameters and attempts to ensure that the tuning parameters are within acceptable ranges. An automatic algorithm would have to balance the need for the tuning parameters to be large enough to obtain a good fit with the need that the tuning parameters not be so large that the computation time is excessive; overfitting is not a concern because it is controlled by λ.

In this section, we propose two methods for choosing the tuning parameters, the myopic and the full-search algorithms, which are similar to algorithms of the same names in Section 5.6.3. The two algorithms are based on the following principle: As the complexity of f increases, each of K, M, N_{iter} should increase. The algorithms use a sequence of values of (K, M, N_{iter}) where each parameter is nondecreasing in the sequence. The myopic algorithm stops when there is no appreciable decrease in GCV between two successive values of (K, M, N_{iter}). Monte Carlo experimentation, discussed in Section 17.5, shows that the values of N_{iter} and M have relatively little effect on the fit, at least within the ranges studied. However, it seems reasonable to increase N_{iter} and M slightly with K. On the other hand, computation time for given K is roughly proportional to $M \times N_{\text{iter}}$, so we avoid $N_{\text{iter}} > 2$ and $M > 6$.

The full-search algorithm computes GCV for all tuning parameter sets and chooses the set that minimizes GCV. Specifically, the sequence of values of (K, M, N_{iter}) that we use are $(10, 2, 1)$, $(20, 3, 2)$, $(40, 4, 2)$, $(80, 6, 2)$, and $(120, 6, 2)$. For each set of tuning parameters, λ is chosen to minimize GCV. The full-search algorithm chooses the set that has the smallest of these minimized GCV values. The myopic algorithm first compares $(10, 2, 1)$ and $(20, 3, 2)$ using GCV. If the value of GCV for $(20, 3, 2)$ is more than a constant C times the GCV value of $(10, 2, 1)$, then we conclude that further increases in the tuning parameters will not appreciably decrease GCV – in the simulations we used $C = 0.98$ and that choice worked well. Therefore, we stop and use $(10, 2, 1)$ or $(20, 3, 2)$ as the final value of the three tuning parameters, whichever has the smallest GCV value. Otherwise, we fit using $(40, 4, 2)$ and compare its GCV value to that of $(20, 3, 2)$. If the value of GCV for $(40, 4, 2)$ is more than C times the GCV value of $(20, 3, 2)$ then we stop and use either $(20, 3, 2)$ or $(40, 4, 2)$, whichever has the smallest GCV, as the final tuning parameters. Otherwise, we continue in this manner, comparing $(40, 4, 2)$ to $(80, 6, 2)$, and so on. If very complex f were contemplated then one could, of course, continue using increasingly larger values of the tuning parameters.

17.4 Bayesian Inference

Bayesian inference for local penalty splines differs little from the Bayesian inference for global penalty splines that was discussed in Section 16.2.2. Suppose that $\varepsilon_1, \ldots, \varepsilon_n$ are independent $N(0, \sigma_\varepsilon^2)$ and that the prior on $\boldsymbol{v}$ is $N\{\boldsymbol{0}, \boldsymbol{\Sigma}(\lambda)\}$, where $\boldsymbol{\Sigma}(\lambda)$ is a covariance matrix that depends on λ. Here $N(\boldsymbol{\mu}, \boldsymbol{\Sigma})$ is the multivariate normal distribution with mean and covariance matrix $\boldsymbol{\mu}$ and $\boldsymbol{\Sigma}$. For now, assume that σ_ε^2 and λ are known. Then the posterior log density of $\boldsymbol{v}$ given $\mathbf{y}$ is, up to an additive function of $\mathbf{y}$ and $(\sigma_\varepsilon^2, \lambda)$, given by

$$-\frac{1}{2}\left(\frac{1}{\sigma_\varepsilon^2}\|\mathbf{y} - \mathbf{C}\boldsymbol{v}\|_2^2 + \boldsymbol{v}^\mathsf{T}\boldsymbol{\Sigma}(\lambda)^{-1}\boldsymbol{v}\right). \tag{17.4}$$

The maximum a posteriori (MAP) estimator of $\boldsymbol{v}$ – that is, the mode of the posterior density – maximizes (17.4). Now let $\beta_0, \ldots, \beta_p$ have improper uniform $(-\infty, \infty)$ priors and let $\{u_k\}_{k=1}^K$ be independent, with β_{p+k} having a $N(0, \sigma_\varepsilon^2/\lambda_k)$ distribution. Then

$$\boldsymbol{\Sigma}^{-1}(\lambda) = \sigma_\varepsilon^2 \,\text{diag}(0, \ldots, 0, \lambda_1, \ldots, \lambda_K). \tag{17.5}$$

More precisely, we let $\beta_0, \ldots, \beta_p$ have a $N(0, \sigma_\beta^2)$ prior and then (17.5) holds in the limit as $\sigma_\beta \to \infty$. The MAP estimator of $\boldsymbol{v}$ minimizes

$$\sum_{i=1}^n \{y_i - f(x_i; \boldsymbol{v})\}^2 + \sum_{k=1}^K \lambda(\kappa_k)u_k^2.$$

Of course, the $\lambda = (\lambda_1^*, \ldots, \lambda_M^*)^\mathsf{T}$ that determines the λ_k will not be known in practice. Empirical Bayes methods replace unknown "hyperparameters" in a prior by estimates. For example, if λ is estimated by GCV and then considered fixed, one is using empirical Bayes methods. Standard calculations show that when λ and σ_ε^2 are known, the posterior distribution of $\boldsymbol{v}$ is

$$N[\hat{\boldsymbol{v}}(\lambda), \sigma_\varepsilon^2\{\mathbf{C}^\mathsf{T}\mathbf{C} + \boldsymbol{\Sigma}(\lambda)\}^{-1}]. \tag{17.6}$$

Also, the posterior distribution of $\mathbf{f} = \{f(x_1), \ldots, f(x_n)\}^\mathsf{T}$ is

$$N[\mathbf{C}\hat{\boldsymbol{v}}(\lambda), \sigma_\varepsilon^2\mathbf{C}\{\mathbf{C}^\mathsf{T}\mathbf{C} + \boldsymbol{\Sigma}(\lambda)\}^{-1}\mathbf{C}^\mathsf{T}]. \tag{17.7}$$

An approximate Bayes posterior replaces λ and σ_ε^2 in (17.6) and (17.7) by estimates. Assuming that λ has been estimated by GCV, one need only estimate σ_ε^2 by $\|\mathbf{y} - \mathbf{C}\hat{\boldsymbol{v}}(\hat{\lambda})\|^2/\{n - df_{\text{fit}}(\hat{\lambda})\}$, where $df_{\text{fit}}(\lambda)$ is defined by (17.3). This gives the approximate posterior distribution for $\mathbf{f}$:

$$\mathbf{f} \sim N[\mathbf{C}\hat{\boldsymbol{v}}(\hat{\lambda}), \hat{\sigma}_\varepsilon^2\mathbf{C}\{\mathbf{C}^\mathsf{T}\mathbf{C} + \boldsymbol{\Sigma}(\hat{\lambda})\}^{-1}\mathbf{C}^\mathsf{T}]. \tag{17.8}$$

The approximate Bayes $100(1 - \alpha)\%$ confidence interval for $f(x_i)$ is

$$\hat{f}(x_i) \pm z(1 - \alpha/2)\,\widehat{\text{st.dev.}}\{\hat{f}(x_i) - f(x_i)\},$$

where $\hat{f}(x_i) = \mathbf{C}_i\hat{\boldsymbol{v}}(\hat{\lambda})$ is the ith element of the posterior mean in (17.8) and $\widehat{\text{st.dev.}}\{\hat{f}(x_i) - f(x_i)\}$ is the square root of the ith diagonal entry of the posterior covariance matrix in (17.8).

Because they estimate hyperparameters yet then pretend that the hyperparameters were known, these approximate Bayesian methods do not account for extra variability in the posterior distribution caused by estimation of hyperparameters in the prior; for discussion see, for example, Morris (1983), Laird and Louis (1987), Kass and Steffey (1989), or Carlin and Louis (2000). Everything else held constant, the underestimation of posterior variance should become worse as M increases, since each λ_m^* ($m = 1, \ldots, M$) will be determined by fewer data and will therefore be more variable. As Nychka (1988) has shown empirically, this underestimation does not appear to be a problem for a global penalty that has only one hyperparameter. However, the local penalty has M hyperparameters. For local penalty splines we have found that the pointwise approximate posterior variance of $\hat{f}$ is too small in the sense that it noticeably underestimates the frequentist MSE.

A simple ad hoc correction to this problem is to multiply the pointwise posterior variances of the local penalty $\hat{f}$ from (17.8) by a constant so that the average pointwise posterior variance of $\hat{f}$ is the same for the global and local penalty estimators. The reasoning behind this correction is as follows. As stated previously, the global penalty approximate posterior variance from (17.8) is nearly equal to the frequentist's MSE on average. The local penalty estimate has an MSE that varies spatially but should be close, on average, to the MSE of the global penalty estimate and hence also close, on average, to the estimated posterior variance of the global penalty estimator. We found that this adjustment is effective in guaranteeing coverage probabilities at least as large as nominal, though in extreme cases of spatial heterogeneity the adjustment can be conservative; see Section 17.5. The reason for the latter is that, in cases of severe spatial heterogeneity, the local penalty MSE will be less, on average, than that of the global penalty estimate. Then, there will be an overcorrection and the local penalty MSE will be overestimated by this adjusted posterior variance. The result is that confidence intervals constructed with this adjustment should be conservative. The empirical evidence in Section 17.5 supports this conjecture. In that section, we refer to these adjusted intervals as *local penalty, conservative* intervals.

Another correction would be to use a fully Bayesian hierarchical model, where the hyperparameters are given a prior. Deely and Lindley (1981) first considered such empirical Bayes methods. An exact Bayesian analysis for penalized splines would require Gibbs sampling or other computationally intensive techniques (discussed in Chapter 16) and would be an interesting area for further research.

There are intermediate positions between the quick, ad hoc, conservative adjustment just proposed and an exact, fully Bayesian analysis. One that we now describe is an approximate fully Bayesian method that uses a small bootstrap experiment and a delta-method correction adopted from Kass and Steffey's (1989) "first-order approximation". Kass and Steffey considered conditionally independent hierarchical models, which are also called empirical Bayes models, but their ideas apply directly to more general hierarchical Bayes models.

The Kass and Steffey approximation is applied to penalized splines as follows. Let $f_i = f(x_i) = \mathbf{C}_i\mathbf{v}$. The posterior variance of f_i is calculated from the joint posterior distribution of $(\mathbf{v}, \lambda)$ and by a standard identity is

$$\mathrm{var}(f_i) = \mathsf{E}\{\mathrm{var}(f_i|\boldsymbol{\lambda})\} + \mathrm{var}\{\mathsf{E}(f_i|\boldsymbol{\lambda})\}.$$

Note that $\mathsf{E}\{\mathrm{var}(f_i|\boldsymbol{\lambda})\}$ is well approximated by the posterior variance of f_i when $\boldsymbol{\lambda}$ is treated as known and fixed at its posterior mode (Kass and Steffey 1989). Thus, $\mathrm{var}\{\mathsf{E}(f_i|\boldsymbol{\lambda})\}$ is the extra variability in posterior distribution of f_i that the approximate posterior variance given by (17.8) does not account for. We estimate $\mathrm{var}\{\mathsf{E}(f_i|\boldsymbol{\lambda})\}$ by the following three steps and add this estimate to the posterior variance given by (17.8).

(1) Use a parametric bootstrap to estimate $\mathrm{Cov}\{\log(\hat{\boldsymbol{\lambda}})\}$. Here the log function is applied elementwise to the vector $\boldsymbol{\lambda}$.
(2) Numerically differentiate $\mathbf{C}\hat{\boldsymbol{v}}(\boldsymbol{\lambda})$ with respect to $\log(\boldsymbol{\lambda})$ at $\boldsymbol{\lambda} = \hat{\boldsymbol{\lambda}}$. We use one-sided numerical derivatives with a step length of 0.1.
(3) Put the results from (1) and (2) into the delta-method formula:

$$\mathrm{var}\{\mathsf{E}(f_i|\boldsymbol{\lambda})\} \simeq \left\{\frac{\partial \mathbf{C}\hat{\boldsymbol{v}}(\hat{\boldsymbol{\lambda}})}{\partial \log(\boldsymbol{\lambda})}\right\}^{\mathsf{T}} \mathrm{Cov}\{\log(\hat{\boldsymbol{\lambda}})\}\left\{\frac{\partial \mathbf{C}\hat{\boldsymbol{v}}(\hat{\boldsymbol{\lambda}})}{\partial \log(\boldsymbol{\lambda})}\right\}. \tag{17.9}$$

When (17.9) is added to the approximate posterior variance from (17.8), we refer to the corresponding confidence intervals as *local penalty, corrected* intervals. Since the correction (17.9) is a relatively small portion of the corrected posterior variance, it need not be estimated by the bootstrap with as great a precision as when a variance is estimated entirely by a bootstrap. In our simulations, we used only 25 bootstrap samples in step (1).

In the simulations of the next section, the local penalty, conservative intervals are close to the more computationally intensive local penalty, corrected intervals. Since the latter have a theoretical justification, this closeness is some justification for the former.

17.5   Simulations

17.5.1   *Effects of the Tuning Parameters*

A Monte Carlo experiment was conducted to learn how the tuning parameters affect the accuracy of the local penalized spline. The regression function

$$f(x; j) = \sqrt{x(1 - x)} \sin\left(\frac{2\pi(1 + 2^{(9-4j)/5})}{x + 2^{(9-4j)/5}}\right) \tag{17.10}$$

was used, where j varied as a factor with levels 3, 4, 5, and 6. Larger values of j imply greater spatial heterogeneity. The sample size was 400, the x were all equally spaced on [0, 1], and σ_ε was 0.2. The three other factors besides j were tuning parameters: K with levels 20, 40, 80, and 120; M with levels 3, 4, 6, and 8; and N_{iter} with levels 1, 2, and 3. A full four-factor design was used with two replications for a total of 384 runs.

The response was $\log(\mathrm{MASE})$, where MASE is as defined in Section 5.6.3. Because of interaction between j and the tuning parameters, quadratic response surfaces in the three tuning parameters were fit with j fixed at each of its four levels. It was found that for $j = 3, 4$, or 5, the tuning parameters had no appreciable effects on log(MASE). For $j = 6$, only the number of knots, K, had an effect on log(MASE). That effect is nonlinear: log(MASE) decreases rapidly as K increases up to about 80, but then log(MASE) levels off.

In summary, for the scenario we simulated, of three tuning parameters only K has a detectable effect on log(MASE). It is important that K be at least a certain minimum value depending on the regression function; however, after K is sufficiently large, further increases in K do not affect accuracy.

17.5.2 *The Automatic Algorithms*

The algorithms in Section 17.3 that choose all tuning parameters automatically were tested on simulated data. As previously mentioned, it is important that the number of knots, K, be sufficiently large that all significant features of the regression function can be modeled. Thus, the main function of the automatic algorithm is to ensure that K is sufficiently large. As reported in Section 17.5.1, the number of subknots and the number of iterations were not noticed to affect accuracy, but in our proposed algorithm we allowed them to increase slightly with K.

There were three simulations that differed in the regression function and the sample size. All three simulations used 300 simulated data sets and a standard deviation of all the ε equal to 0.2. The regression function was always of form (17.10) but with different values of j. The estimators were compared by MASE.

The first simulation used $j = 3$ and $n = 150$. Panel (a) of Figure 17.1 shows a typical data set and the true regression function. Recall that the algorithm can choose as the final value of the tuning parameters (K, M, N_{iter}) one of the vectors $(10, 2, 1)$, $(20, 3, 2)$, $(40, 4, 2)$, $(80, 6, 2)$, or $(120, 6, 2)$. We coded these tuning parameter sets as 1, 2, 3, 4, and 5. Panel (b) shows the MASE for each fixed tuning parameter set and for the myopic and full-search algorithms. Panel (c) gives histograms of the tuning parameter sets chosen by the myopic (left) and full-search (right) algorithms. Finally, panel (d) plots the ASE for tuning parameter set 5 versus ASE for tuning parameter set 1.

In this example, MASE increases with the tuning parameter set number, so that larger values of K, M, and N_{iter} lead to worse estimates. The myopic algorithm chooses the best (i.e., the first) tuning parameter set over two thirds of the time and performs better than the full-search algorithm. In panel (d) one sees that the difference in ASE between the best and the worst tuning parameter sets is quite small for over 90% of the data sets. But there are about 20 out of 300 data sets where the fifth tuning parameter set has a substantially higher ASE than the first tuning parameter set. This behavior explains why one often finds that, for a given data set, the tuning parameters have little effect on the fit – provided that they exceed a certain minimal value – yet MASE still depends on the tuning parameters even when they are above this threshold.

The second simulation used $j = 3$ and $n = 400$. See Figure 17.2. The results are similar to those of the first simulation, except that now the myopic algorithm chooses the first tuning parameter set nearly 100% of the time and has a MASE value nearly as small as the best of the fixed tuning parameter set estimators.

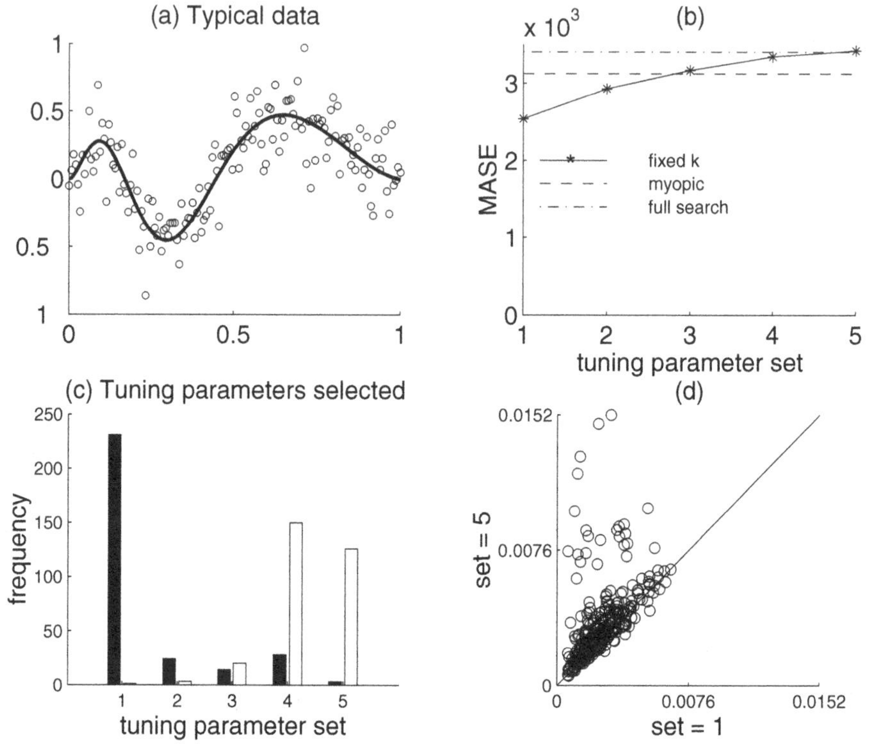

Figure 17.1 Study
of the automatic
algorithm for
choosing the tuning
parameters with
$n = 150$ and $j = 3$
so that there is low
spatial variability.
(a) Typical data set
and true regression
function. (b) MASE
of estimates using
each of the five
sets of fixed tuning
parameter values and
for the myopic and
full-search algorithms.
(c) Histograms of the
tuning parameter sets
chosen by the myopic
(left) and full-search
(right) algorithm.
(d) ASE for tuning
parameter set 5 versus
tuning parameter set 1.

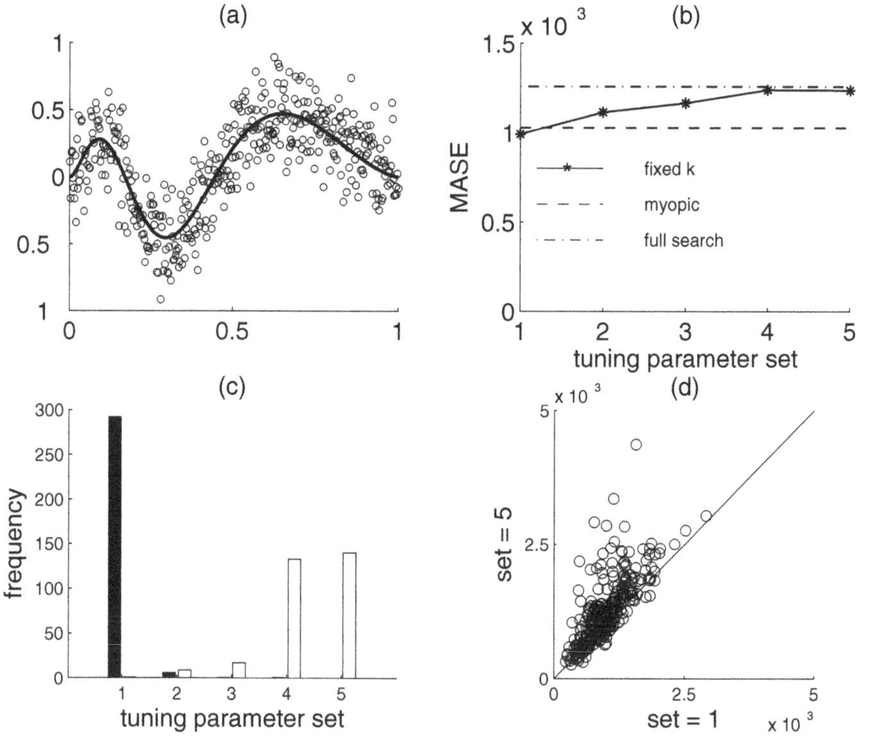

Figure 17.2 Study
of the automatic
algorithm for
choosing the tuning
parameters with
$n = 400$ and $j = 3$ so
that there is low spatial
variability; (a)–(d) as
in Figure 17.1.

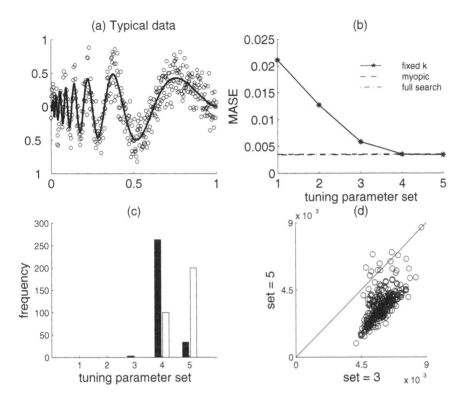

Figure 17.3 Study of the automatic algorithm for choosing the tuning parameters with $n = 400$ and $j = 6$ so that there is substantial spatial variability; (a)–(c) as in Figure 17.1. (d) ASE for tuning parameter set 5 versus tuning parameter set 3.

The third simulation (Figure 17.3) used $n = 400$ and $j = 6$. With this value of j there is serious spatial heterogeneity. The MASE is rather large unless the fourth or fifth tuning parameter set is used. Both the myopic and full-search algorithms choose either the fourth or fifth tuning parameter set in nearly all samples. The bias for tuning parameter sets 1 and 2 is so large that in panel (d) the comparison is between sets 3 and 5, not sets 1 and 5 as in the previous figures. The plot in panel (d) is similar to those in the previous figures except shifted to the right owing to the bias with parameter set 3.

We conclude that the automatic algorithms can supply reasonable, but not optimal, values of the tuning parameters when the user has little idea how to choose them. In particular, for complex functions such as (17.10), both of the automatic algorithms choose K and M large enough to obtain good estimates. However, for less complex functions such as (17.10) with $j = 3$, the MASE is somewhat reduced by choosing K and M small, but the automatic algorithms are not likely to achieve this reduction in MASE.

The myopic algorithm performed better than the full-search algorithm for the examples studied here, but we have seen in Section 5.6.3 that myopic algorithms can stop prematurely if the regression function is particularly nasty. The myopic algorithm can be used provided one exercises caution and checks whether it might have stopped prematurely. The full-search and myopic algorithms can easily be computed together and, in fact, our MATLAB routine does this. It is useful to compute both and to check whether they give vastly different estimates.

Figure 17.4 Typical
data in the Bayesian
inference study: local
and global penalty
splines.

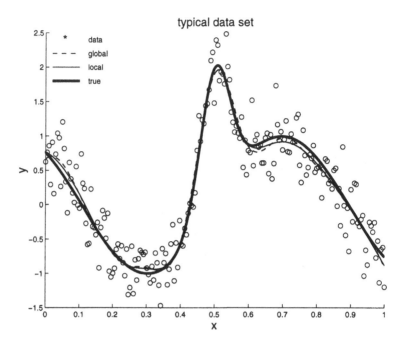

If they do, that is evidence that the myopic algorithm may have stopped prematurely. If the myopic and full-search estimates are similar, we recommend using
the myopic algorithm because it is likely to be slightly closer to the true regression function.

17.5.3 *Bayesian Inference*

To compare the posterior distribution with and without a local penalty, we used a
spatially heterogeneous regression function

$$f(x) = \sin\{8(x - 0.5)\} + 2\exp\{-16^2(x - 0.5)^2\}. \tag{17.11}$$

The x_i were equally spaced on $[0, 1]$, the sample size was $n = 200$, and the ε_i
were normally distributed with $\sigma_\varepsilon = 0.3$. We used quadratic splines with $K = 40$ knots, the number of subknots was $M = 5$, and the number of iterations to
minimize GCV using the local penalty was $N_{\text{iter}} = 1$.

Figure 17.4 shows a typical data set and the global and local penalty estimates.
The global and local penalty estimate are difficult to distinguish visually. In cases
with more extreme spatial heterogeneity, this is not true; see an example in Ruppert and Carroll (2000, p. 215), where the global penalty estimate is noticeably
undersmoothed but where the local penalty estimate is much smoother in a region
where f is nearly constant.

Figure 17.5 shows the pointwise MSE and squared bias of the global penalty
estimator calculated from 300 Monte Carlo samples. Also shown is the pointwise
posterior variance given by (17.8) averaged over the 300 repetitions. The posterior variance should be estimating the MSE. We see that the posterior variance is

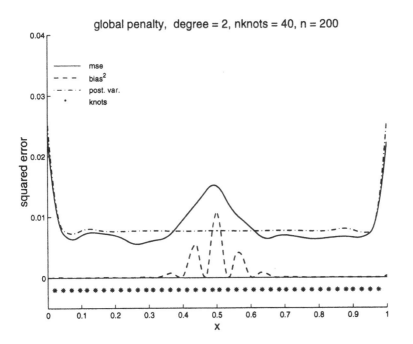

Figure 17.5 Bayesian inference study: behavior of the global penalty estimator. Plots of pointwise MSE, squared bias, and average (over the Monte Carlo trials) posterior variance. The MSE and the posterior variance have been smoothed to reduce the Monte Carlo variance. The posterior variance assumes that λ is known, so the variability in $\hat{\lambda}$ is not taken into account.

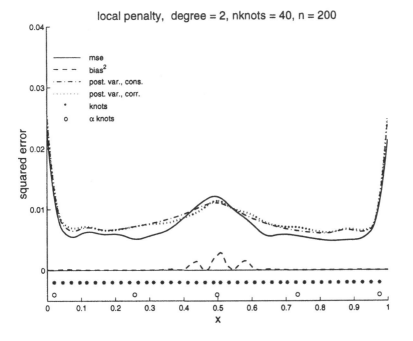

Figure 17.6 Bayesian inference study: behavior of the local penalty estimator. Plots of pointwise MSE, squared bias, and average (over the Monte Carlo trials) posterior variance. The MSE and the posterior variance have been smoothed to reduce the Monte Carlo variance.

constant, except for boundary effects, and cannot detect the spatial heterogeneity in the MSE.

Figure 17.6 is similar to Figure 17.5 but is for the local penalty estimator. Two posterior variances are shown, the conservative adjustment and the Kass–Steffey correction. One can see that the MSE is somewhat different than in Figure 17.5

Figure 17.7 Bayesian inference study using function (17.11). Pointwise coverage probabilities of 95% Bayesian confidence intervals for $f(x_i)$ using global and local penalties. The probabilities have been smoothed to remove Monte Carlo variability. The local penalty intervals use the conservative adjustment to the posterior variance and the Kass–Steffey correction.

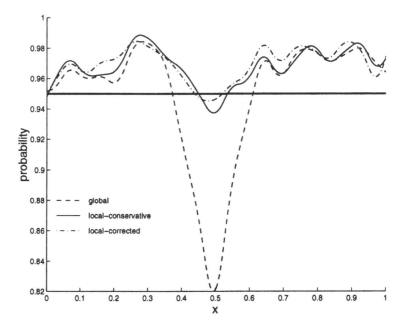

since the estimator reduces bias by adapting to spatial heterogeneity. Also, the posterior variance tracks the MSE better than for the global penalty estimator.

In Figure 17.7 we present the Monte Carlo estimates of the pointwise coverage probabilities of nominal 95% Bayesian confidence intervals based on the global and local penalty estimators. These coverage probabilities have been smoothed by penalized splines to remove some of the Monte Carlo variability. All three confidence interval procedures achieve pointwise coverage probabilities close to 95%. Because the local penalty methods are somewhat conservative, the global penalty method is, on average, the closest to 95%, but the local penalty methods avoid low coverage probabilities around features in f.

17.6 LIDAR Example

As mentioned in Section 6.8.3, an interesting feature of the LIDAR example is that there is more scientific interest in the first derivative (f') than in f itself because $-f'(x)$ is proportional to concentration at range x; see Ruppert et al. (1997) for further discussion. For the estimation of f, a global penalty works satisfactorily. Visually, the local penalty estimate of f is difficult to distinguish from the global penalty fit.

However, for the estimation of f', a local penalty appears to improve upon a global penalty. Figure 6.13 (p. 155) shows the derivatives (times -1) of fitted splines and their confidence intervals using global and local penalties. Notice that the confidence intervals using the local penalty are generally narrower than for the global penalty – except at the peak, where the extra width should be reflecting real uncertainty. The local penalty estimate has a sharper peak and less noise in the flat areas.

17.7   Additive Models

17.7.1   *An Algorithm for Additive Models*

Spatially adaptive penalties can be easily extended to additive models. Suppose we have d predictor variables and that $\mathbf{x}_i = (x_{i,1}, \ldots, x_{i,d})^{\mathsf{T}}$ is the vector of predictor variables for the ith case. The additive model is

$$y_i = \beta_0 + \sum_{j=1}^{d} f_j(x_{i,j}) + \varepsilon_i.$$

Let the jth predictor variable have K_j knots, $\kappa_{1,j}, \ldots, \kappa_{K_j, j}$. Then the additive spline model is

$$f(\mathbf{x}, \mathbf{v}) = \beta_0 + \sum_{j=1}^{d} \left(\beta_{1,j} x_j + \cdots + \beta_{p,j} x_j^p + \sum_{k=1}^{K_j} u_{k,j} (x_j - \kappa_{k,j})_+^p \right).$$

The parameter vector is $\mathbf{v} = [\beta_0, \beta_{1,1}, \ldots, u_{K_1,1}, \ldots, u_{K_d,d}]^{\mathsf{T}}$. Let $\lambda_j(\cdot)$ be the penalty function for the jth predictor. Then the penalized criterion to minimize is

$$\sum_{i=1}^{n} \{y_i - f(\mathbf{x}_i; \mathbf{v})\}^2 + \sum_{j=1}^{d} \lambda_j^2 (\kappa_{k,j}) u_{k,j}^2.$$

Consider three levels of complexity of the penalty:

(1) $\lambda_j(\cdot) \equiv \lambda$ (a common global penalty);
(2) $\lambda_j(\cdot) \equiv \lambda_j$ (separate global penalties);
(3) $\lambda_j(\cdot)$ is a linear spline (separate local penalties).

The following algorithm allows one to fit separate local penalties using only 1-dimensional grid searches for minimizing GCV. First minimize GCV using a common global penalty. For this penalty to be reasonable, one should standardize the predictors so that they have common standard deviations. Then, using the common global penalty as a starting value, minimize GCV over separate global penalties. The d global penalty parameters are varied one at a time during minimization, with the rationale that the optimal value of λ_j depends only slightly on the $\lambda_{j'}$, $j' \neq j$. Finally, using separate global penalties as starting values, minimize GCV over separate local penalties. The jth local penalty has M_j parameters, so there are a total of $M_1 + \cdots + M_d$ penalty parameters. These parameters are then varied in succession to minimize GCV.

17.7.2   *Simulations of an Additive Model*

To evaluate the practicality of this algorithm, we used an example in which we added two spatially homogeneous component functions to a spatially heterogeneous function. Thus, there were three predictor variables, which for each case were independently distributed as Uniform(0, 1) random variables. The components of f were spatially homogeneous, $f_1(x_1) = \sin(4\pi x_1)$ and $f_2(x_2) = x_2^3$, and f_3 was the spatially heterogeneous function

Figure 17.8
Log-log plot of the
computation time for
fitting an additive
model with local
penalties as a function
of the number of
variables, d. A linear
fit (slope = 2.45) is
also shown.

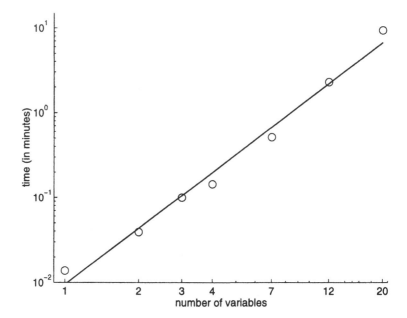

$$f_3(x_3) = \exp\{-400(x_3 - 0.6)^2\}$$
$$+ \tfrac{5}{3}\exp\{-500(x_3 - 0.75)^2\} + 2\exp\{-500(x_3 - 0.9)^2\}.$$

As in Section 17.5.3, $n = 300$ and the ε were independent N(0, 0.25). We used quadratic splines and 10, 10, and 40 knots for f_1, f_2, and f_3, respectively. The local penalty estimate had four subknots for all three functions.

First consider computation time. For a single data set and using our MATLAB program on a SUN Ultra 1 computer, the common global penalty estimate took 2.1 seconds to compute, the separate global penalty estimate took an additional 1.5 seconds, and the separate local penalties estimate took an additional 10.4 seconds. Thus, local penalties are more computationally intensive than global penalties but still feasible for small values of the number of predictor variables d – for example, for $d = 3$ here.

Now consider larger values of d. Everything else held constant, the number of parameters of an additive model grows linearly in d and, since matrix inversion time is cubic in dimension, the time for a single fit should grow cubically in d. Since the number of fits needed for the sequential grid searching just described will grow linearly in d, the total computation time for local penalties should be roughly proportional to d^4. To test this rough calculation empirically, we found the computation time for fitting additive models with 300 data points, 10 knots per variable, and 4 subknots per variable. The number of predictor variables d took seven values from 1 to 20. Figure 17.8 is a log-log plot of computation time versus d. A linear fit on the log scale is also shown; its slope is 2.2, not 4 as the quartic model predicts. The actual data show log-times that are nonlinear in $\log(d)$ with an increasing slope. Thus, a quartic model of time as a function of d may work for large values of d, but a quadratic or cubic model would be better for d in the "usual" range of 1 to 20. A likely reason that the quartic model doesn't fit well for smaller d is that it ignores parts of the computation that are linear, quadratic, and cubic in d. The computation time for 7 variables is about 0.5 minutes but for 20 variables is about 9.4 minutes. It seems clear that local additive fitting is feasible up to at least 8–10 variables and perhaps 20 variables, but it is only "interactive" up to 4 variables.

An important point to keep in mind is that computation times are largely independent of the sample size n. The reason for this is that once $\mathbf{C}^T\mathbf{C}$ and $\mathbf{C}^T\mathbf{y}$ have been computed, all computation times are independent of n, and the computation of $\mathbf{C}^T\mathbf{C}$ and $\mathbf{C}^T\mathbf{y}$ is quite fast unless n is enormous.

Now consider statistical efficiency. The MSEs that were computed over 500 Monte Carlo samples for the separate local penalties estimator were 0.010, 0.0046, and 0.0165 for f_1, f_2, and f_3, respectively. Thus, f_2 is relatively easy to estimate and f_3 is slightly more difficult to estimate than f_1. Ratios of the MSE for common global penalties to separate local penalties were 1.26, 2.36, and 1.23 for f_1, f_2, and f_3, respectively, whereas ratios of the MSE for separate global penalties to separate local penalties were 0.85, 0.88, and 1.20 for f_1, f_2, and f_3, respectively. Thus, for all three component functions, the common global penalty estimator with a single smoothing parameter is less efficient than the fully adaptive estimator with separate local penalties. For the spatially homogeneous functions f_1 and f_2, there is some loss of efficiency when using local penalties rather than separate global penalties, but the spatially heterogeneous f_3 is best estimated by separate local penalties. These results are somewhat different than what we found for univariate regression, where no efficiency loss was noticed when a local penalty was used even though a global penalty would have been adequate. There may be practical situations where one knows that a certain component function is spatially heterogeneous but the other component functions are not. Then greater efficiency should be achievable by using global penalties for the spatially homogeneous component functions and local penalties for the spatially heterogeneous ones.

The results in this section provide evidence that sequential 1-dimensional grid searches to find the smoothing parameter vector are effective. The reason for this is that the optimal value of one tuning parameter depends only weakly upon the other tuning parameters. The result is that searches over a rather large number of tuning parameters (up to 60 when $d = 15$ and there are four subknots per variable) do appear to be feasible.

17.8   Bibliographical Notes

The spatially adaptive local penalty estimators were introduced by Ruppert and Carroll (2000). Other methods for spatially adaptive smoothing include automatic knot selection, local polynomial regression with local bandwidths (e.g. Ruppert 1997b), and wavelets. The automatic knot selection literature is discussed in Section 3.4. For wavelets, see Cai (1999) and the bibliographic notes at the end of Chapter 3.

18

Analyses

In Chapter 1 we introduced several substantive problems that mainly involved scientific studies. In this chapter, we return to these problems. The goal here is not simply to illustrate semiparametric modeling techniques but to show how these techniques can be integrated into scientific studies. Analyses for about half of the studies have recently been published and so, in order to save space, we will simply refer the interested reader to the relevant journal article.

18.1   Cancer Rates on Cape Cod

An analysis of the Cape Cod cancer data is given in French and Wand (2003). In their presentation, a logistic geoadditive model (Section 13.6) leads to maps showing regions of elevated relative cancer risk after accounting for age and smoking status. The model developed there also accounts for missingness (missing values) in the smoking variable.

18.2   Assessing the Carcinogenicity of Phenolphthalein

Parise and colleagues (2001) used semiparametric logistic mixed models to assess the carcinogenicity of phenolphthalein. After adjusting for rodent weight, they were not able to find a significant dose effect for phenolphthalein.

18.3   Salinity and Fishing in North Carolina

Real data sets often illustrate several different statistical principles. The salinity data set is not simply an example of semiparametric modeling; it also shows the differing effects of outliers on parametric and nonparametric modeling.

The salinity data are introduced in Section 1.2. Recall the definitions of the variables: `salinity` is the measured value of salinity in Pamlico Sound, `lagged.sal` is salinity two weeks earlier, and `discharge` is the amount of fresh water flowing into the Sound from rivers. In this example there are two unusual values of `discharge`, and the question naturally arises of whether these data points should be included. We will see that this question is less of an issue if one uses a semiparametric rather than a parametric model.

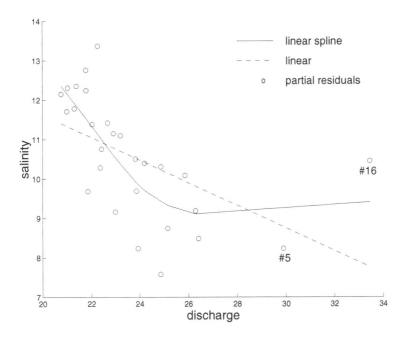

Figure 18.1 Salinity data: plot of $\hat{\beta}_0 + \hat{\beta}_1 \overline{\texttt{lagged.sal}} + \hat{f}(\texttt{discharge})$, modeling f as a penalized linear spline (solid) and modeling f as a linear function (dashed). Also plotted are the salinity partial residuals after regression on lagged.sal.

Figure 1.4 suggests that the effect of lagged.sal on salinity is linear but the effect of discharge on salinity may be nonlinear. Therefore, we used the semiparametric additive model,

$$\texttt{salinity}_i = \beta_0 + \beta_1 \texttt{lagged.sal}_i + f(\texttt{discharge}_i) + \varepsilon_i. \qquad (18.1)$$

To visualize $\hat{f}$, we follow the recommendation in Section 2.5.1.1 by plotting the fitted model versus discharge with lagged.sal fixed at its average value, so in Figure 18.1 we show $\hat{\beta}_0 + \hat{\beta}_1 \overline{\texttt{lagged.sal}} + \hat{f}(\texttt{discharge})$ using a 10-knot penalized linear spline for f and also modeling f as linear. The smoothing parameter for the penalized spline was determined by minimizing GCV. Also plotted are the partial residuals from the spline model. Partial residuals are defined in Section 2.5.1.3, and here they are given by

$$\texttt{salinity}_i - \{\hat{\beta}_1(\texttt{lagged.sal}_i - \overline{\texttt{lagged.sal}})\}.$$

As can be seen in Figure 18.1, all but two of the values of discharge are between 20.8 and 26.4. The anomalous values are case 5 with discharge = 29.9 and case 16 with discharge = 33.4. Most of the curvature in the spline estimate is due to the accommodation of cases 5 and 16.

Figure 18.2 is a plot of the hat diagonals (or leverages) against case number. One can see that case 16 has high leverage. The sum of the hat diagonals is $df_{\text{fit}} = 4.31$ for the linear spline and 3 for the linear model. Therefore, the average hat diagonal is 0.154 (= 4.31/28) for the linear spline model and 0.1071 (= 3/28) for the linear model. For both models, the leverage for case 16 is about five times the average. A case with a leverage value more than two times the average value is often considered a high leverage point – see Belsley et al. (1980, sec. 2.1).

Figure 18.2 Salinity
data: plot of diagonals
of the hat matrices for
linear spline model
and linear model.

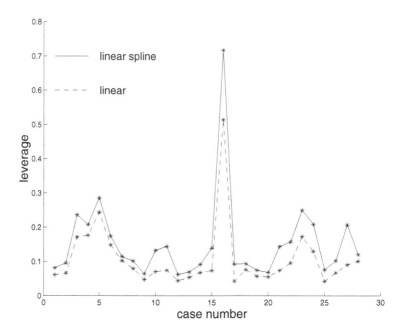

Figure 18.3 Salinity
data: plot of Cook's D
for linear spline model
and linear model.

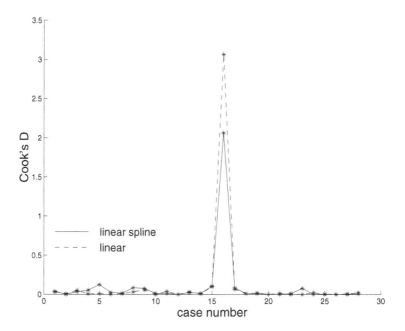

So case 16 is a high-leverage observation, but high leverage implies only a *potential* for being highly influential. What, in fact, is the influence of this case? Figure 18.3 is a plot of Cook's D for both the linear spline and linear models.

The effects of case 16 on the fits is evident in Figure 18.1. The effect on the spline fit is to put the estimated curve up on the right side. For the linear model, the

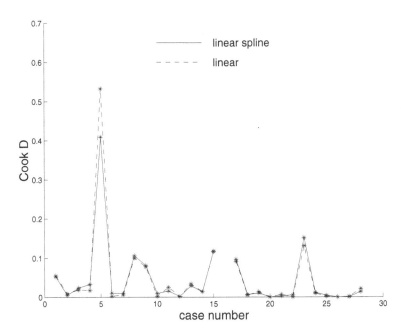

Figure 18.4 Salinity data *without* case 16: plot of Cook's D for linear spline model and linear model.

effect is to tilt the entire estimated curve. The spline fit has a certain robustness, since the effect of case 16 is somewhat localized to large values of discharge. This can be considered an advantage of a semiparametric fit.

Figure 18.4 is a plot of Cook's D for fits without case 16. One can see that with case 16 deleted, case 5 now has a high distance value. One must now wonder whether to delete both cases 5 and 16, though we hesitate to delete any data point unless it is known to be an error. In this case study, there is no evidence that either case 5 or case 16 is in error. It has been noticed that if the value of discharge were exactly 10 units less, 23.4 instead of 33.4, then its value of salinity would be consistent with the rest of the data. However, it could very well be that case 16 is a valid data point, and there are ways to explain case 16 without maintaining that it is erroneous. One hypothesis is that large values of discharge indicate that the rivers flowing into Pamlico Sound are flowing so rapidly that their fresh water flows through the sound into the ocean without mixing with the water in the sound. If true, this hypothesis would explain why very large values of discharge are not associated with low salinity.

Fortunately, the question of which observations, if any, to delete is less of a pressing issue when one uses semiparametric rather than linear models. To appreciate this fact, look at Figure 18.5. This graph shows the plots of the estimated effects of discharge for the semiparametric model with all the data, with case 16 deleted, and with cases 5 and 16 deleted. Notice that the fits are quite similar for discharge less than 26.4, which is the largest value observed besides cases 5 and 16. If one deletes case 16, the result is that the curve is estimated over a shorter interval, but the curve is changed only slightly in the region were the bulk

Figure 18.5 Salinity
data: plot of $\hat{\beta}_0 +$
$\hat{\beta}_1\overline{\texttt{lagged.sal}} +$
$\hat{f}(\texttt{discharge})$ for
penalized linear spline
fits using the full data
set and with selected
cases deleted.

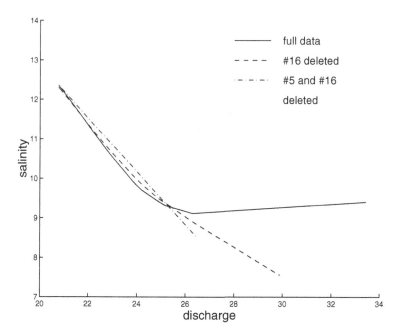

Figure 18.6 Salinity
data: plot of $\hat{\beta}_0 +$
$\hat{\beta}_1\overline{\texttt{lagged.sal}} +$
$\hat{f}(\texttt{discharge})$ for
linear fits using the
full data set and with
selected cases deleted.

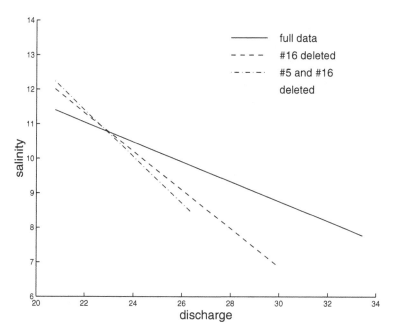

of the data lie. The same is true if both cases 5 and 16 are deleted. With a linear
model, in contrast, if case 16 is deleted or if both 5 and 16 are deleted then the
fitted curves change everywhere – see Figure 18.6.

The intended purpose of our modeling was to estimate salinity at time
points where salinity was not measured but discharge and lagged.sal

were known – or at least where `discharge` was known and `lagged.sal` had been estimated. For example, if `salinity` were known for one week, then it could be estimated for a period two weeks later by using `discharge`. Using that "imputed" value of `salinity`, one could estimate `salinity` forward another two weeks, and so forth.

If `discharge` is less than, say, 27 (as is typically true), then the estimation of `salinity` will not depend heavily on which data points are used to fit the model.

What we have have seen in this case study is an example of a general phenomenon: parametric models have a rigidity, so that an outlier affects the global fit; whereas nonparametric models are flexible, so that an outlier affects only the fit in its locality.

18.4 Management of a Retirement Fund

The response variable

$$\texttt{contribution} \equiv \text{year-end contribution}$$

is heavily skewed. Its logarithm displayed approximate normality, so we worked with log(`contribution`) throughout the analysis.

The spreadsheet contains nine possible predictors of log(`contribution`). An additive model fit containing all nine variables suggested that the variables with the most predictive power are:

> `group` – indicator variable taking the value 1 if the client also has a group life or health insurance policy with Best Retirement Inc. (BRI) and 0 otherwise;
> `susan` – indicator variable taking the value 1 if Susan Shepard sold the policy and 0 otherwise;
> `eligible` – number of employees eligible to participate in retirement plan;
> `salary` – average annual employee salary in dollars.

As mentioned in Section 1.3, Susan Shepard is the only sales representative who has been specifically trained to deal with 401(k) retirement plans. BRI would like to know if Susan's expertise is a factor that influences year-end contributions to 401(k) retirement plans. If so, BRI would like to consider training other sales representatives.

We fit all additive models containing these variables, with `eligible` and `salary` entering either linearly or nonlinearly. Each model fit was obtained using a mixed model representation of penalized spline-based additive models with REML estimation of degrees of freedom. The corresponding corrected AIC value, AIC_C (Section 5.3.4), was recorded. The results are summarized in Table 18.1.

The clear winner according to AIC_C is the model

$$\begin{aligned} \text{E}\{\log(\texttt{contribution})\} &= \beta_0 + \beta_1 \texttt{group} + \beta_2 \texttt{susan} \\ &\quad + \beta_3 \texttt{eligible} + f_2(\texttt{salary}). \end{aligned} \tag{18.2}$$

Table 18.1 Corrected AIC values for four models containing group, susan, eligible, and salary.

Model for E{log(contribution)}	AIC$_C$
$\beta_0 + \beta_1\texttt{group} + \beta_2\texttt{susan} + \beta_3\texttt{eligible} + \beta_4\texttt{salary}$	3.7831
$\beta_0 + \beta_1\texttt{group} + \beta_2\texttt{susan} + f_1(\texttt{eligible}) + \beta_4\texttt{salary}$	3.7832
$\beta_0 + \beta_1\texttt{group} + \beta_2\texttt{susan} + \beta_3\texttt{eligible} + f_2(\texttt{salary})$	3.6896
$\beta_0 + \beta_1\texttt{group} + \beta_2\texttt{susan} + f_1(\texttt{eligible}) + f_2(\texttt{salary})$	3.7831

Figure 18.7 Graphical display of REML fit of model (18.2).

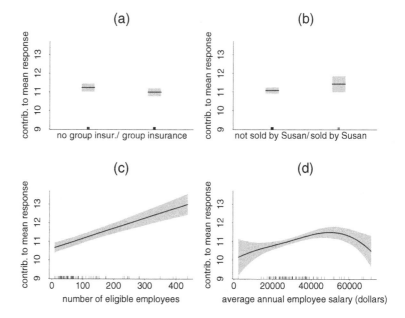

The residuals after fitting were examined and seemed well-behaved. The fit is shown in Figure 18.7. Panel (d) shows salary having a pronounced nonlinear effect. The coefficient corresponding to susan is

$$\hat{\beta}_2 = 0.3467 \quad \text{with} \quad \widehat{\text{st.dev.}}(\hat{\beta}_2) = 0.2248.$$

The corresponding t-ratio is $0.3467/0.2248 = 1.542$ and has a one-sided approximate p-value of 0.062. Hence, there is some evidence that Susan's training positively influences year-end contributions.

18.5   Biomonitoring of Airborne Mercury

Ruppert (1997a) gives a detailed analysis of these data using local polynomial regression. We have also fit a tensor product spline to these data, and the fit is similar to that in Ruppert (1997a). Tensor product methods are successful in this case study because the data are in a roughly rectangular region. Radial basis fitting (as discussed in Chapter 13) could also be used and would be more appropriate in similar problems where the data are in a more irregularly shaped region.

18.6 Term Structure of Interest Rates

The problem of estimating the term structure of interest rates was introduced in Section 1.5. Recall that we have two types of bonds, AT&T bonds, which have coupons, and U.S. STRIPS, which are zero-coupon bonds. There are 118 U.S. STRIPS prices but only five AT&T bond prices. We will start with a nonparametric model for the U.S. STRIPS prices. As in Section 2.8, the response is

$$y_i = 100 \frac{P(0)}{P(T_i)},$$

where $P(t)$ is the price of the ith U.S. STRIPS at time t, 0 is the time at which the prices are given, and T_i is the maturity of the ith bond; $P(T_i)$ is called the par value. Theoretically, the expected value of y_i is

$$100 \exp\left(-\int_0^{T_i} r(x)\, dx\right),$$

where r is the forward rate curve. We will use the linear spline model,

$$r(x) = \beta_0 + \beta_1 x + \sum_{k=1}^{K} u_k (x - \kappa_k)_+,$$

where $\kappa_1, \ldots, \kappa_K$ are equally spaced quantiles of the maturities. This gives us the nonlinear regression model

$$y_i = 100 \exp\left(-\beta_0 T_i - \frac{1}{2}\beta_1 T_i^2 - \frac{1}{2}\sum_{k=1}^{K} u_k (T_i - \kappa_k)_+^2\right) + \varepsilon_i, \qquad (18.3)$$

where $\varepsilon_1, \ldots, \varepsilon_n$ are assumed to be independent $N(0, \sigma^2)$. Note that model (18.3) is a Gaussian generalized regression model but with a log link function rather than the usual identity link function for Gaussian responses. The log link comes from financial theory. Model (18.3) is also a nonlinear regression model of the type discussed in Section 2.8. In that section we saw that the parameters in nonlinear regression models could be estimated by nonlinear least squares. Like other spline models, model (18.3) is overparameterized and so penalized nonlinear least squares is needed.

Define $\theta = (\beta_0, \beta_1, u_1, \ldots, u_K)^{\mathsf{T}}$. The penalized nonlinear least-squares estimator of θ is the minimizer of

$$SS(\theta) = \sum_{i=1}^{n} \left\{ y_i - 100 \exp\left(-\beta_0 T_i - \frac{1}{2}\beta_1 T_i^2 - \frac{1}{2}\sum_{k=1}^{K} u_k (T_i - \kappa_k)_+^2\right)\right\}^2$$

$$+ \lambda^2 \sum_{k=1}^{K} u_k^2. \qquad (18.4)$$

The hat matrix is defined as follows. As in Section 2.8, let $\mathbf{X}$ be the $n \times (2+K)$ matrix whose (i, j)th element is the partial derivative

Figure 18.8 Term structure example: GCV function.

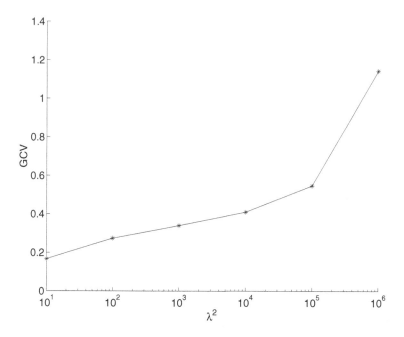

$$\frac{\partial}{\partial \theta_j} 100 \exp\left(-\beta_0 T_i - \frac{1}{2}\beta_1 T_i^2 - \frac{1}{2}\sum_{k=1}^{K} u_k (T_i - \kappa_k)_+^2\right).$$

Let **D** be the diagonal matrix with two zeros and K ones along the diagonal. Then the hat matrix is

$$\mathbf{H} = \mathbf{X}(\mathbf{X}^\mathsf{T}\mathbf{X} + \lambda^2 \mathbf{D})^{-1}\mathbf{X}^\mathsf{T}.$$

The degrees of freedom for the fit is

$$df_{\mathrm{fit}}(\lambda) = \mathrm{tr}(\mathbf{H}) = \mathrm{tr}\{(\mathbf{X}^\mathsf{T}\mathbf{X} + \lambda^2 \mathbf{D})^{-1}(\mathbf{X}^\mathsf{T}\mathbf{X})\}.$$

The GCV function is defined as before – for example, by equation (5.5). Figure 18.8 shows the GCV function for $\log_{10}(\lambda^2) = 1, 2, \ldots, 6$. Notice that GCV is minimized by a small value of λ, which induces the very rough estimate in the upper left panel of Figure 18.9 – this estimate even becomes negative at the right boundary, though negative values are not plotted in the figure. It is puzzling that GCV selects such a rough estimate and, in fact, would select an even rougher one if smaller values of λ^2 had been tried. The problem is likely that the GCV assumption of independent errors is not satisfied. It may be that observations with similar values of T_i are correlated.

To investigate possible correlation of the errors, a plot of the first-order autocorrelation function of the residuals is shown in Figure 18.10. We would like the autocorrelation to be zero. Notice that instead it is rather high and becomes worse as λ increases. An estimated autocorrelation function is considered significant if its absolute value exceeds $2/\sqrt{n} = 2/\sqrt{118} = 0.184$. Here, all six autocorrelations are significant by this criterion.

Figure 18.11 is a plot of the residuals in time order using $\lambda^2 = 10^4$ for the fit; the first-order residual autocorrelation is 0.4 in this case. The dependence among

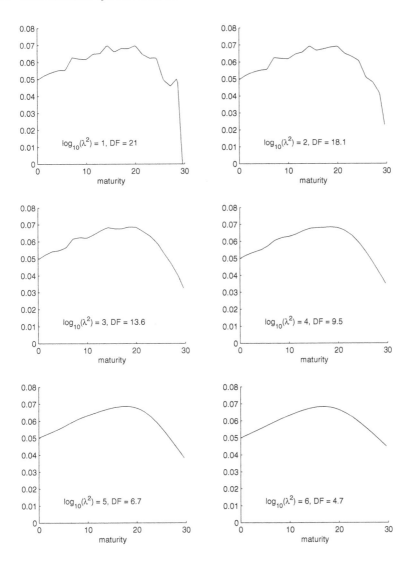

Figure 18.9 Term structure example: spline estimates of the forward rate curve for six values of λ^2.

the residuals is quite clear. Figure 18.12 is the same plot except that $\lambda^2 = 10$, where the residual autocorrelation is 0.2.

Another problem with the data is that the observation with the largest value of T_i has an outlying price – in fact, its price is higher than the observation with the second largest T_i. As mentioned in Chapter 1, we learned during the production of this book that this observation is a contaminant. If this price were correct, then that would imply a negative interest rate when maturity is between these two values. Yet this single outlier is not what drives GCV to select small λ, because GCV continues to select small λ even after this case is dropped.

How should we select λ? Since GCV is selecting a value of λ giving little smoothing and since the assumptions behind GCV are suspect, probably the best way is visual inspection. Use $\lambda^2 = 10^4$ corresponding to $df_{\mathrm{fit}} = 9.5$ and the middle right plot in Figure 18.9. The estimated curve in this plot shows the main

Figure 18.10 Term
structure example:
first-order residual
autocorrelation.

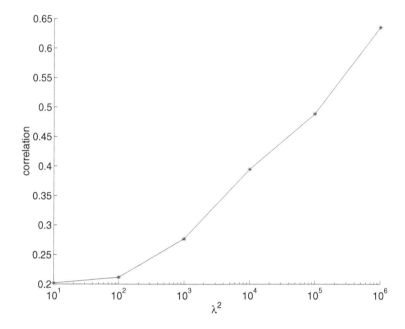

Figure 18.11 Term
structure example:
plot of residuals
against maturity when
$\log_{10}(\lambda^2) = 4$.

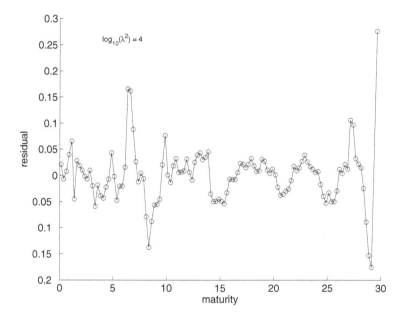

trend without exhibiting too much wiggling. If the estimated curve were used
to price a credit derivative, then it would be important to do a sensitivity analysis to find just where the derivative's price is sensitive to the choice of λ: at the
small GCV value or at the larger value selected by eye. Recently, Jarrow and
colleagues (2003) have developed a new method of selecting λ based upon Ruppert's (1997b) EBBS methodology, which chooses a value of λ similar to the one
we have selected by eye.

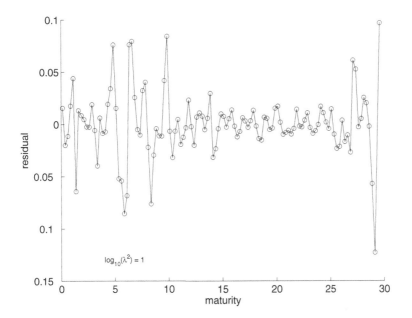

Figure 18.12 Term structure example: plot of residuals against maturity when $\log_{10}(\lambda^2) = 1$.

18.7 Air Pollution and Mortality in Milan: The Harvesting Effect

The analyses of these data required embellishment of generalized additive models (Chapter 11), including *distributed lags* (e.g. Davidson and MacKinnon 1993), in order to account for lagged effects of air pollution and to handle the harvesting effect. Zanobetti et al. (2000) provide full details on this modeling and an analysis of the Milan mortality data. A harvesting effect is seen to be apparent in the Milan mortality data, with the distributed lag component capturing this effect.

19

Epilogue

19.1 Introduction

The final draft of this book was written in 2002 and reflects our priorities and views on semiparametric regression at that time. However, the interplay between statistical methodology and applications is currently in a dynamic state. We hope that our coverage of the main ideas of semiparametric regression will serve as a reasonable basis for whatever new directions semiparametric regression takes. In this closing chapter, we note that the approach to semiparametric regression used throughout most of the book can be distilled into just a few basic ideas. We also mention some notable omissions and comment on future directions.

19.2 Minimalist Statistics

One of the major themes of this book is the use of the mixed model framework to fit and make inferences concerning a wide variety of semiparametric regression models, though we have intentionally used both the mixed model and more classical GCV methods in our examples. This approach has the advantage of requiring little more than familiarity with mixed model methodology, as outlined in Chapter 4 and Section 10.8. In particular, fitting is achieved through just two fundamental and well-established principles:

(1) estimation of parameters via (restricted) maximum likelihood; and
(2) prediction of random effects via best prediction.

If there is an important scientific exception to the basis model – such as a predictor being subject to measurement error – then these principles can still be used for fitting, as demonstrated in Chapter 15. However, as seen there and in Section 10.8, maximum likelihood and best prediction are sometimes hindered by the presence of intractable integrals. Computational schemes such as Laplace approximation, Monte Carlo EM, and Markov chain Monte Carlo algorithms are then important for implementation. If a Bayesian approach is used then similar comments apply, with fitting corresponding to finding the mode or mean of a posterior density. Inference can be made within the mixed model framework using, for example, likelihood ratio tests.

We call this streamlining of statistical methodology *minimalist statistics*. As the field of statistics finds itself increasingly intertwined with other disciplines and as required models become more complicated, we believe that such minimalism is of critical importance. There is only so much time available to educate interdisciplinary researchers and practitioners in statistical theory and methodology. Even in the training of doctoral level statisticians, time is a limiting resource and streamlining may be just as important.

19.3   Some Omitted Topics

The choice of topics covered in the book is partly driven by the case studies of Chapter 1 and partly by what we see as a logical progression of themes for learning semiparametric regression. There are a number of additional topics that we have not been able to cover owing to space and time constraints. It became apparent to us that getting the book to students and applied workers as soon as possible was far more important than covering all topics. We give brief mention to some of the omitted topics here.

19.3.1   *Robustness*

Regression methodology that is resistant to the adverse effects of outlying response values has been the subject of an enormous amount of literature over the past few decades (e.g., Hampel et al. 1986; Rousseeuw and Leroy 1987; Staudte and Sheather 1990; Wilcox 1997). Some of the main approaches to robust regression involve M-estimation (Huber 1983) and the t-distribution likelihood (Lange, Little, and Taylor 1989). Welsh and Richardson (1997) provide a detailed survey of robustness in the linear mixed model, and the approaches described there could be used to "robustify" the semiparametric regression estimators in this book. Kammann, Staudenmayer, and Wand (2002) explored the t-distribution approach for penalized splines in the mixed model framework. Robustness has been built into other smoothers by, for example, Fan, Hu, and Truong (1994) and Smith and Kohn (1996).

19.3.2   *Quantile Regression*

Throughout the book we focused on estimation of conditional means of the response variables. An alternative is to work with conditional quantiles. This is appropriate for response variables that are heavily skewed and prone to outliers. Quantile regression is also a form of robust regression. Formulation, fitting algorithms, and theory for both parametric and nonparametric quantile regression models have received a fair amount of treatment in the recent literature. See, for example, Koenker, Ng, and Portnoy (1994) and Yu and Jones (1998).

19.3.3   *Nonquadratic Penalties*

All penalization in the book involved quadratic forms of certain coefficients. A number of alternatives are possible. One is the *lasso,* an acronym for *least absolute shrinkage and selection operator* (Tibshirani 1996). For the linear regression model

$$y_i = \beta_0 + \sum_{j=1}^{d} \beta_j x_{ji} + \varepsilon_i, \quad 1 \le i \le n,$$

the lasso estimates the β_j by minimizing

$$\sum_{i=1}^{n} \left(y_i - \beta_0 - \sum_{j=1}^{d} \beta_j x_{ji} \right)^2 \quad \text{subject to} \quad \sum_{j=1}^{d} |\beta_j| \le C$$

for some *shrinkage factor C* > 0. One of the advantages is that the least significant coefficients are estimated to be exactly zero, so model selection is more clear-cut. The constraint can also be imposed on $\sum_{j=1}^{d} |\beta_j|^q$ for other values of $q > 0$; see Yu and Ruppert (2002). Hastie, Tibshirani, and Friedman (2001) provide detailed discussion on the advantages of nonquadratic penalties.

19.3.4 *Highly Adaptive Smoothing*

Penalized splines with a single smoothing parameter can model nonlinear relationships quite well, provided that the curvature is not too heterogeneous. The nonlinearity present in most of the examples in this book is adequately handled through smoothers of this type, and the models enjoy the benefits of simplicity and ease of fit. Yet in some application areas, such as speech recognition and neuroscience, discontinuities and varying amounts of curvature in the signal are the norm. A literature in what might be called *highly adaptive* smoothing has emerged to deal with such data. Good performance requires algorithms that are more complex. The spatially adaptive penalties in Chapter 17 are an effective means of achieving an adaptive smoother, but there are other approaches in the literature; some are based on careful knot selection. This was touched on in Section 3.4, where penalized splines were first introduced as an alternative to the more adaptive regression spline approaches. Several references are given there. Others involving spline-type smoothers are Holmes and Mallick (2001), Zhou and Shen (2001), Lee (2002), Shen and Ye (2002), and the forthcoming book by Hansen and colleagues (2003). As discussed in Section 3.15.2, wavelets also offer highly adaptive smoothing; see Ogden (1996) for access to this literature.

19.3.5 *Missing Data*

Missing data is a common problem in many application areas, and careful adjustment is often necessary for trustworthy and efficient analyses. The missing data problem has similarities with the measurement error problem covered in Chapter 15. There are numerous proposed strategies for dealing with missing data; these are surveyed in the books by Little and Rubin (1987) and Schafer (1997). Particularly noteworthy, given Section 19.2, are likelihood-based models that account for missing data – for example, Little and Rubin (1987), Ibrahim (1990), and

Ibrahim et al. (2001). The latter reference deals with generalized mixed models. French and Wand (2003) explore the extension of these ideas to semiparametric regression models with mixed model representation.

19.3.6 *Functional Data Analysis*

These days, data often arrive as curves or images, and methodology for their analysis has received a great deal of treatment in the past decade or so. Ramsay and Silverman (1997) provide a broad overview of functional data analysis. There are a number of connections between semiparametric regression and functional data analysis that we have not discussed in this book.

19.3.7 *Survival Analysis*

Time-to-event data are common in certain application areas, such as reliability and actuarial science – and ubiquitous in others, such as clinical trials and drug development. Survival analysis is concerned with statistical modeling and inference for such data. Popular survival analysis models, such as the Cox proportional hazards model, traditionally cater to parametric functional effects. However, the semiparametric extension can be done analogously to the methods described throughout this book. Therneau and Grambsch (2000) provide a summary of this work and access to related literature.

19.3.8 *Single-Index Models*

Additive models (Chapters 8 and 11) and varying coefficient models (Section 12.4) are both means by which the effects of several predictors can be modeled in a flexible and interpretable fashion. Another such model is the single-index model (Ichimura 1993). If $\mathbf{x}_i$ ($1 \leq i \leq n$) represents a vector of continuous measurements, then a single-index model is one of the form

$$y_i = f(\boldsymbol{\alpha}^\mathsf{T}\mathbf{x}_i) + \varepsilon_i, \qquad (19.1)$$

where f is a smooth but unspecified function and where the $\boldsymbol{\alpha}$ are coefficients in the *single index* $\boldsymbol{\alpha}^\mathsf{T}\mathbf{x}_i$. A common extension of (19.1), called a partially linear single-index model, is the addition of a linear term $\boldsymbol{\beta}^\mathsf{T}\mathbf{z}_i$, where $\mathbf{z}_i$ is another vector of predictors. Contributions to the estimation and theoretical properties of single-index models include Härdle, Hall, and Ichimura (1993), Ichimura (1993), Weisberg and Welsh (1994), Carroll et al. (1997), and Yu and Ruppert (2002). The last reference deals with penalized spline estimation of f.

19.3.9 *Diagnostics*

Because of space and time considerations, this book does not give as much treatment of diagnostics as we would have liked. Of course, this topic has not been omitted entirely. Partial residual plots were discussed in Chapter 8, and the residual analysis in Chapter 2 can be applied to residuals from a semiparametric fit. However, there is a vast literature on regression diagnostics for linear regression. Unfortunately, similar diagnostics for semiparametric regression is a relatively neglected area of research, although it receives some attention in the books of Hastie and Tibshirani (1990) and Loader (1999). Other contributions include Eubank (1985), Thomas (1991), and Gu (1992b). Diagnostics to accompany the methodology described in this book could be assembled by adapting and extending the methods developed in these references, but this is yet to be done.

19.3.10 *Statistical Learning*

Statistical learning is a relatively new branch of statistics (see e.g. Vapnik 1998, 2000) that has grown out of the explosion in amounts of data being collected in various application areas. Some related topics are *data mining and knowledge discovery, machine learning, pattern recognition, artificial intelligence,* and *bioinformatics.* Examples of learning problems include prediction of a loan applicant defaulting based on questionnaire data, identification of handwritten postal codes from a digital image, and inference of which genes cluster together based on their expression profiles on a microarray. The recent book by Hastie et al. (2001) summarizes solutions to learning problems from a statistical perspective, and semiparametric regression is prominently featured. For example, binary response additive models can be used to build very accurate classifiers through the notion of *boosting* (Schapire et al. 1998; Jiang 2002). *Neural networks* (Bishop 1995; Ripley 1996) and *support vector machines* (Cristianini and Shawe-Taylor 2000) are powerful statistical learning tools with strong connections to semiparametric regression.

19.3.11 *Constrained Smoothing*

In the nonparametric regression situation

$$\mathsf{E}(y_i|x_i) = f(x_i), \quad 1 \le i \le n,$$

it is sometimes the case that $f(\cdot)$ should obey certain conditions. For example, if x corresponds to age measurements on a group of children and y corresponds to height measurements (subject to error), then it is reasonable to assume that f is monotonically increasing. Therefore, the estimate f should be computed according to the monotonicity constraint. Other possible constraints include convexity and f having a fixed value at one or more x.

Smoothing under constraints has a substantial literature. See, for example, Ramsay (1988), Tantiyaswasdikul and Woodroofe (1994), Delecroix, Simioni, and Thomas-Agnan (1995), Delecroix and Thomas-Agnan (2000), and Mammen et al. (2001). The last paper shows how to incorporate constraints into smoothers quite generally through projection ideas and can be applied to each of the smoothers described in this book.

19.3.12   *Smoothing Geographical Count Data*

Chapter 13 dealt with the smoothing of geographically referenced "point" data, where each sampling unit has a corresponding longitude–latitude pair. A related problem is the smoothing of geographical "count" data, corresponding to aggregated summaries for "small areas" such as U.K. enumeration districts and U.S. census block groups. Smoothing is also important in this context because (a) the variance of standardized incidence rates is inversely proportional to the expected count and (b) areas with small populations will have high sampling variability (see e.g. Elliott et al. 2000). Methodology for disease mapping has close ties with mixed model fitting (Breslow and Clayton 1993; Leroux, Lei, and Breslow 1999; Wakefield, Best, and Waller 2000; MacNab and Dean 2001). Incorporating the semiparametric regression models for nonlinear confounders can be done straightforwardly (Fahrmeir and Lang 2002).

The standardized incidence rate of a particular occurrence (often a disease) is the ratio of the observed number of occurrences to the expected number according to some reference population.

19.3.13   *Other Topics*

The list of omitted topics in this section is not exhaustive. Many other topics of research overlap with the methodology presented in this book. Examples include image analysis, multilevel models, and spatiotemporal models.

19.4   Future Research

Semiparametric regression has a bright future. In this book we have shown how it can be useful in applications such as environmental epidemiology and finance, two fields that correspond to our recent collaborative research interests. Other areas that make extensive use of semiparametric regression are data mining and knowledge discovery, atmospheric research, hydrology, and econometrics. Emerging application areas such as functional magnetic resonance imaging (in neuroscience) and computational genomics and proteomics are benefiting from flexible regression methodology. Future research in semiparametric regression that is tied to application areas such as these promises to be very fruitful.

A

Technical Complements

A.1 Introduction

Throughout the book, a number of technical definitions and results have been used. This appendix provides some details on each of them. Further details can be found in specialized text books, some of which are listed in Section A.6.

A.2 Matrix Definitions and Results

In this section we assume some familiarity with the fundamentals of matrix algebra and determinants. For notation, we will use bold capital letters (e.g. $\mathbf{A}$) for matrices. Vectors which will be denoted by bold lowercase letters such as $\boldsymbol{a}$ and are assumed to be columns. The transpose of $\mathbf{A}$ is written as $\mathbf{A}^\mathsf{T}$.

A.2.1 *Trace*

The *trace* of a square matrix $\mathbf{A}$ is the sum of its diagonal entries and is usually denoted by $\mathrm{tr}(\mathbf{A})$. For example,

$$\mathrm{tr}\left(\begin{bmatrix} 7 & 4 \\ 3 & 2 \end{bmatrix}\right) = 7 + 2 = 9.$$

A very useful result concerning trace is

$$\mathrm{tr}(\mathbf{AB}) = \mathrm{tr}(\mathbf{BA})$$

for any two matrices $\mathbf{A}$ and $\mathbf{B}$ where $\mathbf{AB}$ is defined and square, which requires that $\mathbf{A}$ and $\mathbf{B}^\mathsf{T}$ have the same dimensions.

A.2.2 *Eigenvalues and Eigenvectors*

Let $\mathbf{A}$ be an $m \times m$ matrix. Then a number λ is an *eigenvalue* of $\mathbf{A}$ if it satisfies

$$\mathbf{A}\mathbf{x} = \lambda\mathbf{x}$$

for some $m \times 1$ vector $\mathbf{x}$, called the *eigenvector* associated with λ. For example,

$$\begin{bmatrix} 1 & 3 \\ 3 & 1 \end{bmatrix}\begin{bmatrix} 1 \\ 1 \end{bmatrix} = 4\begin{bmatrix} 1 \\ 1 \end{bmatrix},$$

so $\lambda = 4$ is an eigenvalue of the matrix

$$\mathbf{A} = \begin{bmatrix} 1 & 3 \\ 3 & 1 \end{bmatrix} \quad \text{with corresponding eigenvector } \mathbf{x} = \begin{bmatrix} 1 \\ 1 \end{bmatrix}.$$

Any nonzero multiple of $\mathbf{x}$ is also an eigenvector corresponding to λ.

The eigenvalues of $\mathbf{A}$ may be found by solving the equation

$$|\mathbf{A} - \lambda\mathbf{I}| = 0.$$

This is an mth-degree polynomial with solutions $\lambda_1, \ldots, \lambda_m$, although some of these may be complex numbers and not all are necessarily distinct. These m solutions are the full set of eigenvalues of $\mathbf{A}$. If $\mathbf{A}$ is symmetric, then all of its eigenvalues are real.

A.2.3   Rank

The *rank* of a matrix $\mathbf{A}$ is defined as the number of linearly independent columns (see Section A.3.3) in $\mathbf{A}$ and is usually denoted by rank$(\mathbf{A})$. If $\mathbf{A}$ is a square matrix, then

$$\text{rank}(\mathbf{A}) = \text{number of eigenvalues of } \mathbf{A} \text{ that are nonzero.} \qquad (A.1)$$

Note that repeated (nondistinct) eigenvalues are included in the right-hand side of (A.1).

A.2.4   Diagonalization

The *diagonalization* of an $m \times 1$ vector $\boldsymbol{a}$ (or $1 \times m$ vector $\boldsymbol{a}^\mathsf{T}$) is the $m \times m$ matrix with ith diagonal entry equal to the ith entry of $\boldsymbol{a}$ and off-diagonal entries equal to zero; this is denoted by diag$(\boldsymbol{a})$. For example,

$$\text{diag}([-1 \ \ 8]) = \text{diag}\left(\begin{bmatrix} -1 \\ 8 \end{bmatrix}\right) = \begin{bmatrix} -1 & 0 \\ 0 & 8 \end{bmatrix}.$$

If $\mathbf{A}$ is an $m \times m$ matrix, then we define diagonal$(\mathbf{A})$ to be the vector $[A_{11}, \ldots, A_{mm}]^\mathsf{T}$ with elements equal to the diagonal entries of $\mathbf{A}$.

A.2.5   Elementwise Function Notation

Let

$$\boldsymbol{a} = \begin{bmatrix} a_1 \\ \vdots \\ a_p \end{bmatrix} \quad \text{and} \quad \mathbf{b} = \begin{bmatrix} b_1 \\ \vdots \\ b_p \end{bmatrix}$$

be general vectors of the same length. Throughout this book we use the following notation:

$$\boldsymbol{a} \odot \mathbf{b} = \begin{bmatrix} a_1 b_1 \\ \vdots \\ a_p b_p \end{bmatrix}, \quad \boldsymbol{a}/\mathbf{b} = \begin{bmatrix} a_1/b_1 \\ \vdots \\ a_p/b_p \end{bmatrix}, \quad s(\boldsymbol{a}) = \begin{bmatrix} s(a_1) \\ \vdots \\ s(a_p) \end{bmatrix},$$

where $s: \mathbb{R} \to \mathbb{R}$ is a scalar function. For example,

$$\begin{bmatrix} 9 \\ 4 \\ 15 \end{bmatrix} \odot \begin{bmatrix} 3 \\ 2 \\ 5 \end{bmatrix} = \begin{bmatrix} 27 \\ 8 \\ 75 \end{bmatrix}, \qquad \begin{bmatrix} 9 \\ 4 \\ 15 \end{bmatrix} \bigg/ \begin{bmatrix} 3 \\ 2 \\ 5 \end{bmatrix} = \begin{bmatrix} 3 \\ 2 \\ 3 \end{bmatrix},$$

and

$$\log_{10} \left(\begin{bmatrix} 10 \\ 1 \\ 1000 \end{bmatrix} \right) = \begin{bmatrix} 1 \\ 0 \\ 3 \end{bmatrix}.$$

We will use $\mathbf{1}$ to denote a vector of ones, with dimension clear from the context.

A.2.6 *Definiteness*

A symmetric matrix $\mathbf{A}$ is said to be *positive definite* if

$$\mathbf{x}^{\mathsf{T}} \mathbf{A} \mathbf{x} > 0 \quad \text{for all } \mathbf{x} \neq \mathbf{0}.$$

A symmetric matrix $\mathbf{A}$ is said to be *positive semidefinite* if

$$\mathbf{x}^{\mathsf{T}} \mathbf{A} \mathbf{x} \geq 0 \quad \text{for all } \mathbf{x}.$$

A symmetric matrix $\mathbf{A}$ is positive definite if and only if each of its eigenvalues is positive; it is positive semidefinite if and only if each of its eigenvalues is nonnegative.

An example of a positive definite matrix is

$$\mathbf{A} = \begin{bmatrix} 7 & 2 \\ 2 & 2 \end{bmatrix},$$

since

$$[x_1 \ x_2] \begin{bmatrix} 7 & 2 \\ 2 & 2 \end{bmatrix} \begin{bmatrix} x_1 \\ x_2 \end{bmatrix} = 7x_1^2 + 4x_1 x_2 + 2x_2^2 = 5x_1^2 + 2(x_1 + x_2)^2 > 0$$

for all $[x_1 \ x_2]^{\mathsf{T}} \neq \mathbf{0}$.

An example of a positive semidefinite matrix is

$$\mathbf{A} = \begin{bmatrix} 1 & -1 \\ -1 & 1 \end{bmatrix},$$

since

$$[x_1 \ x_2] \begin{bmatrix} 1 & -1 \\ -1 & 1 \end{bmatrix} \begin{bmatrix} x_1 \\ x_2 \end{bmatrix} = (x_1 - x_2)^2,$$

which of course is nonnegative but equals zero whenever $x_1 = x_2$.

A.2.7 *Triangular Matrices*

A square matrix $\mathbf{R}$ is an *upper triangular* matrix if every entry below the main diagonal of $\mathbf{R}$ is zero. An example of an upper triangular matrix is

$$\begin{bmatrix} 8 & 11 & -5 \\ 0 & 7 & 2 \\ 0 & 0 & 9 \end{bmatrix}.$$

A square matrix $\mathbf{L}$ is a *lower triangular* matrix if every entry above the main diagonal of $\mathbf{L}$ is zero.

A.2.8 Cholesky Decomposition

If $\mathbf{A}$ is a symmetric positive definite matrix, then there exists an upper triangular matrix $\mathbf{R}$ such that
$$\mathbf{A} = \mathbf{R}^\mathsf{T}\mathbf{R}. \tag{A.2}$$

The right-hand side of (A.2) is called the *Cholesky decomposition* of the matrix $\mathbf{A}$. An example of a Cholesky decomposition is
$$\begin{bmatrix} 9 & 27 \\ 27 & 202 \end{bmatrix} = \begin{bmatrix} 3 & 0 \\ 9 & 11 \end{bmatrix}\begin{bmatrix} 3 & 9 \\ 0 & 11 \end{bmatrix}.$$

There exist fast and numerically stable algorithms for computing the Cholesky decomposition, which is an important tool for matrix computations.

A.2.9 QR Decomposition

For a general $m \times n$ matrix $\mathbf{A}$, a *QR decomposition* of $\mathbf{A}$ is
$$\mathbf{A} = \mathbf{QR},$$

where $\mathbf{Q}$ is an $m \times n$ matrix for which $\mathbf{Q}^\mathsf{T}\mathbf{Q} = \mathbf{I}$ and $\mathbf{R}$ is an $n \times n$ upper triangular matrix. An example of a QR decomposition is
$$\begin{bmatrix} 5/\sqrt{2} & 6\sqrt{2} \\ 0 & 0 \\ 5/\sqrt{2} & 2\sqrt{2} \end{bmatrix} = \begin{bmatrix} 1/\sqrt{2} & 1/\sqrt{2} \\ 0 & 0 \\ 1/\sqrt{2} & -1/\sqrt{2} \end{bmatrix}\begin{bmatrix} 5 & 8 \\ 0 & 4 \end{bmatrix}.$$

A.2.10 Singular Value Decomposition

For a general $m \times n$ matrix $\mathbf{A}$, the *singular value decomposition* of $\mathbf{A}$ is
$$\mathbf{A} = \mathbf{U}\,\mathrm{diag}(\mathbf{d})\mathbf{V}^\mathsf{T},$$

where $\mathbf{U}$ is an $m \times m$ matrix and $\mathbf{V}$ is an $n \times m$ matrix such that $\mathbf{U}^\mathsf{T}\mathbf{U} = \mathbf{V}^\mathsf{T}\mathbf{V} = \mathbf{I}_m$ and where $\mathbf{d}$ is a $m \times 1$ vector with nonnegative entries. The entries of $\mathbf{d}$ are called the *singular values* of $\mathbf{A}$. An example of a singular value decomposition is
$$\begin{bmatrix} 121/\sqrt{2} & 0 & 8\sqrt{2} \\ -121/\sqrt{2} & 0 & 8\sqrt{2} \end{bmatrix} = \begin{bmatrix} -1/\sqrt{2} & 1/\sqrt{2} \\ 1/\sqrt{2} & 1/\sqrt{2} \end{bmatrix}\mathrm{diag}\left(\begin{bmatrix} 121 \\ 16 \end{bmatrix}\right)\begin{bmatrix} -1 & 0 & 0 \\ 0 & 0 & 1 \end{bmatrix}.$$

A.2.11 Matrix Square Root

The *principal square root* of $\mathbf{A}$ is defined to be
$$\mathbf{A}^{1/2} = \mathbf{U}\,\mathrm{diag}(\sqrt{\mathbf{d}})\mathbf{V}^\mathsf{T},$$

where $\mathbf{U}$, $\mathbf{d}$, and $\mathbf{V}$ correspond to the singular value decomposition of $\mathbf{A}$ (Section A.2.10) and $\sqrt{\mathbf{d}}$ is obtained by replacing the entries of $\mathbf{d}$ by their nonnegative square roots.

For example,

$$
\begin{bmatrix} 121/\sqrt{2} & 0 & 8\sqrt{2} \\ -121/\sqrt{2} & 0 & 8\sqrt{2} \end{bmatrix}^{1/2} = \begin{bmatrix} -1/\sqrt{2} & 1\sqrt{2} \\ 1/\sqrt{2} & 1\sqrt{2} \end{bmatrix} \mathrm{diag}\left(\begin{bmatrix} 11 \\ 4 \end{bmatrix}\right)\begin{bmatrix} -1 & 0 & 0 \\ 0 & 0 & 1 \end{bmatrix}
$$

$$
= \begin{bmatrix} 11/\sqrt{2} & 0 & 2\sqrt{2} \\ -11/\sqrt{2} & 0 & 2\sqrt{2} \end{bmatrix}.
$$

The matrices

$$
\mathbf{A}^{1/2}(\mathbf{A}^{1/2})^{\mathsf{T}} \quad \text{and} \quad (\mathbf{A}^{1/2})^{\mathsf{T}}\mathbf{A}^{1/2} \tag{A.3}
$$

are both positive definite provided $\mathbf{A}$ is invertible. If $\mathbf{A}$ is symmetric and positive semidefinite, then the matrices in (A.3) are each equal to $\mathbf{A}$.

A.2.12   *Derivative Vector and Hessian Matrix*

Let f be a scalar-valued function with argument $\mathbf{x} \in \mathbb{R}^p$. The *derivative vector* of f, $\mathsf{D}f(\mathbf{x})$, is the $1 \times p$ vector whose ith entry is

$$
\partial f(\mathbf{x})\partial x_i.
$$

If $\mathbf{f}(\mathbf{x}) \equiv [f_1(\mathbf{x}), \ldots, f_m(\mathbf{x})]^{\mathsf{T}}$ is a vector-valued function of the vector $\mathbf{x}$, then $\mathsf{D}\mathbf{f}(\mathbf{x})$ is a matrix whose ith row is $\mathsf{D}f_i(\mathbf{x})$.

The *Hessian matrix* of a scalar-valued f is

$$
\mathsf{H}f(\mathbf{x}) = \mathsf{D}\{\mathsf{D}f(\mathbf{x})^{\mathsf{T}}\}
$$

and is, alternatively, the $p \times p$ matrix with (i, j) entry equal to

$$
\frac{\partial^2 f(\mathbf{x})}{\partial x_i \partial x_j}.
$$

For example, if

$$
f\left(\begin{bmatrix} x_1 \\ x_2 \end{bmatrix}\right) = x_1^2 x_2 - \cos(x_2 + 7),
$$

then

$$
\mathsf{D}f\left(\begin{bmatrix} x_1 \\ x_2 \end{bmatrix}\right) = [2x_1 x_2 \quad x_1^2 + \sin(x_2 + 7)]
$$

and

$$
\mathsf{H}f\left(\begin{bmatrix} x_1 \\ x_2 \end{bmatrix}\right) = \begin{bmatrix} 2x_2 & 2x_1 \\ 2x_1 & \cos(x_2 + 7) \end{bmatrix}.
$$

Often f has a matrix algebraic expression in terms of its argument $\mathbf{x}$. Then it is usually the case that matrix algebraic expressions exist for $\mathsf{D}f(\mathbf{x})$ and $\mathsf{H}f(\mathbf{x})$. For example, if

$$
f(\mathbf{x}) = \tfrac{1}{2}\mathbf{x}^{\mathsf{T}}\mathbf{x}\log(\mathbf{x}^{\mathsf{T}}\mathbf{x})
$$

then

$$
\mathsf{D}f(\mathbf{x}) = \{\log(\mathbf{x}^{\mathsf{T}}\mathbf{x}) + 1\}\mathbf{x}^{\mathsf{T}},
$$

$$
\mathsf{H}f(\mathbf{x}) = (2/\mathbf{x}^{\mathsf{T}}\mathbf{x})\mathbf{x}\mathbf{x}^{\mathsf{T}} + \{\log(\mathbf{x}^{\mathsf{T}}\mathbf{x}) + 1\}\mathbf{I}.
$$

Details on how to obtain such expressions are given in, for example, Magnus and Neudecker (1999) and Wand (2002).

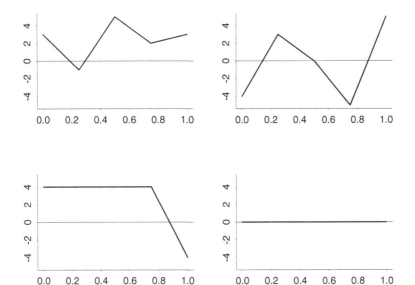

Figure A.1 Four members of the vector space $\mathcal{V}_1$. The last one is the zero vector.

A.3 Linear Algebra

A.3.1 *Vectors and Vector Spaces*

In matrix algebraic parlance (e.g., Section A.2), a vector is an $m \times 1$ matrix such as

$$\mathbf{x} = \begin{bmatrix} 6 \\ 1 \\ 8 \end{bmatrix}.$$

However, in the field of linear algebra the term "vector" is much more general. A vector is an element of a *vector space*. A vector space is a set with certain properties, given that addition and scalar multiplication (i.e., multiplication by a real number) are defined for its elements. The basic properties of a vector space $\mathcal{V}$ are:

$$\text{if } \mathbf{x}, \mathbf{y} \in \mathcal{V} \text{ then } \mathbf{x} + \mathbf{y} \in \mathcal{V};$$
$$\text{if } \mathbf{x} \in \mathcal{V} \text{ and } \alpha \in \mathbb{R} \text{ then } \alpha\mathbf{x} \in \mathcal{V}. \tag{A.4}$$

Several other conditions, such as existence of a zero vector, are also required for $\mathcal{V}$ to be a vector space. The full definition of a vector space can be found in any linear algebra textbook.

It is easy to see that the set of $m \times 1$ matrices satisfy (A.4) and, indeed, they do form a vector space. However, many different types of sets form vector spaces. An example is

$\mathcal{V}_1 =$ set of real-valued piecewise continuous functions f defined on
the interval $[0, 1]$ such that the knots of f are a subset of $\left\{ \frac{1}{4}, \frac{1}{2}, \frac{3}{4} \right\}$.

Four members of $\mathcal{V}_1$ are shown in Figure A.1, including the zero vector of the space: the function that is zero for all arguments in $[0, 1]$.

It is easily checked that the members of $\mathcal{V}_1$ satisfy (A.4) and the other conditions required for a vector space.

A.3.2 Linear Combination and Span

If $\mathbf{x}_1, \ldots, \mathbf{x}_k$ is a set of vectors and $\alpha_1, \ldots, \alpha_k \in \mathbb{R}$ is a set of k numbers, then

$$\alpha_1 \mathbf{x}_1 + \cdots + \alpha_k \mathbf{x}_k$$

is also a vector. The *span* of a set of vectors $\mathcal{S}$ is the set of all linear combinations of the vectors in that set and is often denoted by span($\mathcal{S}$). For example,

$$\text{span}\left(\left\{\begin{bmatrix} 8 \\ 3 \end{bmatrix}, \begin{bmatrix} 5 \\ -7 \end{bmatrix}\right\}\right) = \mathbb{R}^2,$$

the set of all 2×1 matrices. We also say that $\mathcal{S}$ *spans* $\mathbb{R}^2$.

A.3.3 Linear Dependence and Independence

Vectors $\mathbf{x}_1, \ldots, \mathbf{x}_k$ are said to be *linearly dependent* if there exist numbers $\alpha_1, \ldots, \alpha_k$, not all zero, such that

$$\alpha_1 \mathbf{x}_1 + \cdots + \alpha_k \mathbf{x}_k = \mathbf{0}.$$

Otherwise $\mathbf{x}_1, \ldots, \mathbf{x}_k$ are said to be *linearly independent.* For example, the vectors

$$\begin{bmatrix} 5 \\ 2 \end{bmatrix} \quad \text{and} \quad \begin{bmatrix} 10 \\ 4 \end{bmatrix}$$

are linearly dependent because

$$2\begin{bmatrix} 5 \\ 2 \end{bmatrix} + (-1)\begin{bmatrix} 10 \\ 4 \end{bmatrix} = \begin{bmatrix} 0 \\ 0 \end{bmatrix}.$$

On the other hand, the vectors

$$\begin{bmatrix} 5 \\ 2 \end{bmatrix} \quad \text{and} \quad \begin{bmatrix} 10 \\ 3 \end{bmatrix}$$

are linearly independent because

$$\alpha_1\begin{bmatrix} 5 \\ 2 \end{bmatrix} + \alpha_2\begin{bmatrix} 10 \\ 3 \end{bmatrix} = \begin{bmatrix} 0 \\ 0 \end{bmatrix}$$

if and only if $\alpha_1 = \alpha_2 = 0$.

A.3.4 Bases

A set of vectors $\mathcal{B}$ forms a *basis* for the vector space $\mathcal{V}$ if it spans $\mathcal{V}$ and if its members are linearly independent of one another. For example,

$$\left\{\begin{bmatrix} 8 \\ 3 \end{bmatrix}, \begin{bmatrix} 5 \\ -7 \end{bmatrix}\right\}$$

is a basis for $\mathbb{R}^2$. Any other pair of linearly independent 2×1 matrices forms a basis for $\mathbb{R}^2$. A particularly simple one is

$$\left\{\begin{bmatrix} 1 \\ 0 \end{bmatrix}, \begin{bmatrix} 0 \\ 1 \end{bmatrix}\right\},$$

which is sometimes called the *natural basis* for $\mathbb{R}^2$.

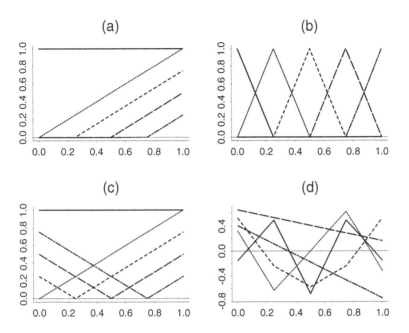

Figure A.2 Four bases of the vector space $\mathcal{V}_1$: (a) truncated line, (b) B-spline, (c) radial, and (d) Demmler–Reinsch.

The number of elements in a basis for a particular vector space $\mathcal{V}$ is always the same; this is called the *dimension* of the vector space. Note that $\mathbb{R}^2$ has dimension equal to 2.

Figure A.2 shows four different bases for $\mathcal{V}_1$: (a) truncated line; (b) B-spline (see Section 3.7.1); (c) radial (see Section 3.7.3); and (d) Demmler–Reinsch (see Section B.1.1.1). Note that the dimension of $\mathcal{V}_1$ is 5.

A.4   Probability Definitions and Results

A random vector $\mathbf{x} = [x_1, \ldots, x_n]^\mathsf{T}$ is a vector whose components are random variables.

A.4.1   *Mean of a Random Vector*

The *mean* or *expectation vector*, $\mathsf{E}(\mathbf{x})$, of $\mathbf{x}$ contains the expected values of the entries of $\mathbf{x}$:

$$\mathsf{E}(\mathbf{x}) = \begin{bmatrix} \mathsf{E}(x_1) \\ \vdots \\ \mathsf{E}(x_n) \end{bmatrix}.$$

A.4.2   *Covariance Matrix of a Random Vector*

The *covariance matrix* of $\mathbf{x}$ is an $n \times n$ matrix, denoted $\mathrm{Cov}(\mathbf{x})$, whose (i, j)th entry is the covariance between x_i and x_j. An equivalent definition is

$$\mathrm{Cov}(\mathbf{x}) = \mathsf{E}[\{\mathbf{x} - \mathsf{E}(\mathbf{x})\}\{\mathbf{x} - \mathsf{E}(\mathbf{x})\}^\mathsf{T}].$$

A.4.3 *Conditional Distribution, Mean, and Covariance Matrix*

Let $\mathbf{x}$ and $\mathbf{y}$ be two random vectors with probability density functions $p_\mathbf{x}$ and $p_\mathbf{y}$ (respectively) and with joint probability density function $p_{\mathbf{y},\mathbf{x}}(\mathbf{y}, \mathbf{x})$. The *conditional distribution of* $\mathbf{y}$ *given* $\mathbf{x}$ is that distribution having probability density function

$$p_{\mathbf{y}|\mathbf{x}}(\mathbf{y}|\mathbf{x}) = p_{\mathbf{y},\mathbf{x}}(\mathbf{y}, \mathbf{x})/p_\mathbf{x}(\mathbf{x})$$

for all $\mathbf{x}$ at which $p_\mathbf{x}(\mathbf{x}) > 0$.

The *conditional expectation of* $\mathbf{y}$ *given* $\mathbf{x}$ is

$$\mathsf{E}(\mathbf{y}|\mathbf{x}) = \int_{\mathbb{R}^{d_y}} p_{\mathbf{y}|\mathbf{x}}(\mathbf{y}|\mathbf{x})\,d\mathbf{y},$$

where d_y is the length of $\mathbf{y}$.

The *conditional covariance matrix of* $\mathbf{y}$ *given* $\mathbf{x}$ is

$$\mathrm{Cov}(\mathbf{y}|\mathbf{x}) = \mathsf{E}[\{\mathbf{y} - \mathsf{E}(\mathbf{y}|\mathbf{x})\}\{\mathbf{y} - \mathsf{E}(\mathbf{y}|\mathbf{x})\}^\mathsf{T}|\mathbf{x}].$$

A.4.4 *Bayes Theorem*

Bayes theorem is the following result, which expresses $p_{\mathbf{x}|\mathbf{y}}(\mathbf{x}|\mathbf{y})$ in terms of $p_{\mathbf{y}|\mathbf{x}}(\mathbf{y}|\mathbf{x})$:

$$p_{\mathbf{x}|\mathbf{y}}(\mathbf{x}|\mathbf{y}) = \frac{p_{\mathbf{y}|\mathbf{x}}(\mathbf{y}|\mathbf{x})p_\mathbf{x}(\mathbf{x})}{\int_{\mathbb{R}^{d_x}} p_{\mathbf{y}|\mathbf{x}}(\mathbf{y}|\mathbf{x})p_\mathbf{x}(\mathbf{x})\,d\mathbf{x}}.$$

Here d_x is the length of $\mathbf{x}$.

A.4.5 *Results Concerning Mean Vector and Covariance Matrix*

If $\mathbf{x}$ is a random vector, $\mathbf{A}$ is a constant matrix, and $\mathbf{c}$ is a constant vector whose dimensions are such that $\mathbf{Ax} + \mathbf{c}$ is defined, then

$$\mathsf{E}(\mathbf{Ax} + \mathbf{c}) = \mathbf{A}\mathsf{E}(\mathbf{x}) + \mathbf{c} \tag{A.5}$$

and

$$\mathrm{Cov}(\mathbf{Ax} + \mathbf{c}) = \mathbf{A}\,\mathrm{Cov}(\mathbf{x})\mathbf{A}^\mathsf{T}. \tag{A.6}$$

The mean of a *quadratic form*, $\mathbf{x}^\mathsf{T}\mathbf{Ax}$, is given by

$$\mathsf{E}(\mathbf{x}^\mathsf{T}\mathbf{Ax}) = \mathsf{E}(\mathbf{x})^\mathsf{T}\mathbf{A}\mathsf{E}(\mathbf{x}) + \mathrm{tr}\{\mathbf{A}\,\mathrm{Cov}(\mathbf{x})\}. \tag{A.7}$$

The following relationships hold between conditional and unconditional means and covariance matrices:

$$\mathsf{E}(\mathbf{y}) = \mathsf{E}\{\mathsf{E}(\mathbf{y}|\mathbf{x})\};$$
$$\mathrm{Cov}(\mathbf{y}) = \mathsf{E}\{\mathrm{Cov}(\mathbf{y}|\mathbf{x})\} + \mathrm{Cov}\{\mathsf{E}(\mathbf{y}|\mathbf{x})\}. \tag{A.8}$$

A.4.6 *Multivariate Normal Distribution*

The $d \times 1$ random vector $\mathbf{x}$ has a multivariate normal distribution with mean $\boldsymbol{\mu}$ and invertible covariance matrix $\boldsymbol{\Sigma}$ if its probability density function is

$$p_\mathbf{x}(\mathbf{x}) = (2\pi)^{-d/2}|\mathbf{\Sigma}|^{-1/2}\exp\{-\tfrac{1}{2}(\mathbf{x}-\boldsymbol{\mu})^\mathsf{T}\mathbf{\Sigma}^{-1}(\mathbf{x}-\boldsymbol{\mu})\};$$

we denote this by

$$\mathbf{x} \sim N(\boldsymbol{\mu}, \mathbf{\Sigma}).$$

If

$$\begin{bmatrix} \mathbf{x}_1 \\ \mathbf{x}_2 \end{bmatrix} \sim N\left(\begin{bmatrix} \boldsymbol{\mu}_1 \\ \boldsymbol{\mu}_2 \end{bmatrix}, \begin{bmatrix} \mathbf{\Sigma}_{11} & \mathbf{\Sigma}_{12} \\ \mathbf{\Sigma}_{21} & \mathbf{\Sigma}_{22} \end{bmatrix}\right)$$

is a general partitioned normal random vector, then the marginal distribution of $\mathbf{x}_2$ is $N(\boldsymbol{\mu}_2, \mathbf{\Sigma}_{22})$ and the conditional distribution of $\mathbf{x}_2$ given $\mathbf{x}_1$ is

$$\mathbf{x}_2|\mathbf{x}_1 \sim N\{\boldsymbol{\mu}_2 + \mathbf{\Sigma}_{21}\mathbf{\Sigma}_{11}^{-1}(\mathbf{x}_1 - \boldsymbol{\mu}_1), \mathbf{\Sigma}_{22} - \mathbf{\Sigma}_{21}\mathbf{\Sigma}_{11}^{-1}\mathbf{\Sigma}_{12}\}.$$

Analogous results hold for $\mathbf{x}_1$. Alternatively, $\mathbf{x}_2|\mathbf{x}_1$ may be expressed in terms of the submatrices of the *inverse* covariance matrix of $[\mathbf{x}_1^\mathsf{T}\ \mathbf{x}_2^\mathsf{T}]^\mathsf{T}$. If

$$\begin{bmatrix} \mathbf{x}_1 \\ \mathbf{x}_2 \end{bmatrix} \sim N\left(\begin{bmatrix} \boldsymbol{\mu}_1 \\ \boldsymbol{\mu}_2 \end{bmatrix}, \begin{bmatrix} \mathbf{Q}_{11} & \mathbf{Q}_{12} \\ \mathbf{Q}_{21} & \mathbf{Q}_{22} \end{bmatrix}^{-1}\right),$$

then

$$\mathbf{x}_2|\mathbf{x}_1 \sim N\{\boldsymbol{\mu}_2 - \mathbf{Q}_{22}^{-1}\mathbf{Q}_{21}(\mathbf{x}_1 - \boldsymbol{\mu}_1), \mathbf{Q}_{22}^{-1}\}.$$

A.5   Maximum Likelihood Estimation

Maximum likelihood estimation is a general technique for estimation of parameters in a statistical model. Maximum likelihood estimators generally have good statistical properties and, in many circumstances, are optimal among competing estimators (see e.g. Huber 1967).

Let $\mathbf{y}$ denote the vector of observed data, and suppose that it is modeled to have probability density function

$$p_\mathbf{y}(\mathbf{y}; \boldsymbol{\theta}).$$

Here $\boldsymbol{\theta}$ is a vector of parameters requiring estimation. Let Θ be the set of permissible values of $\boldsymbol{\theta}$. Then the *likelihood of* $\boldsymbol{\theta}$ is

$$\mathcal{L}(\boldsymbol{\theta}) \equiv p_\mathbf{y}(\mathbf{y}; \boldsymbol{\theta})$$

considered as a function of $\boldsymbol{\theta} \in \Theta$. The *maximum likelihood estimator* of $\boldsymbol{\theta}$ is

$$\hat{\boldsymbol{\theta}} = \text{the } \boldsymbol{\theta} \text{ that maximizes } \mathcal{L}(\boldsymbol{\theta}) \text{ over } \Theta.$$

A.6   Bibliographical Notes

Detailed references on matrix algebra geared toward statistics are Searle (1982) and Harville (2000). There are numerous elementary linear algebra books; for example, Strang (1998), Halmos (1995), Anton and Rorres (2000), and Nicholson (2001). Elementary probability and statistics books are also quite numerous. Good examples are Hoel (1984), Hogg and Craig (1995), and Casella and Berger (2002).

B

Computational Issues

B.1 Fast Computation of Penalized Spline Smooths

A penalized spline smooth involves fitting a ridge regression. Automatic smooth-ing parameter selection may involve several such ridge regressions. Their direct implementation can be both slow and numerically unstable. In this section we describe algorithms that lead to large improvements in both facets.

B.1.1 Demmler–Reinsch Orthogonalization

Suppose that

$$\hat{\mathbf{f}}_\alpha = \mathbf{C}(\mathbf{C}^\mathsf{T}\mathbf{C} + \alpha\mathbf{D})^{-1}\mathbf{C}^\mathsf{T}\mathbf{y} \tag{B.1}$$

for some symmetric matrix $\mathbf{D}$. Such is the case for Gaussian data for all low-rank spline smoothers described in the book. Algorithm A.1 allows for fast and stable calculation of (B.1).

Algorithm A.1 Inputs: $\mathbf{y}$, $\mathbf{C}$, $\mathbf{D}$, α.

(1) Obtain the Cholesky decomposition of $\mathbf{C}^\mathsf{T}\mathbf{C}$:

$$\mathbf{C}^\mathsf{T}\mathbf{C} = \mathbf{R}^\mathsf{T}\mathbf{R},$$

where $\mathbf{R}$ is square and invertible.

(2) Form the symmetric matrix $\mathbf{R}^{-\mathsf{T}}\mathbf{D}\mathbf{R}^{-1}$ and obtain its singular value de-composition:

$$\mathbf{R}^{-\mathsf{T}}\mathbf{D}\mathbf{R}^{-1} = \mathbf{U}\,\mathrm{diag}(\mathbf{s})\mathbf{U}^\mathsf{T}.$$

(3) Compute the matrix and vector,

$$\mathbf{A} \equiv \mathbf{C}\mathbf{R}^{-1}\mathbf{U} \quad\text{and}\quad \mathbf{b} \equiv \mathbf{A}^\mathsf{T}\mathbf{y}.$$

(4) The fitted values are then

$$\hat{\mathbf{f}}_\alpha = \mathbf{A}\left(\frac{\mathbf{b}}{1 + \alpha\mathbf{s}}\right)$$

with corresponding degrees of freedom

$$df_{\mathrm{fit}}(\alpha) = \mathbf{1}^\mathsf{T}\left(\frac{1}{1 + \alpha\mathbf{s}}\right).$$

The Cholesky decomposition applies only to nonsingular matrices. If $\mathbf{C}$ is ill-conditioned then it is advisable to add a small multiple of $\mathbf{D}$ to $\mathbf{C}^\mathsf{T}\mathbf{C}$ before applying the Cholesky decomposition, so that

$$\mathbf{C}^\mathsf{T}\mathbf{C} + \delta\mathbf{D} = \mathbf{R}^\mathsf{T}\mathbf{R},$$

where δ is small (e.g., $\delta = 10^{-10}$).

Once the matrix $\mathbf{A}$ and vectors $\mathbf{b}$ and $\mathbf{s}$ have been computed, the vector of fits (for different values of α) reduces to a matrix multiplication. Therefore, $\hat{\mathbf{f}}_\alpha$ and $df_{\mathrm{fit}}(\alpha)$ can be computed cheaply for several α-values. This is particularly useful when solving for the α corresponding to a prespecified number of degrees of freedom.

Automatic smoothing parameter selection also benefits greatly from cheap calculation for several smooths. This is because common methods, such as all those described in Section 5.3, depend only on $df_{\mathrm{fit}}(\alpha)$ and

$$\mathrm{RSS}(\alpha) = \|\mathbf{y} - \hat{\mathbf{f}}_\alpha\|^2.$$

For example,

$$\mathrm{GCV}(\alpha) = \mathrm{RSS}(\alpha)/\{1 - df_{\mathrm{fit}}(\alpha)/n\}^2,$$

$$\mathrm{AIC}(\alpha) = \log\{\mathrm{RSS}(\alpha)\} + 2df_{\mathrm{fit}}(\alpha)/n,$$

$$\mathrm{AIC}_C(\alpha) = \log\{\mathrm{RSS}(\alpha)\} + \frac{2\{df_{\mathrm{fit}}(\alpha) + 1\}}{n - df_{\mathrm{fit}}(\alpha) - 2}.$$

Grid searches for the smoothing parameter that minimizes each of these criteria can be done very rapidly.

If the length of $\mathbf{y}$ is very large, then direct computation of $\mathrm{RSS}(\alpha)$ can be costly. A faster alternative is to compute $\mathbf{y}^\mathsf{T}\mathbf{y}$, $\mathbf{b} = \mathbf{A}^\mathsf{T}\mathbf{y}$, and $\mathbf{A}^\mathsf{T}\mathbf{A}$ (which will be low-dimensional) and then use

$$\mathrm{RSS}(\alpha) = \mathbf{y}^\mathsf{T}\mathbf{y} - 2\mathbf{y}^\mathsf{T}\hat{\mathbf{f}}_\alpha + \hat{\mathbf{f}}_\alpha^\mathsf{T}\hat{\mathbf{f}}_\alpha$$

$$= \mathbf{y}^\mathsf{T}\mathbf{y} - 2\mathbf{b}^\mathsf{T}\left(\frac{\mathbf{b}}{1 + \alpha\mathbf{s}}\right) + \left(\frac{\mathbf{b}}{1 + \alpha\mathbf{s}}\right)^\mathsf{T}\mathbf{A}^\mathsf{T}\mathbf{A}\left(\frac{\mathbf{b}}{1 + \alpha\mathbf{s}}\right).$$

Confidence bands that adjust for bias as described in Section 6.4 can be formed using

$$\mathrm{st.dev.}(\hat{\mathbf{f}}_\alpha - \mathbf{f}) = \sigma_\varepsilon\sqrt{\mathrm{diagonal}\left\{\mathbf{A}\,\mathrm{diag}\left(\frac{1}{1 + \alpha\mathbf{s}}\right)\mathbf{A}^\mathsf{T}\right\}},$$

assuming that $\mathrm{Cov}(\mathbf{y} - \mathbf{f}) = \sigma_\varepsilon^2\mathbf{I}$. When computing the diagonal in this expression, one does not need to first compute the entire $n \times n$ matrix. Rather, one can use

$$\mathrm{diagonal}\left\{\mathbf{A}\,\mathrm{diag}\left(\frac{1}{1 + \alpha\mathbf{s}}\right)\mathbf{A}^\mathsf{T}\right\} = \left[\left\{\mathbf{A}\,\mathrm{diag}\left(\frac{1}{1 + \alpha\mathbf{s}}\right)\right\} \odot \mathbf{A}\right]\mathbf{1},$$

where $\mathbf{1}$ is a vector of ones of length equal to the number of basis elements – that is, the number of columns of $\mathbf{C}$.

If $\mathbf{x}$ is a vector, then $\mathrm{diag}(\mathbf{x})$ is the diagonal matrix with the elements of $\mathbf{x}$ on its diagonal. The inverse operation is $\mathrm{diagonal}(\mathbf{A})$, which converts a square matrix $\mathbf{A}$ to the vector containing the diagonal elements of $\mathbf{A}$.

A reasonable estimate of σ_ε^2 is

$$\hat{\sigma}_\varepsilon^2 = \text{RSS}(\alpha)/df_{\text{res}}(\alpha),$$

where $df_{\text{res}}(\alpha)$ can be computed as

$$df_{\text{res}}(\alpha) = n - 2\mathbf{1}^\mathsf{T}\left(\frac{1}{1+\alpha\mathbf{s}}\right) + \left\|\frac{1}{1+\alpha\mathbf{s}}\right\|^2.$$

If computations over a grid

$$\mathbf{g} = [g_1, \ldots, g_M]^\mathsf{T}$$

are required then, in the definition of $\hat{\mathbf{f}}_\alpha$, $\mathbf{A} = \mathbf{C}\mathbf{R}^{-1}\mathbf{U}$ should be replaced by $\mathbf{C}_g\mathbf{R}^{-1}\mathbf{U}$, where the columns of $\mathbf{C}_g$ have the same form as those in $\mathbf{C}$ but with the x_i replaced by the g_ℓ.

B.1.1.1   *Justification of Algorithm A.1*

Now

$$\mathbf{R}^{-\mathsf{T}}\mathbf{D}\mathbf{R}^{-1} = \mathbf{U}\,\text{diag}(\mathbf{s})\mathbf{U}^\mathsf{T} \quad \text{with } \mathbf{U}^\mathsf{T}\mathbf{U} = \mathbf{I}.$$

Since $\mathbf{U}$ is a square matrix, $\mathbf{U}^\mathsf{T} = \mathbf{U}^{-1}$ and so

$$\mathbf{D} = \mathbf{R}^\mathsf{T}\mathbf{U}\,\text{diag}(\mathbf{s})\mathbf{U}^{-1}\mathbf{R}.$$

Also,

$$\mathbf{C}^\mathsf{T}\mathbf{C} = \mathbf{R}^\mathsf{T}\mathbf{R} = \mathbf{R}^\mathsf{T}\mathbf{U}\mathbf{U}^{-1}\mathbf{R}$$

and consequently

$$\mathbf{C}^\mathsf{T}\mathbf{C} + \alpha\mathbf{D} = \mathbf{R}^\mathsf{T}\mathbf{U}\{\mathbf{I} + \alpha\,\text{diag}(\mathbf{s})\}\mathbf{U}^{-1}\mathbf{R}.$$

Hence

$$\hat{\mathbf{f}}_\alpha = \mathbf{C}[\mathbf{R}^\mathsf{T}\mathbf{U}\{\mathbf{I} + \alpha\,\text{diag}(\mathbf{s})\}\mathbf{U}^{-1}\mathbf{R}]^{-1}\mathbf{C}^\mathsf{T}\mathbf{y}$$

$$= (\mathbf{C}\mathbf{R}^{-1}\mathbf{U})\{\text{diag}(1+\alpha\mathbf{s})\}^{-1}(\mathbf{C}\mathbf{R}^{-1}\mathbf{U})^\mathsf{T}\mathbf{y} = \mathbf{A}\left(\frac{\mathbf{b}}{1+\alpha\mathbf{s}}\right),$$

where $\mathbf{A} \equiv \mathbf{C}\mathbf{R}^{-1}\mathbf{U}$ and $\mathbf{b} \equiv \mathbf{A}^\mathsf{T}\mathbf{y}$.

The degrees-of-freedom expression follows quickly from the result

$$\mathbf{A}^\mathsf{T}\mathbf{A} = \mathbf{U}^\mathsf{T}\mathbf{R}^{-\mathsf{T}}\mathbf{C}^\mathsf{T}\mathbf{C}\mathbf{R}^{-1}\mathbf{U} = \mathbf{U}^{-1}\mathbf{R}^{-\mathsf{T}}\mathbf{R}^\mathsf{T}\mathbf{R}\mathbf{R}^{-1}\mathbf{U} = \mathbf{I}.$$

The new expression for $\hat{\mathbf{f}}_\alpha$ is thus of the form

$$\hat{\mathbf{f}}_\alpha = \mathbf{A}\{\mathbf{A}^\mathsf{T}\mathbf{A} + \alpha\,\text{diag}(\mathbf{s})\}^{-1}\mathbf{A}^\mathsf{T}\mathbf{y}.$$

Comparison with (B.1) shows that we have effectively replaced the basis functions in $\mathbf{C}$ with those in $\mathbf{A}$ where this design matrix has the *orthogonality* property $\mathbf{A}^\mathsf{T}\mathbf{A} = \mathbf{I}$. The columns of $\mathbf{A}$ correspond to the *Demmler–Reinsch* basis for the vector space spanned by $\mathbf{C}$ (see Section A.3.4). The orthogonality property is crucial for fast computation over several smoothing parameters.

B.1.1.2 S-PLUS *Implementation of Algorithm A.1*

We will now give some S-PLUS code for implementation of Algorithm A.1 as well as code for automatic smoothing parameter selection via GCV and variability bars.

The basic inputs are the **x** and **y** scatterplot vectors. Here we will take these to correspond to some components of the S-PLUS data set `air`:

```
x <- $radiation
y <- air$ozone^(1/3)
```
(B.2)

Now set the default number and location of the knots:

```
num.knots <- max(5,min(floor(length(
                unique(x))/4),35))
knots <- quantile(unique(x),seq(0,1,length=
             (num.knots+2))[-c(1,(num.knots+2))])
```
(B.3)

Set up a logarithmic grid of alpha values:

```
alpha.low <- 1
alpha.upp <- 10000000
num.alpha <- 25
alpha.vec <- 10^(seq(log10(alpha.low),
                 log10(alpha.upp),length=num.alpha))
```

Set up design matrices, for linear splines in this case:

```
n <- length(x)
X <- cbind(rep(1,n),x)
Z <- outer(x,knots,"-")
Z <- Z*(Z>0)
```
(B.4)

Set up input matrices for Algorithm A.1:

```
C.mat <- cbind(X,Z)
CTC <- t(C.mat)%*%C.mat
D.mat <- diag(c(rep(0,ncol(X)),rep(1,ncol(Z))))
```

Carry out steps (1) and (2) of Algorithm A.1:

```
R.mat <- chol(CTC)
svd.out <- svd(t(solve(t(R.mat),
                 t(solve(t(R.mat),D.mat)))))
s.vec <- svd.out$d
U.mat <- svd.out$u
```

Obtain a suite of fits for a vector of smoothing parameters, along with corresponding GCV values, and determine the minimum smoothing parameter:

```
A.mat <- C.mat%*%backsolve(R.mat,U.mat)
b.vec <- as.vector(t(A.mat)%*%y)
r.mat <- 1/(1+outer(s.vec,alpha.vec))
f.hats <- A.mat%*%(b.vec*r.mat)
y.vecs <- matrix(rep(y,num.alpha),n,num.alpha)
RSS.vec <- apply((y.vecs-f.hats)^2,2,sum)
df.vec <- apply(r.mat,2,sum)
GCV.vec <- RSS.vec/((1-df.vec/n)^2)
ind.min <- order(GCV.vec)[1]
if (ind.min==1)
    stop("Hit left boundary; make alpha.low smaller.")
if (ind.min==num.alpha)
    stop("Hit right boundary; make alpha.upp bigger.")
alpha.GCV <- alpha.vec[ind.min]
```

Compute lower and upper limits of variability bars:

```
df.res <- n-2*df.vec[ind.min]
            +sum((1/(1+alpha.GCV*s.vec))^2)
sig.eps.hat <- sqrt(RSS.vec[ind.min]/df.res)
st.dev.hat <- sig.eps.hat*sqrt(diag(A.mat%*%
                              (r.mat[,ind.min]*t(A.mat)))))
var.bar.upp <- f.hats[,ind.min]+2*st.dev.hat
var.bar.low <- f.hats[,ind.min]-2*st.dev.hat
```

Plot GCV curve and minimum GCV fit, along with variability bars:

```
par(mfrow=c(1,2))
plot(log10(alpha.vec),GCV.vec,type="l",bty="l",
    xlab="log10(alpha)",ylab="GCV")
points(log10(alpha.GCV),min(GCV.vec),pch=4,
      lwd=2,cex=1.2)
lines(rep(log10(alpha.GCV),2),c(0,min(GCV.vec)),
      err=-1)
plot(x,y,pch=1,bty="l")
lines(x[order(x)],f.hats[order(x),ind.min])
lines(x[order(x)],var.bar.low[order(x)],lty=3)
lines(x[order(x)],var.bar.upp[order(x)],lty=3)
```

The result is shown in Figure B.1.

B.1.1.3   MATLAB *Implementation of Algorithm A.1*

Algorithm A.1 is implemented by the MATLAB program PsplineDR04.m that is available at the book's website along with other MATLAB programs for smoothing. PsplineDR04.m can be called as in the following example, which fits a 25-knot linear spline to the LIDAR data.

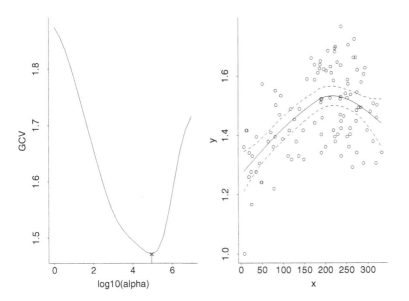

Figure B.1 Plots obtained by running the S-PLUS code listed in Section B.1.1.2.

The first two arguments in the call are the x- and y-variables. The third argument, which is optional and can be omitted, is a MATLAB "structure" that contains values for tuning parameters. Any tuning parameter not defined in this structure will be given a default value.

```
load lidar.dat ;
x = lidar(:,1) ;
y = lidar(:,2) ;
fit = PsplineDR04(x,y,struct('degree',1,'nknots',25)) ;
plot(fit.xgrid,fit.mhat,x,y,'o') ;
```

PsplineDR04.m calls the program quantileknots01.m as well as the program powerbasis01.m. quantileknots01.m finds the sample quantiles of the unique values of the x-variable and is as follows:

```
function knots = quantileknots01(x,nknots) ;
x = unique(x) ;
n = length(x) ;
xsort = sort(x) ;
loc = n*(1:nknots)' ./ (nknots+1) ;
knots = xsort(round(loc)) ;
```

powerbasis01.m creates the polynomial and truncated power basis functions that form the matrix **C**:

```
function xm = powerbasis01(x,degree,knots,der) ;
%    Returns the power basis functions of a
%    spline of given degree
```

```
%
%    USAGE: xm=powerbasis(x,degree,knots)
%
%    Last edit:   3/22/99

if nargin < 4 ;
       der = 0 ;
end ;

if der > degree ;
disp('*************************************') ;
disp('*************************************') ;
disp('WARNING: der > degree --- xm not') ;
disp('returned by powerbasis') ;
disp('*************************************') ;
disp('*************************************') ;
return ;
end ;


n = size(x,1) ;
nknots = length(knots);
mx = mean(x) ;
stdx = std(x) + 100*eps ;

if der == 0 ;
   xm = ones(n,1);
   else ;
   xm = zeros(n,1) ;
end ;

    for i=1:degree ;
        if i < der ;
            xm = [xm zeros(n,1)] ;
            else ;
            xm = [xm prod((i-der+1):i) * x.^(i-der)] ;
        end ;
    end ;

    if nknots > 0 ;
        for i=1:(nknots) ;
        xm = [xm prod((degree-der+1):degree) * ...
          (x-knots(i)).^(degree-der).*(x > knots(i))] ;
        end ;
    end ;
```

To explain how PsplineDR04.m works, the program will be listed in a sequence
of small sections of code with explanatory text between them. The beginning of
the program consists of comments:

```
function fit = PsplineDR04(x,y,param);
%    Fits a P-spline to univariate x's
%    with Demmler-Reinsch algorithm.
%
%    Copied from PsolineDR02 - like that program but
%    optional input enters as a structure.
%
%    Also allows a quadratic integral penalty
%    on the 2nd derivative
%
%    USAGE: fit=PsplineDR03(x,y,param);
%
%            INPUT - REQUIRED
%    x = independent variable (univariate)
%    y = response (same length as x)
%
%            INPUT - OPTIONAL (put in a structure
%                      that is the third argument in)
%    degree = degree of the spline (default is 2)
%            (changed to 3 if smooth_spline_penalty=1)
%    nknots = number of knots (default min of
%            floor(.3*n) and 20)
%    extrapen: if 1 then the x^degree term is
%            penalized; if 0 then not (default is 0)
%    penwt = trial values of the penalty weight
%            (one is chosen by minimizing gcv
%            (default is logspace(-12,12,100))
%    boundstab = parameter passed to quantileknots
%                (see that program)
%    knots (default is to generate nknots using the
%            program quantileknots)
%    istd = 1 if x and y are to be standardized
%                before computation (default=1)
%    smooth_spline_penalty: if 1 then penalty is on the
%            integral from min(x) to max(x) of the
%            square of the second derivative
%
%            OUTPUT
%    Returns a structure "fit" with the following
%    components
%
```

```
%    CALLS: powerbasis01, quantileknots01
%
%    Last edit: 9/17/2002
```

Next, any tuning constants not specified in the third input argument are set to default values.

```
n = size(x,1) ;

if nargin < 3 ;
        param = '' ;
end ;

if isfield(param,'gcvfact') == 0 ;
        gcvfact = 1 ;
else ;
        gcvfact = param.gcvfact ;
end ;

if isfield(param,'istd') == 0 ;
        istd = 1 ;
else ;
        istd = param.istd ;
end ;

if isfield(param,'penwt') == 0 ;
        penwt = logspace(-12,12,100) ;
else ;
        penwt = param.penwt ;
end ;

if isfield(param,'nknots') == 0 ;
        nknots = min([floor(.3*n), 20]) ;
else ;
        nknots = param.nknots ;
end ;

if isfield(param,'knots') == 0 ;
        knots = quantileknots01(x,nknots) ;
else ;
        knots = param.knots ;
        nknots = length(knots) ;
end ;

if isfield(param,'degree') == 0 ;
        degree = 2 ;
```

```
else ;
        degree = param.degree ;
end ;

if isfield(param,'smooth_spline_penalty') == 0 ;
        smooth_spline_penalty = 0 ;
else ;
        smooth_spline_penalty
        = param.smooth_spline_penalty ;
end ;
```

Here the *x*-variable is standardized if istd is equal to 1.

```
originalx = x ;
if istd == 1 ;
        stdx = std(x) ;
        meanx = mean(x) ;
        x = (x - meanx) ./ stdx ;
        meany = mean(y) ;
        y = y - meany ;
        knots = (knots - meanx) ./ stdx ;
end ;

n = length(x) ;
```

If smooth_spline_penalty equals 1, then a cubic smoothing spline penalty
is used and so degree is set equal to 3.

```
if smooth_spline_penalty == 1 ;
        degree = 3 ;
end ;
```

Next, knots and the matrix **C** (which is called xm in the program) are created.
Then the penalty matrix **D** is created.

```
xm = powerbasis01(x,degree,knots) ;
xx = xm'*xm ;
id = [zeros(1,degree+1)   ones(1,nknots)] ;
D = diag(id) ;
if smooth_spline_penalty == 1 ;
maxx = max(x) ;
knots2 = [min(x);knots] ;

for i = 1:4+nknots ;

if i < 3 ;
D(i,:) = zeros(1,4+nknots) ;
elseif i == 3 ;
```

```
D(i,1) = 0 ;
D(i,2) = 0 ;
D(i,3) = 4 ;

for i2 = 4:4+nknots ;
D(i,i2) = 6*(maxx-knots2(i2-3))^2 ;
end ;   %          end "for i2 = 4:4+nknots"

elseif i > 3 ;
D(i,1) = 0 ;
D(i,2) = 0 ;
D(i,3) = 6*(maxx-knots2(i-3))^2 ;

for i2 = 4:4+ nknots ;
knotmax = knots2( max([i-3 i2-3]) ) ;
knotmin = knots2( min([i-3 i2-3]) ) ;
D(i,i2) = 36*( (maxx-knotmax)^3/3 + ...
            (knotmax-knotmin) * (maxx-knotmax)^2/2) ;
end ;   %          end "for i2 = 4:4+nknots"
end ;   %          end "if i < 3"
end ;   %          end "for i:4+nknots"
end ;   %          end "if smooth_spline_penalty == 1"
```

Next, Algorithm A.1 is started. First, the matrix **R** is created using MATLAB's built-in Cholesky factorization command.

```
R = chol(xx + 10e-10*D) ;
B = inv(R') ;
[U,C] = eig(B*D*B') ;
Z = xm*B'*U ;

Zy = Z'*y ;
ZZ = Z'*Z ;
```

In the code that follows, penwt is the vector of values of the penalty λ, while beta, asr, dfres, and gcv are (respectively) the spline basis coefficients, average squared residual, degrees of freedom for residuals, and generalized cross-validation statistics, all computed at each value of penwt. These are initialized as zero matrices to set up storage.

```
m = length(penwt) ;
beta = zeros(size(xm,2),m) ;
asr = zeros(m,1) ;
gcv = asr ;
trsd = asr ;
trsdsd = asr ;
dfres = asr ;
```

```
aic = asr ;
ssy = y'*y ;
          for i=1:m ;
          oneld = 1 ./ (1 + penwt(i)*diag(C)) ;
          trsd(i) = sum(oneld) ;
          alpha(:,i) = (Zy) .* oneld ;
          asr(i) = (ssy - 2*Zy'*alpha(:,i) ...
                     + alpha(:,i)'*ZZ*alpha(:,i))/n ;
          trsdsd(i) = sum(oneld.*oneld) ;
          dfres(i) = n - 2*trsd(i) + trsdsd(i) ;
          gcv(i) = asr(i) / (1 - gcvfact*trsd(i)/n)^2 ;
          sigma2 = asr(i) / (1 - trsd(i)/n) ;
          aic(i) = n*log(sigma2) + 2*trsd(i) ;
          end ;
```

Next we have imin, the index of penwt where gcv is minimized; a, the value of penwt at this minimum; and dffit, degrees of freedom for the fit. The fitted curve at the GCV selected value of λ is computed.

```
imin = min(find( (gcv==min(gcv)) ) ) ;
a = penwt(imin) ;

dffit = trsd ;
alpha = alpha(:,imin) ;
beta = B'* U * alpha ;
dbeta = length(beta) ;
yhat = Z*alpha ;
res = y - yhat ;
sigma2hat = n*asr(imin) ./ dfres(imin) ;
sigmahat = sqrt(sigma2hat) ;

oneld = 1 ./ (1 + penwt(imin)*diag(C)) ;
postvaralpha = sigma2hat*diag(oneld);
varalpha = sigma2hat * diag(oneld.^2) ;
postvaryhat = (Z.*(Z*postvaralpha))*ones(dbeta,1) ;
postvarbeta = B'*U*postvaralpha*U'*B ;
varbeta = B'*U*varalpha*U'*B ;
varyhat = (Z.*(Z*varalpha))*ones(dbeta,1) ;
```

Now some diagnostics, Cook's D, the hat diagonals, and the studentized residuals are computed.

```
[m1,m2] = size(Z) ;
hi = Z.*(Z*diag(oneld))*ones(m2,1) ;
cookD = (res.^2) .* hi ./ ( dffit(imin)*(1-hi) ) ;
studres = res .* (sqrt(1-hi)) ./ sigmahat ;
Z = [] ;
```

Now the estimate of the derivative is computed.

```
yhatder = '' ;
postvaryhatder = '' ;

if degree > 0 ;

xmder = xm ;
xmder(:,1) = 0*xmder(:,1) ;

for i = 2:degree+1 ;
        xmder(:,i) = (i-1)*(abs(xm(:,i))).^( ...
            (i-2)/(i-1) ) ... .* (sign(x)).^(i-2) ;
end ;

for i=degree+2:degree+1+nknots ;
        xmder(:,i) = degree*(abs(xm(:,i))).^( ...
            (degree-1)/degree ) ... .* (xm(:,i) > 0) ;
end ;

yhatder = xmder*beta ;
postvaryhatder = ((xmder*postvarbeta).*xmder) ...
                *ones(dbeta,1) ;
clear xmder ;
end ;    %        end "if degree > 0"

if istd == 1 ;
        yhat = yhat + meany ;
end ;
```

Confidence limits for the fitted curve and its derivative are then computed.

```
ulimit = yhat + 2*sqrt(postvaryhat) ;
llimit = yhat - 2*sqrt(postvaryhat) ;
ulimitder = yhatder + 2*sqrt(postvaryhatder) ;
llimitder = yhatder - 2*sqrt(postvaryhatder) ;
```

Next, the fitted curve and its derivative are computed on an equally spaced grid. The estimates are called mhat and mhatder, and the grid is called xgrid; xgridm is the matrix of spline basis functions evaluated at xgrid.

```
xgrid = linspace(min(x),max(x),200)' ;
xgridm = powerbasis01(xgrid,degree,knots) ;

if istd == 1 ;

        xgrid = meanx + stdx*xgrid ;
        mhat = meany + xgridm*beta ;
```

```
        knots = meanx + stdx*knots ;
        yhatder = yhatder./stdx ;
        ulimitder = ulimitder./stdx ;
        llimitder = llimitder./stdx ;

else ;
        mhat = xgridm*beta ;
end ;


[B,I] = unique(originalx) ;
```

Next, confidence limits are evaluated at xgrid.

```
ulimit_xgrid = interp1(originalx(I),ulimit(I), ...
                       xgrid,'cubic') ;
llimit_xgrid = interp1(originalx(I),llimit(I), ...
                       xgrid,'cubic') ;

ulimitder_xgrid = interp1(originalx(I), ...
                          ulimitder(I),xgrid,'cubic') ;
llimitder_xgrid = interp1(originalx(I), ...
                          llimitder(I),xgrid,'cubic') ;

mhatder = interp1(originalx(I),yhatder(I),
                  xgrid,'cubic') ;
```

Various statistics are collected into a structure called fit, which is the output. Exactly which statistics are included in fit depends on whether or not the sample size n is large (i.e., over 5000).

```
if n < 5001 ;

fit=struct('yhat',yhat,'beta',beta,'gcv',gcv, ...
        'imin',imin,'dffit',dffit, ...
        'knots',knots,'postvarbeta',postvarbeta, ...
        'postvaryhat',postvaryhat,'xm',xm,'xx',xx, ...
        'a',a,'penwt',penwt,'sigma2hat',sigma2hat, ...
        'dfres',dfres,'yhatder',yhatder, ...
        'postvaryhatder',postvaryhatder, ...
        'ulimit',ulimit,'llimit',llimit, ...
        'ulimitder',ulimitder, ...
        'llimitder',llimitder,'asr',asr, ...
        'xgrid',xgrid,'mhat',mhat, ...
        'x',meanx+stdx*x,'degree',degree, ...
        'varbeta',varbeta,'varyhat',varyhat, ...
        'ulimit_xgrid',ulimit_xgrid, ...
        'llimit_xgrid',llimit_xgrid, ...
```

```
            'ulimitder_xgrid',ulimitder_xgrid,  ...
            'llimitder_xgrid',llimitder_xgrid,  ...
            'mhatder',mhatder,'hi',hi,'res',res,  ...
            'cookD',cookD,'studres',studres)  ;
else ;
fit=struct('yhat',yhat,'beta',beta,'gcv',gcv,  ...
            'imin',imin,'dffit',dffit,'knots',knots,  ...
            'postvarbeta',postvarbeta,  ...
            'postvaryhat',postvaryhat,'xx',xx,'a',a,  ...
            'penwt',penwt,'sigma2hat',sigma2hat,  ...
            'dfres',dfres,'yhatder',yhatder,  ...
            'postvaryhatder',postvaryhatder,  ...
            'ulimit',ulimit,'llimit',llimit,  ...
            'ulimitder',ulimitder,  ...
            'llimitder',llimitder,'asr',asr,  ...
            'xgrid',xgrid,'mhat',mhat,  ...
            'x',meanx+stdx*x,'degree',degree,  ...
            'varbeta',varbeta,'varyhat'varyhat)  ;
end ;
```

B.1.1.4 *Multipredictor Extension*

Demmler–Reinsch orthogonalization can be extended to fit semiparametric models involving several predictors. For example, penalized spline fitting of the additive model

$$y_i = f(s_i) + g(t_i) + \varepsilon_i$$

involves solving the ridge regression

$$(\mathbf{C}^{\mathsf{T}}\mathbf{C} + \alpha_s \mathbf{D}_s + \alpha_t \mathbf{D}_t)^{-1}\mathbf{C}^{\mathsf{T}}\mathbf{y}$$

for smoothing parameters α_s and α_t. For fixed α_s, a suite of estimates of g for several α_t values can be obtained by replacing $\mathbf{C}^{\mathsf{T}}\mathbf{C}$ in step (2) of Algorithm A.1 by $\mathbf{C}^{\mathsf{T}}\mathbf{C} + \alpha_s \mathbf{D}_s$ and vice versa.

B.1.2 *QR Decomposition*

An alternative approach to handling the ridge regressions that arise in penalized spline models is through QR decomposition (see e.g. Golub and Van Loan 1983; Hastie 1996). Algorithm A.2 provides the fitting procedure. Here $\mathbf{D}$ is assumed to be positive semidefinite.

Algorithm A.2   Inputs: $\mathbf{y}$, $\mathbf{C}$, $\mathbf{D}$, α.

(1) Form the augmented matrices

$$\mathbf{C}_a = \begin{bmatrix} \mathbf{C} \\ \sqrt{\alpha}\mathbf{D}^{1/2} \end{bmatrix} \quad \text{and} \quad \mathbf{y}_a = \begin{bmatrix} \mathbf{y} \\ \mathbf{0} \end{bmatrix}.$$

(2) Obtain the QR decomposition of $\mathbf{C}_a$,

$$\mathbf{C}_a = \mathbf{QR},$$

and set

$$\mathbf{Q}_1 = \text{matrix consisting of first } n \text{ rows of } \mathbf{Q}.$$

(3) The fitted values are then

$$\hat{\mathbf{f}}_\alpha = \mathbf{CR}^{-1}\mathbf{Q}_1^\mathsf{T}\mathbf{y}.$$

Note that the degrees of freedom and standard error estimates can be computed from

$$df_{\text{fit}}(\alpha) = \text{tr}(\mathbf{R}^{-1}\mathbf{Q}_1^\mathsf{T}\mathbf{C})$$

and

$$\text{st.dev.}(\hat{\mathbf{f}}_\alpha - \mathbf{f}) = \sigma_\varepsilon\sqrt{\text{diagonal}(\mathbf{CR}^{-1}\mathbf{Q}_1^\mathsf{T})}.$$

Other quantities described in Section B.1.1 can be computed straightforwardly using $\mathbf{Q}_1$ and $\mathbf{R}$. Hastie (1996) provides some of the details.

B.1.2.1 *Justification of Algorithm A.2*
Note that

$$\mathbf{C}_a^\mathsf{T}\mathbf{C}_a = \mathbf{C}^\mathsf{T}\mathbf{C} + \alpha\mathbf{D} \quad \text{and} \quad \mathbf{C}_a^\mathsf{T}\mathbf{y}_a = \mathbf{C}^\mathsf{T}\mathbf{y},$$

so

$$\hat{\mathbf{f}}_\alpha = \mathbf{C}\hat{\boldsymbol{\beta}}_a,$$

where

$$\hat{\boldsymbol{\beta}}_a \equiv (\mathbf{C}_a^\mathsf{T}\mathbf{C}_a)^{-1}\mathbf{C}_a^\mathsf{T}\mathbf{y}_a. \tag{B.5}$$

However, (B.5) corresponds to ordinary least-squares estimation, for which the QR approach is standard (Golub and Van Loan 1983). The solution is

$$\hat{\boldsymbol{\beta}}_a = \mathbf{R}^{-1}\mathbf{Q}^\mathsf{T}\mathbf{y}_a = \mathbf{R}^{-1}\mathbf{Q}_1^\mathsf{T}\mathbf{y}.$$

B.1.2.2 *Multipredictor Extension*
Algorithm A.2 can be easily extended to generalized ridge regressions of the form

$$(\mathbf{C}^\mathsf{T}\mathbf{C} + \mathbf{B})^{-1}\mathbf{C}^\mathsf{T}\mathbf{y},$$

where $\mathbf{B}$ is positive semidefinite, and hence allow fitting of the various multipredictor models considered in this book. Hastie (1996) gives some of the details.

B.2 Computation of Covariance Matrix Estimators

Many of the models described in the book are special cases of the linear mixed model

$$\mathbf{y} = \mathbf{X}\boldsymbol{\beta} + \mathbf{Z}\mathbf{u} + \boldsymbol{\varepsilon}, \quad \text{where } \text{Cov}\begin{bmatrix} \mathbf{u} \\ \boldsymbol{\varepsilon} \end{bmatrix} = \begin{bmatrix} \mathbf{G} & \mathbf{0} \\ \mathbf{0} & \mathbf{R} \end{bmatrix}.$$

As described in Section 4.5.4, the covariance matrices $\mathbf{G}$ and $\mathbf{R}$ can be estimated by assuming normality of $\mathbf{y}$ and maximizing either the log-likelihood or restricted log-likelihood. We will focus on the latter:

$$\ell_R(\mathbf{V}) = -\tfrac{1}{2}[\log|\mathbf{V}| + \mathbf{y}^\mathsf{T}\mathbf{V}^{-1}\{\mathbf{I} - \mathbf{X}(\mathbf{X}^\mathsf{T}\mathbf{V}^{-1}\mathbf{X})^{-1}\mathbf{X}^\mathsf{T}\mathbf{V}^{-1}\}\mathbf{y}$$
$$- \log|\mathbf{X}^\mathsf{T}\mathbf{V}^{-1}\mathbf{X}|] - \tfrac{n}{2}\log(2\pi), \tag{B.6}$$

with

$$\mathbf{V} \equiv \mathrm{Cov}(\mathbf{y}) = \mathbf{Z}\mathbf{G}\mathbf{Z}^\mathsf{T} + \mathbf{R}.$$

However, direct computation of (B.6) and its derivatives is hindered by the presence of the determinant and inverse of the $n \times n$ matrix $\mathbf{V}$. The identities

$$\mathbf{V}^{-1} = \mathbf{R}^{-1} - \mathbf{R}^{-1}\mathbf{Z}\mathbf{G}(\mathbf{I} + \mathbf{Z}^\mathsf{T}\mathbf{R}^{-1}\mathbf{Z}\mathbf{G})^{-1}\mathbf{Z}^\mathsf{T}\mathbf{R}^{-1},$$
$$|\mathbf{V}| = |\mathbf{R}||\mathbf{I} + \mathbf{Z}^\mathsf{T}\mathbf{R}^{-1}\mathbf{Z}\mathbf{G}| \tag{B.7}$$

(Harville 1977) help overcome this problem. Assuming that $\mathbf{R}$ has a simple form (e.g., is diagonal), the matrices requiring inverses and determinants are $q \times q$, where q is the number of columns in $\mathbf{Z}$.

Most of the examples considered in the book involve the fitting of variance component models, which take the form

$$\mathbf{G} = \underset{1 \le \ell \le c}{\mathrm{blockdiag}}(\sigma_{u\ell}^2\mathbf{I}), \quad \mathbf{R} = \sigma_\varepsilon^2\mathbf{I}.$$

In this case it is useful to introduce the vector

$$\boldsymbol{\alpha} = [\alpha_1, \ldots, \alpha_c]^\mathsf{T} \equiv [\sigma_\varepsilon^2/\sigma_{u1}^2, \ldots, \sigma_\varepsilon^2/\sigma_{uc}^2]^\mathsf{T}$$

and the matrices

$$\mathbf{A}(\boldsymbol{\alpha}) \equiv \underset{1 \le \ell \le c}{\mathrm{blockdiag}}(\alpha_\ell\mathbf{I}) \quad \text{and} \quad \boldsymbol{\Psi}(\boldsymbol{\alpha}) \equiv \sigma_\varepsilon^2\mathbf{V}^{-1}.$$

Then, using the identities (B.7) and letting p be the number of columns in $\mathbf{X}$, the following set of computing formulas can be derived:

$$\ell_R(\sigma_\varepsilon^2, \boldsymbol{\alpha}) = -\tfrac{1}{2}[(n - p)\log(\sigma_\varepsilon^2) + \{\mathbf{y} - \mathbf{X}\boldsymbol{\beta}(\boldsymbol{\alpha})\}^\mathsf{T}\boldsymbol{\Psi}(\boldsymbol{\alpha})\{\mathbf{y} - \mathbf{X}\boldsymbol{\beta}(\boldsymbol{\alpha})\}/\sigma_\varepsilon^2$$
$$+ \log|\mathbf{I} + \mathbf{Z}^\mathsf{T}\mathbf{Z}\mathbf{A}(\boldsymbol{\alpha})^{-1}| + \log|\mathbf{X}^\mathsf{T}\boldsymbol{\Psi}(\boldsymbol{\alpha})\mathbf{X}|] - \tfrac{n}{2}\log(2\pi);$$
$$\boldsymbol{\beta}(\boldsymbol{\alpha}) = \{\mathbf{X}^\mathsf{T}\boldsymbol{\Psi}(\boldsymbol{\alpha})\mathbf{X}\}^{-1}\mathbf{X}^\mathsf{T}\boldsymbol{\Psi}(\boldsymbol{\alpha})\mathbf{y};$$
$$\boldsymbol{\Psi}(\boldsymbol{\alpha}) = \mathbf{I} - \mathbf{Z}\{\mathbf{A}(\boldsymbol{\alpha}) + \mathbf{Z}^\mathsf{T}\mathbf{Z}\}^{-1}\mathbf{Z}^\mathsf{T}.$$

For fixed $\boldsymbol{\alpha}$, $\ell_R(\sigma_\varepsilon^2, \boldsymbol{\alpha})$ is maximized over $\sigma_\varepsilon^2 > 0$ by

$$\hat{\sigma}_\varepsilon^2(\boldsymbol{\alpha}) = \{\mathbf{y} - \mathbf{X}\boldsymbol{\beta}(\boldsymbol{\alpha})\}^\mathsf{T}\boldsymbol{\Psi}(\boldsymbol{\alpha})\{\mathbf{y} - \mathbf{X}\boldsymbol{\beta}(\boldsymbol{\alpha})\}/(n - p).$$

Therefore, all that remains is maximization of $\ell_R(\hat{\sigma}_\varepsilon^2(\boldsymbol{\alpha}), \boldsymbol{\alpha})$ over $\boldsymbol{\alpha} \in \mathbb{R}_+^c$, and this must be done numerically. Further details on this numerical maximization – including expressions for partial derivatives – are given in Lindstrom and Bates (1988), Searle et al. (1992), and Pinheiro and Bates (2000).

B.3 Software

B.3.1 *Smoothing Software*

The support of smoothing functionality in commercial software packages has increased quite markedly in recent years. This section provides a summary of developments known to us at the time of writing.

B.3.1.1 S-PLUS *Functions*

The commercial version of S-PLUS comes equipped with a number of functions for scatterplot smoothing as well as a function for fitting generalized additive models. Scatterplot smoothing functions include ksmooth(), loess(), smooth.spline(), and supsmu(). Of these, smooth.spline() is the one closest to the penalized spline smoothers used throughout this book. Indeed, for sample sizes greater than 50, smooth.spline() uses a low-rank approximation to the smoothing spline that can be thought of as a penalized spline with cubic radial basis functions.

Suppose that commands (B.2) have been issued. Then the fitted values, with GCV smoothing parameter selection, can be obtained using:

```
fit <- smooth.spline(x,y)
f.hat <- predict.smooth.spline(fit,x)$y
```

The degrees-of-freedom value for the fit may also be specified by the user. For example, a smooth with 12 degrees of freedom is obtained by replacing the first line above with:

```
fit <- smooth.spline(x,y,df=12)
```

To illustrate the use of additive models, suppose that three predictors x1, x2, x3 and a response y are set. An example, based on the S-PLUS data set air, is:

```
x1 <- air$radiation
x2 <- air$temperature
x3 <- air$wind
y  <- air$ozone^(1/3)
```
(B.8)

Then an additive model fit based on low-rank smoothing splines can be obtained and plotted as follows:

```
fit <- gam(y~s(x1)+s(x2)+s(x3))
par(mfrow=c(2,2))
plot(fit,ask=F,se=T,bty="l")
```

Note that, by default, each smooth uses 3 degrees of freedom (not counting the intercept). The user may also specify the number of degrees of freedom through commands such as:

```
fit <- gam(y~s(x1,df=7)+s(x2,df=6)+s(x3,df=8))
```

The function `step.gam()` implements the model selection algorithm as described in Section 8.6.2.

B.3.1.2 `S-PLUS` *and* `R` *Modules*

Some researchers have written their own smoothing "modules" and made them available on the Internet. Those known to us are either in the `S` (`S-PLUS`) language or the closely related `R` language. Some examples are `KernSmooth` and `LOCFIT`, described (respectively) in the books by Wand and Jones (1995) and Loader (1999). Typically such modules are placed on the developer's personal website, but both `S-PLUS` and `R` have designated Internet repositories for modules. Each have several modules for smoothing. At the time of writing, `S-PLUS` modules could be downloaded from the site ⟨`lib.stat.cmu.edu/S/`⟩, which is part of the `Statlib` system maintained by the Department of Statistics, Carnegie Mellon University. Modules in the `R` language may be downloaded from the Comprehensive R Archive Network (CRAN). This site has several "mirrors" but at the time of writing is accessible via the `Statlib` system at ⟨`lib.stat.cmu.edu/R/CRAN/`⟩. If you use `S-PLUS` or R, it is worth browsing these sites from time to time for new and updated modules.

B.3.1.3 *The* `SemiPar` *Module*

An `S-PLUS` module named `SemiPar` has been developed in parallel to the writing of this book. It handles many of the models discussed in the book. The central function of the `SemiPar` module is named `spm()`, an acronym for semiparametric model, and has similarities with the `S-PLUS` function `gam()`. However, `spm()` has the following features not available in `gam()`:

- penalized splines of arbitrary rank and degree (the knots may be input or chosen by default);
- the option to have the degrees of freedom for all smooth functions in additive models chosen automatically via REML or ML;
- bivariate smoothing, including incorporation into additive models (called *geoadditive models* in Section 13.6).

Information on `SemiPar` is posted on the website corresponding to this book. The URL is ⟨`http://www.cup.org/titles/0521785162.htm`⟩.

B.3.1.4 `SAS` *Procedures*

The `SAS` package has several procedures for smoothing and generalized additive model fitting, for example, `PROC GAM` and `PROC TPSLINE`. Note, in particular, that `PROC GAM` facilitates the choice of multiple smoothing parameters via GCV.

The code given in Section B.3.2 corresponds to the `nlme` library developed by José C. Pinheiro and Douglas M. Bates. This library is not available in earlier releases of `S-PLUS`.

B.3.2 `S-PLUS` *Mixed Model Functions*

The `S-PLUS` function `lme()` fits linear mixed models and therefore can aid the computation of many of the estimators described in this book. For example, if `X` and `Z` are computed using the `S-PLUS` code in equations (B.4), then automatic scatterplot smoothing can be achieved through:

```
fit <- lme(y~-1+X,random=pdIdent(~-1+Z))
beta.hat <- fit$coef$fixed
u.hat <- unlist(fit$coef$random)
f.hat <- X%*%beta.hat + Z%*%u.hat
```

The estimated variance components are:

```
sig.sq.eps.hat <- fit$sigma^2
sig.sq.u.hat <- sig.sq.eps.hat*exp(
                          2*unlist(fit$modelStruct))
```

The REML estimate of α is then:

```
alpha.REML <- sig.sq.eps.hat/sig.sq.u.hat
```

Next, we illustrate how an additive model may be fit using lme(). Suppose that (a) the data are defined by (B.8) and (b) knots corresponding to each of the three predictors are obtained and stored as knots.1, knots.2, and knots.3. First set up the block structure for the random effects covariance matrix:

```
K.1 <- length(knots.1)
K.2 <- length(knots.2)
K.3 <- length(knots.3)
Z.1 <- outer(x1,knots.1,"-")
Z.1 <- Z.1*(Z.1>0)
Z.2 <- outer(x2,knots.2,"-")
Z.2 <- Z.2*(Z.2>0)
Z.3 <- outer(x3,knots.3,"-")
Z.3 <- Z.3*(Z.3>0)
Z <- cbind(Z.1,Z.2,Z.3)
re.block.inds <- list(1:K.1,(K.1+1):(K.1+K.2),
                      (K.1+K.2+1):(K.1+K.2+K.3))
Z.block <- list()
for (i in 1:length(re.block.inds))
    Z.block[[i]] <- as.formula(paste("~Z[,c(",paste(
            re.block.inds[[i]],collapse=","),")]-1"))
```

The additive model can now be fit as follows:

```
fit <- lme(y~-1+X,random=pdBlocked(Z.block,
                          pdClass="pdIdent"))
beta.hat <- fit$coef$fixed
u.hat <- unlist(fit$coef$random)
```

The variance component estimates are given by:

```
sig.sq.eps.hat <- fit$sigma^2
sig.sq.u.hat <- sig.sq.eps.hat*exp(
                          2*unlist(fit$modelStruct))
```

B.3.3   SAS *Mixed Model Procedures*

The following macro computes the default set of knots from a vector x, as done
by the S-PLUS code (B.3).

```
%macro default_knots(librefknots=,data=,
                     varknots=,numknots=);
proc sort data=&data (keep=&varknots) out=q1;
    by &varknots;
run;
data q2;
    set q1;
    by &varknots;
    if first.&varknots;
run;
data &librefknots..knots;
    set q2 nobs=n;
    knotsp=int(n/5);
    if knotsp>=35 then kmx=35; else
    if knotsp<35 then kmx=knotsp;
    %if &numknots ne %then %do;
        ktemp=&numknots;
        if 1 <= ktemp <= 35 then kmx=ktemp;
    %end;
    kintrvl=int(n/kmx);
    knotsok=mod(_n_,kintrvl);
    knots=&varknots;
    if knotsok=0 then output;
    keep knots;
run;
%mend;
```

Given the scatterplot vectors x and y and a set of knots, the following macro
uses PROC MIXED to perform a penalized linear spline regression with REML
estimation of the amount of smoothing.

```
%macro scatter_smooth(libref=,data=,x=,y=,
                      knotdata=,knots=);
data dataw;
    set &data (keep=&y &x);
    m=1;
run;
data kt1;
    set &knotdata nobs=nk;
    call symput('nkt',nk);
run;
```

```
proc transpose data=&knotdata prefix=knots out=knotst;
    var &knots;
run;
data &libref..knotst;
    set knotst;
    m=1;
run;
data dataw;
    merge dataw &libref..knotst;
    by m;
    %let nk=&nkt;
    array Z (&nk) Z1-Z&nk;
    array knots (&nk) knots1-knots&nk;
    do k=1 to &nk;
        Z(k)=&x-knots(k);
        if Z(k)<0 then Z(k)=0;
    end;
    drop knots1-knots&nk _name_;
run;
ods output CovParms=&libref..varcomp;
proc mixed;
    model &y = &x / solution outp=&libref..yhat;
    random Z1-Z&nk / type=toep(1) s;
run;
%mend;
```

B.3.4 WinBugs

WinBugs (Bayesian analysis using Gibbs sampling for Windows) is a flexible software package that implements MCMC simulations and runs under Microsoft Windows. As discussed in Chapter 16, the result of using MCMC is a sample from the joint posterior distribution of parameters. Writing code in WinBugs is similar to summarizing the model, which makes the language intuitive and easy to use. Currently, WinBugs is distributed free of charge over the Internet. In this section, we assume the reader is already familiar with WinBugs. We found that WinBugs produces results quite similar to those in Chapter 16, which were obtained from our own MATLAB programs.

We first describe the logistic semiparametric model used for the union–wages example analyzed previously in Section 16.5.1.1. Then we show how this program can easily be transformed for binomial, Poisson, and Gaussian responses.

B.3.4.1 *Logistic Semiparametric Models*

Specifying the likelihood. For the union–wages example in Section 16.5.1.1, response[k] is the indicator of whether or not the kth individual is a member of the union. The logit of the probability that the kth individual belongs to the union is assumed to have a linear mixed structure. The matrices X and Z are the design matrices for fixed effects (e.g., polynomials) and random effects (e.g., truncated power functions), respectively. beta is the vector of fixed effects parameters and u contains the random effects. The function inprod is the inner product of two vectors.

The program begins with the following code:

```
model
{
for (k in 1:n)
    {
    response[k]~dbern(p[k])
    logit(p[k])<-inprod(x[k,],beta[])
                +inprod(z[k,],u[])
    }
```

In WinBugs code, dbern is the Bernoulli distribution.

Specifying the priors.   We take the fixed effects parameters beta to have independent and very diffuse normal priors. Here 1.0E-6 represents the precision (i.e., the inverse of the variance) of the normal distribution. WinBugs does not allow improper priors, so we cannot specify a uniform prior on beta as in Chapter 16.

The program continues with the following code for the priors:

```
for (l in 1:degree+1)
    {
    beta[l]~dnorm(0,1.0E-6)
    }
```

We assume the random effects u are exchangeable with mean zero, precision tauu, and a normal distribution. They are specified by the code:

```
for (i in 1:nknots)
    {
    u[i]~dnorm(0,tauu)
    }
```

We assume that the prior distribution of the precision tauu of the random effects u is gamma with mean 1 and variance 1000. sigmau is the standard deviation for the distribution of u[i]. The program continues as:

```
tauu~dgamma(1.0E-3,1.0E-3)
sigmau<-1/sqrt(tauu)
```

Here, dgamma(alpha,beta) means the gamma distribution with mean $\alpha\beta^{-1}$ and variance $\alpha\beta^{-2}$.

Constructing the design matrix for the fixed effects.   The fixed effects design matrix can be constructed outside WinBugs, but we prefer to do it here so that the program can easily be used for other applications. For the union–wages example, covariate[k] is the wage per hour for the kth individual. The ℓth column of matrix x corresponds to the ℓth-order power function. The function pow is the power function, whose second argument is the exponent. The number of observations is n and the degree of the spline used for smoothing is degree. The code for constructing this design matrix is:

```
for (k in 1:n)
    {
    for (l in 1:degree+1)
        {
        x[k,l]<-pow(covariate[k],l-1)
        }
    }
```

Constructing the design matrix Z *of random effects.*   The ith column of matrix Z contains the truncated power funtion corresponding to the ith knot. The step function is an indicator of the argument being positive. Here nknots is the number of knots and knot is a vector of knots (usually fixed-sample quantiles for covariate). The program continues:

```
for (k in 1:n)
    {
    for (i in 1:nknots)
        {
        u[k,i]<-(covariate[k]-knot[i])*step(
                                covariate[k]-knot[i])
        z[k,i]<-pow(u[k,i],degree)
        }
    }
}
```

Inputting arguments.   The input arguments containing data and user-specified parameters include:

response – an n-dimensional vector of 0–1 responses (e.g., union membership);

covariate – an n-dimensional vector used for the regression of response (e.g., wages/hour);

n – number of observations;

nknots – number of knots;

degree – degree of the spline;

knots – an nknots-dimensional vector of knots.

B.3.4.2 *Binomial Responses*
This model can be extended to a logistic model simply by replacing the line

```
response[k]~dbern(p[k])
```

with

```
response[k]~dbin(p[k],N[k])
```

where p[k] continues to represent the probability of success and N[k] is the number of repeated Bernoulli experiments. Here N is an n-dimensional vector that needs to be input.

B.3.4.3 *Poisson Semiparametric Models*
To change the model to a Poisson model, we simply replace the likelihood part of the model with

```
for (k in 1:n)
    {
    response[k]~dpois(p[k])
    log(p[k])<-inprod(x[k,],beta[])+inprod(z[k,],u[])
    }
```

B.3.4.4 *Gaussian Semiparametric Models*
To change the model to spline smoothing with normal errors, one needs to replace the likelihood part of the model with

```
for (k in 1:n)
    {
    response[k]~dnorm(m[k],taueps)
    m[k]<-inprod(beta[],x[k,])+inprod(u[],z[k,])
    }
```

where the conditional mean, m[k], of the response given the random effects is modeled as a spline. We also need to add a line specifying the prior distribution of the error precision taueps:

```
taueps~dgamma(1.0E-3,1.0E-3)
```

B.3.4.5 *Other Models*
Because of the flexibility of WinBugs, the programs just described can be extended easily to GAMs, models with measurement error, and variance function models. The beauty of WinBugs is that one need only describe the model – estimation is automatic.

Bibliography

Abelson, P. (1995). Flaws in risk assessment. *Science* 270: 215.

Abramowitz, M., and Stegun, I. (1974). *Handbook of Mathematical Functions, with Formulas, Graphs, and Mathematical Tables.* New York: Dover.

Aerts, M., Claeskens, G., and Wand, M. P. (2002). Some theory for penalized spline generalized additive models. *Journal of Statistical Planning and Inference* 103: 455–70.

Aitkin, M., Anderson, D., Francis, B., and Hinde, J. (1989). *Statistical Modeling in GLIM* (Oxford Statistical Science Series, vol. 4). Oxford University Press.

Akaike, H. (1973). Maximum likelihood identification of Gaussian autoregressive moving average models. *Biometrika* 60: 255–65.

Ansley, C. F., and Kohn, R. (1985). Estimation, filtering, and smoothing in state space models with incompletely specified initial conditions. *Annals of Statistics* 13: 1286–1316.

Anton, H., and Rorres, C. (2000). *Elementary Linear Algebra: Applications Version.* New York: Wiley.

Atkinson, A. C. (1985). *Plots, Transformations, and Regression.* Oxford: Clarendon.

Bates, D. M., and Watts, D. G. (1988). *Nonlinear Regression Analysis and Its Applications.* New York: Wiley.

Beck, N., and Jackman, S. (1998). Beyond linearity by default: Generalized additive models. *American Journal of Political Science* 42: 596–627.

Belsley, D. A., Kuh, E., and Welsch, R. E. (1980). *Regression Diagnostics: Identifying Influential Data and Sources of Collinearity.* New York: Wiley.

Berger, J. O. (1985). *Statistical Decision Theory and Bayesian Analysis,* 2nd ed. New York: Springer-Verlag.

Berndt, E. R. (1991). *The Practice of Econometrics: Classical and Contemporary.* Reading, MA: Addison-Wesley.

Berry, S. M., Carroll, R. J., and Ruppert, D. (2002). Bayesian smoothing and regression splines for measurement error problems. *Journal of the American Statistical Association* 97: 160–9.

Bishop, C. M. (1995). *Neural Networks for Pattern Recognition.* Oxford University Press.

Booth, J. G., and Hobert, J. P. (1999). Maximizing generalized linear mixed model likelihoods with an automated Monte Carlo EM algorithm. *Journal of the Royal Statistical Society, Series B* 61: 265–85.

Bowman, A. W., and Azzalini, A. (1997). *Applied Smoothing Techniques for Data Analysis.* Oxford: Clarendon.

Box, G. E. P., and Cox, D. R. (1964). An analysis of transformations. *Journal of the Royal Statistical Society, Series B* 26: 211–46.

Breslow, N. E., and Clayton, D. G. (1993). Approximate inference in generalized linear mixed models. *Journal of the American Statistical Association* 88: 9–25.

Breslow, N. E., and Lin, X. (1995). Bias correction in generalised linear mixed models with a single component of dispersion. *Biometrika* 82: 81–91.

Brockwell, P. J., and Davis, R. A. (1996). *Introduction to Time Series and Forecasting,* 2nd ed. New York: Springer-Verlag.

Brumback, B. A., and Rice, J. A. (1998). Smoothing spline models for the analysis of nested and crossed samples of curves (with discussion). *Journal of the American Statistical Association* 93: 961–94.

Brumback, B. A., Ruppert, D., and Wand, M. P. (1999). Comment on Shively, Kohn, and Wood. *Journal of the American Statistical Association* 94: 794–7.

Brumback, B. A., Ryan, L. M., Schwartz, J. D., Neas, L. M., Stark, P. C., and Burge, H. A. (2000). Transitional regression models, with application to environmental time series. *Journal of the American Statistical Association* 95: 16–27.

Bryant, P. G., and Smith, M. A. (1995). *Practical Data Analysis: Case Studies in Business Statistics.* Chicago: Irwin.

Buja, A., Hastie, T., and Tibshirani, R. (1989). Linear smoothers and additive models. *Annals of Statistics* 17: 453–510.

Burnham, K. P., and Anderson, D. (2002). *Model Selection and Multi-Model Inference,* 2nd ed. New York: Springer-Verlag.

Cai, T. T. (1999). Adaptive wavelet estimation: A block thresholding and oracle inequality approach. *Annals of Statistics* 27: 898–924.

Cai, Z., Fan, J., and Li, R. Z. (2000). Efficient estimation and inferences for varying-coefficient models. *Journal of the American Statistical Association* 95: 888–902.

Cantoni, E., and Hastie, T. (2002). Degrees of freedom tests for smoothing splines. *Biometrika* 89: 251–65.

Cao, R., Cuevas, A., and González-Manteiga, W. (1994). A comparative study of several smoothing methods in density estimation. *Computational Statistics and Data Analysis* 17: 153–76.

Carlin, B. P., and Louis, T. A. (2000). *Bayes and Empirical Bayes Methods for Data Analysis,* 2nd ed. New York: Chapman & Hall.

Carroll, R. J., Fan, J., Gijbels, I., and Wand, M. P. (1997). Generalized partially linear single-index models. *Journal of the American Statistical Association* 92: 477–89.

Carroll, R. J., Roeder, K., and Wasserman, L. (1999). Flexible parametric measurement error models. *Biometrics* 55: 44–54.

Carroll, R. J., and Ruppert, D. (1982). Robust estimation in heteroscedastic linear models. *Annals of Statistics* 10: 429–41.

Carroll, R. J., and Ruppert, D. (1988). *Transformation and Weighting in Regression.* New York: Chapman & Hall.

Carroll, R. J., Ruppert, D., and Stefanski, L. A. (1995). *Measurement Error in Nonlinear Models.* London: Chapman & Hall.

Carroll, R. J., Ruppert, D., and Welsh, A. H. (1998). Local estimating equations. *Journal of the American Statistical Association* 93: 214–27.

Casella, G., and Berger, R. L. (2002). *Statistical Inference,* 2nd ed. Pacific Grove, CA: Thomson Learning.

Chambers, J. M., and Hastie, T. J. (1993). *Statistical Models in S.* New York: Chapman & Hall.

Chatterjee, S., and Hadi, A. (1998). *Sensitivity Analysis in Linear Regression.* New York: Wiley.

Chatterjee, S., Handcock, M. S., and Simonoff, J. S. (1995). *A Casebook for a First Course in Statistics and Data Analysis.* New York: Wiley.

Chaudhuri, P., and Marron, J. S. (1999). SiZer for exploration of structures in curves. *Journal of the American Statistical Association* 94: 807–23.

Chen, H. (1988). Convergence rates for parametric components in a partly linear model. *Annals of Statistics* 16: 136–46.

Chen, Z. (1993). Fitting multivariate regression functions by interaction spline models. *Journal of the Royal Statistical Society, Series B* 55: 473–91.

Chib, S., and Greenberg, E. (1995). Understanding the Metropolis–Hastings algorithm. *American Statistician* 49: 327–35.

Chiu, S. T. (1996). A comparative review of bandwidth selection for kernel density estimation. *Statistica Sinica* 6: 129–46.

Clayton, D. (1996). Generalized linear mixed models. In W. R. Gilks, S. Richardson, and D. J. Spiegelhalter (Eds.), *Markov Chain Monte Carlo in Practice,* pp. 275–301. London: Chapman & Hall.

Cleveland, W. S. (1979). Robust locally weighted regression and smoothing scatterplots. *Journal of the American Statistical Association* 74: 829–36.

Cleveland, W. S., and Grosse, E. (1991). Computational methods for local regression. *Statistics and Computing* 1: 47–62.

Cook, R. D., and Weisberg, S. (1982). *Residuals and Influence in Regression.* New York: Chapman & Hall.

Coull, B. A., Ruppert, D., and Wand, M. P. (2001). Simple incorporation of interactions into additive models. *Biometrics* 57: 539–45.

Coull, B. A., Schwartz, J., and Wand, M. P. (2001). Respiratory health and air pollution: Additive mixed model analyses. *Biostatistics* 2: 337–49.

Crainiceanu, C., and Ruppert, D. (2002). Asymptotic distribution of likelihood ratio tests in linear mixed models. Unpublished manuscript.

Crainiceanu, C., Ruppert, D., and Vogelsang, T. (2002). Probability that the MLE of a variance component is zero with applications to likelihood ratio tests. Unpublished manuscript.

Craven, P., and Wahba, G. (1979). Smoothing noisy data with spline functions: Estimating the correct degree of smoothing by the method of generalized cross-validation. *Numerische Mathematik* 31: 377–403.

Cressie, N. (1990). Reply to Wahba's letter. *American Statistician* 44: 256–8.

Cressie, N. (1993). *Statistics for Spatial Data.* New York: Wiley.

Cristianini, N., and Shawe-Taylor, J. (2000). *An Introduction to Support Vector Machines.* Cambridge University Press.

Cummins, D. J., Filloon, T. G., and Nychka, D. (2001). Confidence intervals for nonparametric curve estimates: Toward more uniform pointwise coverage. *Journal of the American Statistical Association* 96: 233–46.

Davidian, M., and Carroll, R. J. (1997). Variance function estimation. *Journal of the American Statistical Association* 82: 1079–91.

Davidian, M., and Gallant, R. A. (1993). The nonlinear mixed effects model with a smooth random effects density. *Biometrika* 80: 475–88.

Davidian, M., and Giltinan, D. M. (1995). *Nonlinear Models for Repeated Measurement Data.* New York: Chapman & Hall.

Davidson, R., and MacKinnon, J. G. (1993). *Estimation and Inference in Econometrics.* New York: Oxford University Press.

Davis, J. M., and Speckman, P. (1999). A model for predicting maximum and 8-hr average ozone in Houston. *Atmospheric Environment* 33: 2487–2500.

Deely, J. J., and Lindley, D. V. (1981). Bayes empirical Bayes. *Journal of the American Statistical Association* 76: 833–41.

Delecroix, M., Simioni, M., and Thomas-Agnan, C. (1995). Functional estimation under shape constraints. *Journal of Nonparametric Statistics* 6: 69–89.

Delecroix, M., and Thomas-Agnan, C. (2000). Spline and kernel regression under shape restriction. In M. G. Schimek (Ed.), *Statistical Theory and Computational Aspects of Smoothing,* pp. 109–33. Heidelberg: Physica-Verlag.

Dempster, A. P., Laird, N. M., and Rubin, D. B. (1977). Maximum likelihood from incomplete data via the EM algorithm. *Journal of the Royal Statistical Society, Series B* 39: 1–22.

Dempster, A. P., Rubin, D. B., and Tsutakawa, R. K. (1981). Estimation in covariance components models. *Journal of the American Statistical Association* 76: 341–53.

Denison, D. G. T., Mallick, B. K., and Smith, A. F. M. (1997). Automatic Bayesian curve fitting. *Journal of the Royal Statistical Society, Series B* 60: 333–50.

Dierckx, P. (1995). *Curve and Surface Fitting with Splines* (Monographs on Numerical Analysis). Oxford University Press.

Diggle, P., Heagerty, P., Liang, K.-L., and Zeger, S. (2002). *Analysis of Longitudinal Data,* 2nd ed. Oxford University Press.

Diggle, P. J., Tawn, J. A., and Moyeed, R. A. (1998). Model-based geostatistics (with discussion). *Applied Statistics* 47: 299–350.

DiMatteo, I., Genovese, C. R., and Kass, R. E. (2001). Bayesian curve-fitting with free-knot splines. *Biometrika* 88: 1055–72.

Dobson, A. (2002). *An Introduction to Generalized Linear Models,* 2nd ed. London / Boca Raton, FL: Chapman & Hall / CRC Press.

Dominici, F., Zeger, S. L., and Samet, J. M. (2000). A measurement error model for time-series studies of air pollution and mortality. *Biostatistics* 1: 157–75.

Donnelly, C. A., Laird, N. M., and Ware, J. H. (1995). Prediction and creation of smooth curves for temporally correlated longitudinal data. *Journal of the American Statistical Association* 90: 984–9.

Draper, N. R., and Smith, H. (1998). *Applied Regression Analysis,* 3rd ed. New York: Wiley.

Duchon, J. (1976). Fonctions-spline et esperances conditionnelles de champs gaussiens. *Ann. Sci. Univ. Clermont Ferrand II Math.* 14: 19–27.

Ecker, M. D., and Heltshe, J. F. (1994). Geostatistical estimates of scallop abundance. In N. Lange et al. (Eds.), *Case Studies in Biometry,* pp. 107–24. New York: Wiley.

Efromovich, S. (1999). *Nonparametric Curve Estimation.* New York: Springer-Verlag.

Eilers, P. H. C., and Marx, B. D. (1996). Flexible smoothing with B-splines and penalties (with discussion). *Statistical Science* 11: 89–121.

Elliott, P., Wakefield, J. C., Best, N. G., and Briggs, D. J. (2000). *Spatial Epidemiology.* Oxford University Press.

Engels, E. A., Rosenberg, P. S., and Biggar, R. J. (1999). Zoster incidence in human immunodeficiency virus-infected hemophiliacs and homosexual men, 1984–1997. *Journal of Infectious Diseases* 180: 1784–9.

Engle, R. F., Granger, C. W. J., Rice, J., and Weiss, A. (1986). Semiparametric estimates of the relation between weather and electricity sales. *Journal of the American Statistical Association* 81: 310–20.

Eubank, R. L. (1984). The hat matrix for smoothing splines. *Statistics and Probability Letters* 2: 9–14.

Eubank, R. L. (1985). Diagnostics for smoothing splines. *Journal of the Royal Statistical Society, Series B* 47: 332–41.

Eubank, R. L. (1988). *Spline Smoothing and Nonparametric Regression.* New York: Marcel Dekker.

Eubank, R. L. (1994). A simple smoothing spline. *American Statistician* 48: 103–6.

Eubank, R. L. (1999). *Nonparametric Regression and Spline Smoothing.* New York: Marcel Dekker.

Eubank, R. L., and Gunst, R. (1986). Diagnostics for penalized least-squares estimators. *Statistics and Probability Letters* 4: 265–72.

Ezekiel, M. (1924). A method for handling curvilinear correlation for any number of variables. *Journal of the American Statistical Association* 19: 431–53.

Fahrmeir, L., and Lang, S. (2002). Bayesian inference for generalized additive mixed models based on Markov random field priors. Unpublished manuscript.

Fahrmeir, L., and Tutz, G. (2002). *Multivariate Statistical Modelling Based on Generalized Linear Models,* 2nd ed. New York: Springer-Verlag.

Fan, J. (1992). Design-adaptive nonparametric regression. *Journal of the American Statistical Association* 87: 1273–94.

Fan, J. (1993). Local linear regression smoothers and their minimax efficiencies. *Annals of Statistics* 21: 196–216.

Fan, J., and Gijbels, I. (1996). *Local Polynomial Modelling and Its Applications.* London: Chapman & Hall.

Fan, J., Hu, I.-C., and Truong, Y. K. (1994). Robust nonparametric function estimation. *Scandinavian Journal of Statistics* 21: 433–46.

Fan, J., and Marron, J. S. (1994). Fast implementations of nonparametric curve estimators. *Journal of Computational and Graphical Statistics* 3: 35–56.

Fan, J., and Truong, Y. K. (1993). Nonparametric regression with errors in variables. *Annals of Statistics* 21: 1900–25.

Fisher, M., Nychka, D., and Zervos, D. (1994). Fitting the term structure of interest rates with smoothing splines. Economics Division Series, Division of Research and Statistics, Division of Monetary Affairs, Federal Reserve Board, Washington, DC (January 1995).

Fox, J. (2000). *Nonparametric Simple Regression: Smoothing Scatterplots.* Thousand Oaks, CA: Sage.

French, J. L., Kammann, E. E., and Wand, M. P. (2001). Comment on paper by Ke and Wang. *Journal of the American Statistical Association* 96: 1285–8.

French, J. L., and Wand, M. P. (2004). Generalized additive models for cancer mapping with incomplete covariates. *Biostatistics* 5: 177–91.

Friedman, J. H. (1991). Multivariate adaptive regression splines (with discussion). *Annals of Statistics* 19: 1–141.

Friedman, J. H., and Silverman, B. W. (1989). Flexible parsimonious smoothing and additive modeling. *Technometrics* 31: 3–21.

Friedman, J. H., and Stuetzle, W. (1981). Projection pursuit regression. *Journal of the American Statistical Association* 76: 817–23.

Fuller, W. A. (1987). *Measurement Error Models.* New York: Wiley.

Ganguli, B., Staudenmayer, J., and Wand, M. P. (2001). Additive models with predictors subject to measurement error. Unpublished manuscript.

Gelfand, A. E., Sahu, S. K., and Carlin, B. P. (1995). Efficient parametrisations for normal linear mixed models. *Biometrika* 82: 479–88.

Gelman, A., Carlin, J. B., Stern, H. S., and Rubin, D. B. (1995). *Bayesian Data Analysis.* Boca Raton, FL: Chapman & Hall.

Gilks, W. R., Richardson, S., and Spiegelhalter, D. J. (1996). *Markov Chain Monte Carlo in Practice.* London: Chapman & Hall.

Gilks, W. R., and Wild, P. (1992). Adaptive rejection sampling for Gibbs sampling. *Applied Statistics* 41: 337–48.

Gill, J. (2001). *Generalized Linear Models: A Unified Approach.* Thousand Oaks, CA: Sage.

Gilmour, A. R., Anderson, R. D., and Rae, A. L. (1985). The analysis of binomial data by a generalized linear mixed model. *Biometrika* 72: 593–9.

Goldstein, H., and Rasbash, J. (1996). Improved approximations for multilevel models with binary responses. *Journal of the Royal Statistical Society, Series A* 159: 505–13.

Golub, G. H., and Van Loan, C. F. (1983). *Matrix Computations,* 3rd ed. Baltimore: Johns Hopkins University Press.

Gourieroux, C., Monfort, A., and Trognon, A. (1984). Pseudo maximum likelihood methods: Applications to Poisson models. *Econometrica* 52: 701–20.

Gray, R. J. (1992). Flexible methods for analyzing survival data using splines, with application to breast cancer prognosis. *Journal of the American Statistical Association* 87: 942–51.

Gray, R. J. (1994). Spline-based tests in survival analysis. *Biometrics* 50: 640–52.

Green, P. J., and Silverman, B. W. (1994). *Nonparametric Regression and Generalized Linear Models.* London: Chapman & Hall.

Gu, C. (1992a). Cross-validating non-Gaussian data. *Journal of Computational and Graphical Statistics* 1: 169–79.

Gu, C. (1902b). Diagnostics for nonparametric regression models with additive terms. *Journal of the American Statistical Association* 87: 1051–8.

Gu, C. (2000). Multivariate spline regression. In M. G. Schimek (Ed.), *Smoothing and Regression: Approaches, Computation and Application,* pp. 329–55. New York: Wiley.

Gu, C. (2002). *Smoothing Spline ANOVA Models.* New York: Springer-Verlag.

Guo, W. (2002). Functional mixed effects models. *Biometrics* 58: 121–8.

Halmos, P. (1995). *Linear Algebra Problem Book.* Washington, DC: Mathematical Association of America.

Hampel, F. R., Ronchetti, E. M., Rousseeuw, P. J., and Stahel, W. A. (1986). *Robust Statistics: The Approach Based on Influence Functions.* New York: Wiley.

Handcock, M. S., and Wallis, J. R. (1994). An approach to statistical spatial-temporal modeling of meteorological fields. *Journal of the American Statistical Association* 89: 368–78.

Hansen, M. H., and Kooperberg, C. (2002). Spline adaptation in extended linear models (with discussion). *Statistical Science* 17: 2–51.

Hansen, M. H., Huang, J. Z., Kooperberg, C., Stone, C. J., and Truong, Y. K. (2003). *Statistical Modeling with Spline Functions: Methodology and Theory.* New York: Springer-Verlag.

Hardin, J., and Hilbe, J. (2001). *Generalized Linear Models and Extensions.* College Station, TX: Stata Press.

Härdle, W. (1990). *Applied Non-parametric Regression* (Econometric Society Monographs, vol. 19). Cambridge University Press.

Härdle, W. (1991). *Smoothing Techniques, with Implementations in S.* New York: Springer-Verlag.

Härdle, W., and Hall, P. (1993). On the backfitting algorithm for additive regression models. *Statistica Neerlandica* 47: 43–57.

Härdle, W., Hall, P., and Ichimura, H. (1993). Optimal smoothing in single-index models. *Annals of Statistics* 21: 157–78.

Härdle, W., Hall, P., and Marron, J. S. (1988). How far are automatically chosen regression smoothing parameters from their optimum? *Journal of the American Statistical Association* 83: 86–101.

Härdle, W., Kerkyacharian, G., Picard, D., and Tsybakov, A. (1998). *Wavelets, Approximation, and Statistical Applications* (Lecture Notes in Statistics, vol. 129). New York: Springer-Verlag.

Härdle, W., and Korostelev, A. (1996). Search for significant variables in nonparametric additive regression. *Biometrika* 83: 541–9.

Härdle, W., Liang, H., and Gao, J. (2000). *Partially Linear Models.* Heidelberg: Physica-Verlag.

Härdle, W., and Scott, D. W. (1992). Smoothing by weighted averaging of rounded points. *Computational Statistics* 7: 97–128.

Hart, J. D. (1997). *Nonparametric Smoothing and Lack-of-Fit Tests.* New York: Springer-Verlag.

Hart, R. W., Keenan, K., Turtorro, A., Abdo, K. M., Leakey, J., and Lyn-Cook, B. (1995). Symposium overview: Caloric restriction and toxicity. *Fundamental and Applied Toxicology* 25: 184–95.

Harville, D. A. (1977). Maximum likelihood approaches to variance component estimation and to related problems. *Journal of the American Statistical Association* 72: 320–38.

Harville, D. A. (2000). *Matrix Algebra from a Statistician's Perspective.* New York: Springer-Verlag.

Harville, D. A., and Mee, R. W. (1984). A mixed-model procedure for analyzing ordered categorical data. *Biometrics* 40: 393–408.

Haseman, J. K., Bourbina, J., and Eustis, S. L. (1994). The effect of individual housing and other experimental design factors on tumor incidence in B6C3F1 mice. *Fundamental and Applied Toxicology* 23: 44–52.

Hastie, T. J. (1996). Pseudosplines. *Journal of the Royal Statistical Society, Series B* 58: 379–96.

Hastie, T. J., and Tibshirani, R. J. (1990). *Generalized Additive Models.* London: Chapman & Hall.

Hastie, T. J., and Tibshirani, R. J. (1993). Varying-coefficients models. *Journal of the Royal Statistical Society, Series B* 55: 757–96.

Hastie, T., and Tibshirani, R. J. (2000). Bayesian backfitting. *Statistical Science* 15: 196–223.

Hastie, T., Tibshirani, R., and Friedman, J. (2001). *The Elements of Statistical Learning.* New York: Springer-Verlag.

Hastings, W. K. (1970). Monte Carlo sampling methods using Markov chains and their applications. *Biometrika* 57: 97–109.

Hayes, K., and Haslett, J. (1999). Simplifying general least squares. *American Statistician* 53: 376–81.

Heckman, N. E. (1986). Spline smoothing in a partly linear model. *Journal of the Royal Statistical Society, Series B* 48: 244–8.

Henderson, C. (1950). Estimation of genetic parameters. *Annals of Mathematical Statistics* 21: 309–10.

Hoel, P. G. (1984). *Introduction to Mathematical Statistics,* 5th ed. New York: Wiley.

Hogg, R. V., and Craig, A. T. (1995). *Introduction to Mathematical Statistics,* 5th ed. Englewood Cliffs, NJ: Prentice-Hall.

Holmes, C. C., and Mallick, B. K. (2001). Bayesian regression with multivariate linear splines. *Journal of the Royal Statistical Society, Series B* 63: 3–18.

Holst, U., Hössjer, O., Björklund, C., Ragnarson, P., and Edner, H. (1996). Locally weighted least squares kernel regression and statistical evaluation of LIDAR measurements. *Environmetrics* 7: 401–16.

Huber, P. (1967). The behavior of maximum likelihood estimates under nonstandard conditions. *Proceedings of the Fifth Berkeley Symposium on Mathematical Statistics and Probability,* vol. 1, pp. 221–33. Berkeley: University of California Press.

Huber, P. (1983). *Robust Statistics.* Chichester: Wiley.

Hurvich, C. M., Simonoff, J. S., and Tsai, C. (1998). Smoothing parameter selection in nonparametric regression using an improved Akaike information criterion. *Journal of the Royal Statistical Society, Series B* 60: 271–93.

Hutchinson, M. F., and de Hoog, F. R. (1985). Smoothing noisy data with spline functions. *Numerische Mathematik* 47: 99–106.

Ibrahim, J. G. (1990). Incomplete data. *Journal of the American Statistical Association* 85: 765–9.

Ibrahim, J. G., Chen, M. H., and Lipsitz, S. R. (2001). Missing responses in generalized linear mixed models when the missing data mechanism is nonignorable. *Biometrika* 88: 551–64.

Ichimura, H. (1993). Semiparametric least squares (SLS) and weighted SLS estimation of single-index models. *Journal of Econometrics* 58: 71–120.

James, G. M., and Hastie, T. J. (2001). Functional linear discriminant analysis for irregularly sampled curves. *Journal of the Royal Statistical Society, Series B* 63: 533–50.

James, G. M., Hastie, T. J., and Sugar, C. A. (2000). Principal component models for sparse functional data. *Biometrika* 87: 587–602.

Jarrow, R., Ruppert, D., and Yu, Y. (2003). Estimating the term structure of corporate debt with a semiparametric penalized spline model. Technical report, School of Operations Research, Cornell University, Ithaca, NY.

Jiang, W. (2002). On weak base hypotheses and their implications for boosting regression. *Annals of Statistics* 30: 51–73.

Johnson, M. E., Moore, L. M., and Ylvisaker, D. (1990). Minimax and maximin distance designs. *Journal of Statistical Planning and Inference* 26: 131-48.

Jones, M. C., Marron, J. S., and Sheather, S. J. (1996a). A brief survey of bandwidth selection for density estimation. *Journal of the American Statistical Association* 91: 401–7.

Jones, M. C., Marron, J. S., and Sheather, S. J. (1996b). Progress in data-based bandwidth selection for kernel density estimation. *Computational Statistics* 11: 337–81.

Kackar, R. N., and Harville, D. A. (1984). Approximations for standard errors of estimators of fixed and random effects in mixed linear models. *Journal of the American Statistical Association* 79: 853–62.

Kammann, E. E., Staudenmayer, J., and Wand, M. P. (2002). Robustness for general design mixed models using the t-distribution. Unpublished manuscript.

Kammann, E. E., and Wand, M. P. (2003). Geoadditive models. *Applied Statistics* 52: 1–18.

Kass, R. E., and Steffey, D. (1989). Approximate Bayesian inference in conditionally independent hierarchical models (parametric empirical Bayes models). *Journal of the American Statistical Association* 84: 717–26.

Kauermann, G., and Tutz, G. (1999). On model diagnostics using varying coefficient models. *Biometrika* 86: 119–28.

Kelly, C., and Rice, J. (1990). Monotone smoothing with application to dose–response curves and the assessment of synergism. *Biometrics* 46: 1071–85.

Kent, J. T. (1998). Comment on paper by Diggle, Tawn, and Moyeed. *Applied Statistics* 47: 330–1.

Khuri, A. I., Mathew, T., and Sinha, B. K. (1998). *Statistical Tests for Mixed Linear Models.* New York: Wiley.

Kimeldorf, G., and Wahba, G. (1971). Some results on Tchebycheffian spline functions. *Journal of Mathematical Analysis and Applications* 33: 82–95.

Kitanidis, P. K. (1997). *Introduction to Geostatistics: Applications in Hydrogeology.* Cambridge University Press.

Knafl, G., Sacks, J., and Ylvisaker, D. (1985). Confidence bands for regression functions. *Journal of the American Statistical Association* 80: 683–91.

Koenker, R., Ng, P., and Portnoy, S. (1994). Quantile smoothing splines. *Biometrika* 81: 673–80.

Kolmogorov, A. N. (1941). Interpolirovanie ekstrapolivanie statsionarnykh sluchainykh posledovatel 'nostei' [Interpolated and extrapolated random sequences]. *Izv. Akad. Nauk SSSR Ser. Mat.* 5: 3–14.

Krige, D. G. (1966). Two-dimensional weighted moving average trend surfaces for ore evaluation. *Journal of the South African Institute of Mining and Metallurgy* 66: 13–38.

Laird, N. M., and Louis, T. A. (1987). Empirical Bayes confidence intervals based on bootstrap samples (with discussion). *Journal of the American Statistical Association* 82: 739–54.

Laird, N. M., and Ware, J. H. (1982). Random-effects models for longitudinal data. *Biometrics* 38: 963–74.

Lange, K. L., Little, R. J. A., and Taylor, J. M. G. (1989). Robust statistical modeling using the t-distribution. *Journal of the American Statistical Association* 84: 881–96.

Lange, N., Ryan, L., Billard, L., Brillinger, D., Conquest, L., and Greenhouse, J. (Eds.) (1994). *Case Studies in Biometry.* New York: Wiley.

Lee, T. C. M. (2002). Automatic smoothing for discontinuous regression functions. *Statistica Sinica* 13: 823–42.

Leroux, B. G., Lei, X., and Breslow, N. (1999). Estimation of disease rates in small areas: A new mixed model for spatial dependence. In M. E. Halloran and D. Berry (Eds.), *Statistical Models in Epidemiology, the Environment and Clinical Trials,* pp. 179–92. New York: Springer-Verlag.

Liang, K., and Zeger, S. L. (1986). Longitudinal data analysis using generalized linear models. *Biometrika* 73: 13–22.

Lin, X., and Breslow, N. E. (1996). Bias correction in generalized linear mixed models with multiple components of dispersion. *Journal of the American Statistical Association* 91: 1007–16.

Lin, X., and Zhang, D. (1999). Inference in generalized additive mixed models by using smoothing splines. *Journal of the Royal Statistical Society, Series B* 61: 381–400.

Lindsey, J. K. (1997). *Applying Generalized Linear Models.* New York: Springer-Verlag.

Lindstrom, M. J., and Bates, D. M. (1988). Newton–Raphson and EM algorithms for linear mixed-effects models for repeated-measures data. *Journal of the American Statistical Association* 83: 1014–22.

Lindstrom, M. J., and Bates, D. M. (1990). Nonlinear mixed effects models for repeated measures data. *Biometrics* 46: 673–87.

Linhart, H., and Zucchini, W. (1986). *Model Selection.* New York: Wiley.

Linton, O. B., and Härdle, W. (1996). Estimation of additive regression models with known links. *Biometrika* 83: 529–40.

Linton, O., and Nielsen, J. P. (1995). A kernel method of estimating structured nonparametric regression based on marginal integration. *Biometrika* 82: 93–100.

Little, R. J., and Rubin, D. B. (1987). *Statistical Analysis with Missing Data*. New York: Wiley.

Liu, J. (2001). *Monte Carlo Strategies in Scientific Computing*. New York: Springer-Verlag.

Loader, C. (1999). *Local Regression and Likelihood*. New York: Springer-Verlag.

Louis, A. K., Maass, D., and Rieder, A. (1997). *Wavelets: Theory and Applications*. New York: Wiley.

Louis, T. A. (1982). Finding the observed information matrix when using the EM algorithm. *Journal of the Royal Statistical Society, Series B* 44, 226–33.

MacNab, Y. C., and Dean, C. B. (2001). Autoregressive spatial smoothing and temporal spline smoothing for mapping rates. *Biometrics* 57: 949–56.

Magnus, J. R., and Neudecker, H. (1999). *Matrix Differential Calculus with Applications in Statistics and Econometrics,* 2nd ed. Chichester: Wiley.

Mallows, C. L. (1973). Some comments on C_p. *Technometrics* 15: 661–75.

Mammen, E., Marron, J. S., Turlach, B. A., and Wand, M. P. (2001). A general framework for constrained smoothing. *Statistical Science* 16: 232–48.

Mardia, K. V., and Marshall, R. J. (1984). Maximum likelihood estimation of models for residual covariance in spatial regression. *Biometrika* 72: 135–46.

Marron, J. S. (1996). A personal view of smoothing and statistics. In M. G. Schimek (Ed.), *Statistical Theory and Computational Aspects of Smoothing,* pp. 1–9. Heidelberg: Physica-Verlag.

Marschner, I. C., and Bosch, R. J. (1998). Flexible assessment of trends in age-specific HIV incidence using two-dimensional penalized likelihood. *Statistics in Medicine* 17: 1017–31.

Marx, B. D., and Eilers, P. H. C. (1998). Direct generalized additive modeling with penalized likelihood. *Computational Statistics and Data Analysis* 28: 193–209.

Matheron, G. (1965). *Les variables régionalisées et leur estimation*. Paris: Masson.

Maybeck, P. S. (1979). *Stochastic Models, Estimation, and Control*. New York: Academic Press.

McCullagh, P., and Nelder, J. A. (1989). *Generalized Linear Models,* 2nd ed. London: Chapman & Hall.

McCulloch, C. E. (1997). Maximum likelihood algorithms for generalized linear mixed models. *Journal of the American Statistical Association* 92: 162–70.

McCulloch, C. E., and Searle, S. R. (2001). *Generalized, Linear, and Mixed Models*. New York: Wiley.

McGilchrist, C. A., and Aisbett, C. W. (1991). Regression with frailty in survival analysis. *Biometrics* 47: 461–6.

McLachlan, G. J., and Krishnan, T. (1997). *The EM Algorithm and Extensions*. New York: Wiley.

Metropolis, N., Rosenbluth, A. W., Rosenbluth, M. N., Teller, A. H., and Teller, E. (1953). Equations of state calculations by fast computing machines. *Journal of Chemical Physics* 21: 1087–91.

Miller, J. J. (1977). Asymptotic properties of maximum likelihood estimates in the mixed model of the analysis of variance. *Annals of Statistics* 5: 746–62.

Moore, D. S., and McCabe, G. P. (1998). *Introduction to the Practice of Statistics,* 3rd ed. New York: Freeman.

Morris, C. N. (1983). Parametric empirical Bayes inference: Theory and applications (with discussion). *Journal of the American Statistical Association* 78: 47–65.

Müller, H. G. (1988). *Nonparametric Regression Analysis of Longitudinal Data* (Lecture Notes in Statistics, vol. 46). New York: Springer-Verlag.

Müller, P., and Vidakovic, B. (1999). *Bayesian Inference in Wavelet-based Models*. New York: Springer-Verlag.

Myers, R. H., Montgomery, D. C., and Vining, G. G. (2001). *Generalized Linear Models: With Applications in Engineering and the Sciences*. New York: Wiley.

Nadaraya, E. A. (1964). On estimating regression. *Theory of Probability and Its Applications* 9: 141–2.

Nadaraya, E. A. (1989). *Nonparametric Estimation of Probability Densities and Regression Curves* [translated by S. Kotz]. Boston: Kluwer.

Nason, G. P., and Silverman, B. W. (2000). Wavelets for regression and other statistical problems. In M. G. Schimek (Ed.), *Smoothing and Regression. Approaches, Computation and Application,* pp. 159–91. New York: Wiley.

Neter, J., Kutner, M., Nachtsheim, C., and Wasserman, W. (1996). *Applied Linear Statistical Models,* 4th ed. Chicago: Irwin.

Newey, W. K. (1994). The asymptotic variance of semiparametric estimators. *Econometrica* 62: 1349–82.

Nicholson, W. K. (2001). *Elementary Linear Algebra.* New York: McGraw-Hill.

Nychka, D. W. (1988). Confidence intervals for smoothing splines. *Journal of the American Statistical Association* 83: 1134–43.

Nychka, D. W. (2000). Spatial process estimates as smoothers. In M. Schimek (Ed.), *Smoothing and Regression.* Heidelberg: Springer-Verlag.

Nychka, D., and Cummins, D. J. (1996). Comment on paper by Eilers and Marx. *Statistical Science* 11: 104–5.

Nychka, D., Haaland, P., O'Connell, M., and Ellner, S. (1998). FUNFITS, data analysis and statistical tools for estimating functions. In D. Nychka, W. W. Piegorsch, and L. H. Cox (Eds.), *Case Studies in Environmental Statistics* (Lecture Notes in Statistics, vol. 132), pp. 159–79. New York: Springer-Verlag.

Nychka, D., and Saltzman, N. (1998). Design of air quality monitoring networks. In D. Nychka, W. W. Piegorsch, and L. H. Cox (Eds.), *Case Studies in Environmental Statistics* (Lecture Notes in Statistics, vol. 132), pp. 51–76. New York: Springer-Verlag.

O'Connell, M. A., and Wolfinger, R. D. (1997). Spatial regression models, response surfaces, and process optimization. *Journal of Computational and Graphical Statistics* 6: 224–41.

Ogden, R. T. (1996). *Essential Wavelets for Statistical Applications and Data Analysis.* Boston: Birkhäuser.

Opsomer, J. D., and Ruppert, D. (1997). Fitting a bivariate additive model by local polynomial regression. *Annals of Statistics* 25: 186–211.

Opsomer, J. D., Ruppert, D., Wand, M. P., Holst, U. and Hössjer, O. (1999). Kriging with nonparametric variance function estimation. *Biometrics* 55: 704–10.

O'Sullivan, F. (1986). A statistical perspective on ill-posed inverse problems (with discussion). *Statistical Science* 1: 505–27.

O'Sullivan, F. (1988). Fast computation of fully automated log-density and log-hazard estimators. *SIAM Journal on Scientific and Statistical Computing* 9: 363–79.

Pagan, A., and Ullah, A. (1999). *Nonparametric Econometrics* (Themes in Modern Econometrics). Cambridge University Press.

Pagano, M., and Gauvreau, K. (1993). *Principles of Biostatistics.* Florence, KY: Duxbury.

Parise, H., Wand, M. P., Ruppert, D., and Ryan, L. M. (2001). Incorporation of historical controls using semiparametric mixed models. *Journal of the Royal Statistical Society, Series C* 50: 31–42.

Park, B. U., and Turlach, B. A. (1992). Practical performance of several data driven bandwidth selectors. *Computational Statistics* 7: 251–70.

Parker, R. L., and Rice, J. A. (1985). Discussion of "Some aspects of the spline smoothing approach to nonparametric curve fitting" by B. W. Silverman. *Journal of the Royal Statistical Society, Series B* 47: 40–2.

Patterson, H. D., and Thompson, R. (1971). Recovery of inter-block information when block sizes are unequal. *Biometrika* 58: 545–54.

Peixoto, J. L. (1990). A property of well-formulated polynomial regression models. *American Statistician* 44: 26–30.

Pinheiro, J. C., and Bates, D. M. (2000). *Mixed-Effects Models in S and S-PLUS.* New York: Springer-Verlag.

Pope, C. A., Dockery, D. W., Spengler, J. D., and Raizenne, M. E. (1991). Respiratory health and PM_{10} pollution: A daily time series analysis. *American Review of Respiratory Disease* 144: 668–74.

Prasad, N. G. N., and Rao, J. N. K. (1990). The estimation of the mean squared error of small-area estimators. *Journal of the American Statistical Association* 85: 163–71.

Ramsay, J. O. (1988). Monotone regression splines in action. *Statistical Science* 3: 425–41.

Ramsay, J. O., and Silverman, B. W. (1997). *Functional Data Analysis.* New York: Springer-Verlag.

Rao, C. R. (1973). *Linear Statistical Inference and Its Applications.* New York: Wiley.

Ratkowsky, D. A. (1983). *Nonlinear Regression Modeling: A Unified Practical Approach.* New York: Marcel Dekker.

Raudenbush, S. W., Yang, M.-L., and Yosef, M. (2000). Maximum likelihood for generalized linear models with nested random effects via high-order, multivariate Laplace approximation. *Journal of Computational and Graphical Statistics* 9: 141–57.

Rice, J. A., and Wu, C. O. (2001). Nonparametric mixed effects models for unequally sampled noisy curves. *Biometrics* 57: 253–9.

Rice, S. O. (1939). The distribution of the maxima of a random curve. *American Journal of Mathematics* 61: 409–16.

Ripley, B. D. (1996). *Pattern Recognition and Neural Networks.* Cambridge University Press.

Robert, C. P. (1995). Simulation of truncated normal variables. *Statistics and Computing* 5: 121–5.

Robert, C. P., and Casella, G. (1999). *Monte Carlo Statistical Methods.* New York: Springer-Verlag.

Robinson, G. K. (1991). That BLUP is a good thing: The estimation of random effects. *Statistical Science* 6: 15–51.

Roland, J., Keyghobadi, N., and Fownes, S. (2000). *Alpine parnassius* butterfly dispersal: Effects of landscape and population size. *Ecology* 81: 1642–53.

Rousseeuw, P. J., and Leroy, A. M. (1987). *Robust Regression and Outlier Detection.* New York: Wiley.

Rubinstein, R. Y. (1981). *Simulation and the Monte Carlo Method.* New York: Wiley.

Ruppert, D. (1997a). Local polynomial regression and its applications in environmental statistics. In V. Barnett and F. Turkman (Eds.), *Statistics for the Environment,* vol. 3. Chichester: Wiley.

Ruppert, D. (1997b). Empirical-bias bandwidths for local polynomial nonparametric regression and density estimation. *Journal of the American Statistical Association* 92: 1049–62.

Ruppert, D. (2002). Selecting the number of knots for penalized splines. *Journal of Computational and Graphical Statistics* 11: 735–57.

Ruppert, D., and Carroll, R. J. (2000). Spatially-adaptive penalties for spline fitting. *Australian and New Zealand Journal of Statistics* 42: 205–24.

Ruppert, D., Carroll, R. J., and Cressie, N. (1989). A transformation/weighting model for estimating Michaelis–Menten parameters. *Biometrics* 46: 637–56.

Ruppert, D., Carroll, R. J., and Cressie, N. (1991). Response to "Generalized linear models for enzyme-kinetic data" by J. A. Nelder. *Biometrics* 47: 1610–15.

Ruppert, D., and Wand, M. P. (1994). Multivariate locally weighted least squares regression. *Annals of Statistics* 22: 1346–70.

Ruppert, D., Wand, M. P., Holst, U., and Hössjer, O. (1997). Local polynomial variance function estimation. *Technometrics* 39: 262–73.

Ryan, T. P. (1997). *Modern Regression Methods.* New York: Wiley.

Schafer, J. L. (1997). *Analysis of Incomplete Multivariate Data.* New York: Chapman & Hall.

Schall, R. (1991). Estimation in generalized linear models with random effects. *Biometrika* 78: 719–27.

Schapire, R., Freund, Y., Bartlett, P., and Lee, W. (1998). Boosting at the margin: A new explanation for the effectiveness of voting methods. *Annals of Statistics* 26: 1651–86.

Schuster, E. F. (1972). Joint asymptotic distribution of the estimated regression function at a finite number of distinct points. *Annals of Mathematical Statistics* 43: 84–8.

Schwartz, J. (1994). Air pollution and hospital admissions for the elderly in Birmingham, Alabama. *American Journal of Epidemiology* 139: 589–98.

Searle, S. R. (1982). *Matrix Algebra Useful for Statistics*. New York: Wiley.

Searle, S. R., Casella, G., and McCulloch, C. E. (1992). *Variance Components*. New York: Wiley.

Seber, G. A. F., and Wild, C. J. (1989). *Nonlinear Regression*. New York: Wiley.

Seifert, B., Brockmann, M., Engel, J., and Gasser, T. (1994). Fast algorithms for nonparametric curve estimation. *Journal of Computational and Graphical Statistics* 3: 192–213.

Seilkop, S. K. (1995). The effect of body weight on tumor incidence and carcinogenicity testing in B6C3F1 mice and F344 rats. *Fundamental and Applied Toxicology* 24: 247–59.

Self, S. G., and Liang, K.-Y. (1987). Asymptotic properties of maximum likelihood estimators and likelihood ratio tests under nonstandard conditions. *Journal of the American Statistical Association* 82: 605–10.

Shen, X., and Ye, J. (2002). Adaptive model selection. *Journal of the American Statistical Association* 97: 210–21.

Shively, T. S., Kohn, R., and Wood, S. (1999). Variable selection and function estimation in additive nonparametric regression using a data-based prior. *Journal of the American Statistical Association* 94: 777–94.

Shun, Z. (1997). Another look at the salamander mating data: A modified Laplace approximation approach. *Journal of the American Statistical Association* 92: 341–9.

Shun, Z., and McCullagh, P. (1995). Laplace approximation of high dimensional integrals. *Journal of the Royal Statistical Society, Series B* 57: 749–60.

Sigrist, M. (Ed.) (1994). *Air Monitoring by Spectroscopic Techniques* (Chemical Analysis Series, vol. 197). New York: Wiley.

Silvapulle, M. J. (1996). A test in the presence of nuisance parameters. *Journal of the American Statistical Association* 91: 1690–3.

Silverman, B. W. (1984). Spline smoothing: The equivalent variable kernel method. *Annals of Statistics* 12: 898–916.

Simonoff, J. S. (1996). *Smoothing Methods in Statistics*. New York: Springer-Verlag.

Simonoff, J. S., and Tsai, C. (1999). Semiparametric and additive model selection using an improved Akaike information criterion. *Journal of Computational and Graphical Statistics* 8: 22–40.

Smith, M., and Kohn, R. (1996). Nonparametric regression using Bayesian variable selection. *Journal of Econometrics* 75: 317–44.

Smith, M., Wong, C., and Kohn, R. (1998). Additive nonparametric regression with autocorrelated errors. *Journal of the Royal Statistical Society, Series B* 60: 311–31.

Smith, P. L. (1982). Hypothesis testing in *B*-spline regression. *Communications in Statistics, Part B* 11: 143–57.

Spall, J. C. (1991). The Kalman filter and BLUP (Comment on paper by Robinson). *Statistical Science* 6: 39–41.

Speckman, P. (1988). Kernel smoothing in partial linear models. *Journal of the Royal Statistical Society, Series B* 50: 413–36.

Stark, P. C., Ryan, L. M., McDonald, J. L., and Burge, H. A. (1997). Using meteorologic data to model and predict daily ragweed pollen levels. *Aerobiologia* 13: 177–84.

Staudte, R. G., and Sheather, S. J. (1990). *Robust Estimation and Testing*. New York: Wiley.

Steele, B. M. (1996). A modified EM algorithm for estimation in generalized mixed models. *Biometrics* 52: 1295–1310.

Stein, M. L. (1999). *Interpolation of Spatial Data: Some Theory for Kriging*. New York: Springer-Verlag.

Stone, C. J., Hansen, M. H., Kooperberg, C., and Truong, Y. K. (1997). Polynomial splines and their tensor products in extended linear modeling. *Annals of Statistics* 25: 1371–1425.

Stram, D. O., and Lee, J. W. (1994). Variance components testing in the longitudinal mixed effects model. *Biometrics* 50: 1171–7.

Strang, G. (1998). *Introduction to Linear Algebra*, 2nd ed. Wellesley, MA: Wellesley-Cambridge Press.

Tanner, M. A. (1996). *Tools for Statistical Inference: Methods for the Exploration of Posterior Distributions and Likelihood Functions,* 3rd ed. New York: Springer-Verlag.

Tantiyaswasdikul, C., and Woodroofe, M. B. (1994). Isotonic smoothing splines under sequential designs. *Journal of Statistical Planning and Inference* 38: 75–87.

Tarter, M. E., and Lock, M. D. (1993). *Model-free Curve Estimation.* New York: Chapman & Hall.

Therneau, T. M., and Grambsch, P. (2000). *Modeling Survival Data: Extending the Cox Model.* New York: Springer-Verlag.

Thomas, W. (1991). Influence diagnostics for the cross-validated smoothing parameter in spline smoothing. *Journal of the American Statistical Association* 86: 693–8.

Thompson, J. R., and Tapia, R. A. (1990). *Nonparametric Function Estimation, Modeling, and Simulation.* Philadelphia: SIAM.

Tibshirani, R. (1996). Regression shrinkage and selection via the lasso. *Journal of the Royal Statistical Society, Series B* 58: 267–88.

Tierney, L. (1994). Markov chains for exploring posterior distributions. *Annals of Statistics* 22: 1701–28.

Tierney, L., Kass, R. E., and Kadane, J. B. (1989). Fully exponential Laplace approximations to expectations and variances of nonpositive functions. *Journal of the American Statistical Association* 84: 710–16.

Tjøstheim, D., and Auestad, B. H. (1994). Nonparametric identification of nonlinear time series: Projections. *Journal of the American Statistical Association* 89: 1398–1409.

Tsybakov, A. (1986). Robust reconstruction of functions by the local approximation method. *Problems of Information Transmission* 22: 133–46.

Ullah, A. (1985). Specification analysis of econometric models. *Journal of Quantitative Economics* 2: 187–209.

van Dyk, D. A. (2000). Nesting EM algorithms for computational efficiency. *Statistica Sinica* 10: 203–26.

Vapnik, V. N. (1998). *Statistical Learning Theory.* New York: Wiley.

Vapnik, V. N. (2000). *The Nature of Statistical Learning Theory,* 2nd ed. New York: Springer-Verlag.

Verbeke, G., and Molenberghs, G. (1997). *Linear Mixed Models in Practice.* New York: Springer-Verlag.

Verbeke, G., and Molenberghs, G. (2000). *Linear Mixed Models for Longitudinal Data.* New York: Springer-Verlag.

Verbyla, A. P., Cullis, B. R., Kenward, M. G., and Welham, S. J. (1999). The analysis of designed experiments and longitudinal data by using smoothing splines (with discussion). *Journal of the Royal Statistical Society, Series C* 48: 269–312.

Vidakovic, B. (1999). *Statistical Modeling by Wavelets.* New York: Wiley.

Vonesh, E. F., and Chinchilli, V. M. (1997). *Linear and Nonlinear Models for the Analysis of Repeated Measures.* New York: Marcel Dekker.

Wahba, G. (1983). Bayesian "confidence intervals" for the cross-validated smoothing spline. *Journal of the Royal Statistical Society, Series B* 45: 133–50.

Wahba, G. (1985). A comparison of GCV and GML for choosing the smoothing parameter in the generalized spline smoothing problem. *Annals of Statistics* 13: 1378–1402.

Wahba, G. (1986). Partial interaction spline models for the semiparametric estimation of functions of several variables. In T. J. Boardman (Ed.), *Computer Science and Statistics: Proceedings of the 18th Symposium on the Interface,* pp. 75–80. Washington, DC: American Statistical Association.

Wahba, G. (1988). Partial and interaction spline models (with discussion). In J. M. Bernado et al. (Eds.), *Bayesian Statistics 3,* pp. 479–91. New York: Oxford University Press.

Wahba, G. (1990). *Spline Models for Observational Data.* Philadelphia: SIAM.

Wakefield, J. C., Best, N. G., and Waller, L. (2000). Bayesian approaches to disease mapping. In P. Elliott et al. (Eds.), *Spatial Epidemiology,* pp. 104–27. Oxford University Press.

Walter, G. G., and Shen, X. (2001). *Wavelets and Other Orthogonal Systems,* 2nd ed. Boca Raton, FL: Chapman & Hall / CRC Press.

Wand, M. P. (2000). A comparison of regression spline smoothing procedures. *Computational Statistics* 15: 443–62.

Wand, M. P. (2002). Vector differential calculus in statistics. *American Statistician* 56: 55–62.

Wand, M. P., and Jones, M. C. (1995). *Kernel Smoothing.* London: Chapman & Hall.

Wang, N., Lin, X., Gutierrez, R., and Carroll, R. J. (1998). Bias analysis and SIMEX inference in generalized linear mixed measurement error models. *Journal of the American Statistical Association* 93: 249–61.

Wang, Y. (1998). Mixed effects smoothing spline analysis of variance. *Journal of the Royal Statistical Society, Series B* 60: 159–74.

Watson, G. S. (1964). Smooth regression analysis. *Sankhya* A26: 359–72.

Webster, R. (1998). Comment on paper by Diggle, Tawn, and Moyeed. *Applied Statistics* 47: 326–7.

Wei, G. C. G., and Tanner, M. A. (1990). A Monte Carlo implementation of the EM algorithm and the poor man's data augmentation algorithms. *Journal of the American Statistical Association* 85: 699–704.

Weisberg, S. (1985). *Applied Linear Regression,* 2nd ed. New York: Wiley.

Weisberg, S., and Welsh, A. H. (1994). Adapting for the missing link. *Annals of Statistics* 22: 1674–1700.

Welsh, A. H., and Richardson, A. M. (1997). Approaches to the robust estimation of mixed models. In G. S. Maddala and C. R. Rao (Eds.), *Handbook of Statistics,* vol. 15. Amsterdam: Elsevier.

Wilcox, R. R. (1997). *Introduction to Robust Estimation and Hypothesis Testing.* San Diego: Academic Press.

Williams, E. J. (1959). *Regression Analysis.* New York: Wiley.

Wolfinger, R. and O'Connell, M. (1993). Generalized linear mixed models: A pseudo-likelihood approach. *Journal of Statistical Computation and Simulation* 48: 233–43.

Wolfinger, R., Tobias, R., and Sall, J. (1994). Computing Gaussian likelihoods and their derivatives for general linear mixed models. *SIAM Journal on Scientific and Statistical Computing* 15: 1294–1310.

Wood, S. N. (2000). Modelling and smoothing parameter estimation with multiple quadratic penalties. *Journal of the Royal Statistical Society, Series B* 62: 413–28.

Xiang, D., and Wahba, G. (1986). A generalized approximate cross validation for smoothing splines with non-Gaussian data. *Statistica Sinica* 6: 675–92.

Young, S. G., and Bowman, A. W. (1995). Non-parametric analysis of covariance. *Biometrics* 51: 920–31.

Yu, K., and Jones, M. C. (1998). Local linear quantile regression. *Journal of the American Statistical Association* 93: 228–37.

Yu, Y., and Ruppert, D. (2002). Penalized spline estimation in partially linear single index models. *Journal of the American Statistical Association* 97: 1042–54.

Zanobetti, A., Wand, M. P., Schwartz, J., and Ryan, L. M. (2000). Generalized additive distributed lag models. *Biostatistics* 1: 279–92.

Zeger, S. L., and Karim, M. R. (1991). Generalized linear models with random effects: A Gibbs sampling approach. *Journal of the American Statistical Association* 86: 79–86.

Zhang, D., Lin, X., Raz, J., and Sowers, M. (1998). Semi-parametric stochastic mixed models for longitudinal data. *Journal of the American Statistical Society* 93: 710–19.

Zhou, S., and Shen, X. (2001). Spatially adaptive regression splines and accurate knot selection schemes. *Journal of the American Statistical Association* 96: 247–59.

Author Index

Notation Index

Example Index

Subject Index

Printed in the United States
By Bookmasters